The Global Information Technology Report

2004–2005

Efficiency in an Increasingly Connected World

Soumitra Dutta
Augusto Lopez-Claros

palgrave
macmillan

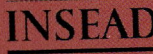
INSEAD

WORLD
ECONOMIC
FORUM

COMMITTED TO
IMPROVING THE STATE
OF THE WORLD

The Global Information Technology Report 2004–2005 is a special project within the framework of the Global Competitiveness Programme. The Global Information Technology Report is the result of a collaboration between the World Economic Forum and INSEAD, France.

At the World Economic Forum:

Professor Klaus Schwab
Executive Chairman

Dr Augusto Lopez-Claros
Director

Jennifer Blanke

Simone Droz

Margareta Drzeniek

Kerry Jaggi

Emma Loades

Irene Mia

Saadia Zahidi

At INSEAD:

Professor Soumitra Dutta
Roland Berger Professor of Business and Technology
Dean for Executive Education

Amit Jain
Research Program Manager

A special thank you to Michael James for his superb editing work and Scott Graham for his excellent graphic design and layout.

The terms *country* and *nation* as used in this report do not in all cases refer to a territorial entity that is a state as understood by international law and practice. The term covers well-defined, geographically self-contained economic areas that may not be states but for which statistical data are maintained on a separate and independent basis.

Positions in these articles and papers are in no way endorsed by Cisco Systems. Cisco is pleased to co-sponsor this report to further the debate on issues that are important to high-tech. To view Cisco's positions on public policy matters, please visit Cisco's Government Affairs homepage. http://www.cisco.com/gov.

First published 2005 by
PALGRAVE MACMILLAN
Houndmills, Basingstoke, Hampshire RG21 6XS and
175 Fifth Avenue, New York, N.Y. 10010
Companies and representatives throughout the world.

PALGRAVE MACMILLAN is the global academic imprint of the Palgrave Macmillan division of St. Martin's Press, LLC and of Palgrave Macmillan Ltd. Macmillan® is a registered trademark in the United States, United Kingdom and other countries. Palgrave is a registered trademark in the European Union and other countries.

ISBN-13: 978-1-4039-4800-7
ISBN-10: 1-4039-4800-3

This book is printed on paper suitable for recycling and made from fully managed and sustained forest sources.

A catalogue record for this book is available from the British Library.
A catalogue record for this book is available from the Library of Congress.

Printed and bound in Great Britain by
Ashford Colour Press Ltd, Gosport

Contents

Preface

Professor Klaus Schwab

World Economic Forum

The global economy is being changed in profound ways by the onward march of science and technology. Technological change has, of course, always been a central engine of economic growth, but what is significant about the past decade is the acceleration in the pace of change and, as more and more countries have made efforts to improve their macroeconomic and policy environments, technology and technological innovation appear to have entered a "golden age", a time when they are emerging as the key drivers of growth and development.

There are, to be sure, still many basic battles to be won in the developing world, addressing fundamental issues of development from reducing poverty levels and the incidence of disease to enhancing opportunity and the quality of life for large segments of the world's population. But, as economists are prone to point out, what matters most is what happens "at the margin", and at the margin technologies today—particularly information and communication technologies (ICT)—are increasingly playing the central catalytic role in pushing the development process forward.

The release of the *Global Information Technology Report 2004–2005* comes at a time of guarded optimism about the near-term evolution of the global economy. World GDP has entered a phase of strong expansion nearly everywhere, and the technology sector, having gone through a period of streamlining and consolidation in the early part of the decade, is making its contribution increasingly felt. Whether this period of economic expansion will be sustained over the next few years, creating the conditions for improvements in per capita income worldwide, or whether it will falter under the weight of uncertainties associated with existing macroeconomic imbalances is, perhaps, the key question confronting businesses and governments everywhere.

But, regardless of how this issue is resolved—and we count ourselves among those who feel that a combination of good macroeconomic policies and ambitious structural reforms will go a long way toward ensuring a sustained recovery—it is clear that ICT will continue to play a growing role in boosting the efficiency of the increasingly integrated global economy and enabling countries to improve resource allocation and boost growth prospects.

The *Global Information Technology Report* series of the World Economic Forum aims to monitor the progress of networked readiness in more than 100 countries, highlighting the policy, institutional, and structural obstacles that prevent countries from fully capturing the benefits of ICT. It is thus a benchmarking tool that also gauges the extent to which best practices are spreading all over the world. Beyond the mere provision of an annual international cross-section of networked readiness, the publication of the Report also may be seen as a vehicle whereby governments, businesses and individuals can assess progress on a regular basis.

We commend the contributors to this Report for their energy and commitment to producing a valuable resource for policy makers and business leaders engaged in the task of promoting networked readiness. We especially thank the editors of the Report, Soumitra Dutta of INSEAD and Augusto Lopez-Claros of the World Economic Forum, for their leadership roles in this project. Appreciation also goes to Irene Mia—a member of the Global Competitiveness Programme—who played a key role at the Forum in day-to-day management of the project, and other members of the program: Jennifer Blanke, Margareta Drzeniek, Simone Droz, Kerry Jaggi, Emma Loades, and Saadia Zahidi.

Forewords

John Chambers, Cisco Systems
Rob Lloyd, Cisco Systems

The theme of the World Economic Forum's 2005 Annual Meeting is *Taking Responsibility for Tough Choices*. One of the tough choices that companies and countries take responsibility for is where to apply their investments. As the global marketplace for products and services becomes more integrated and more competitive, the number of variables involved in these decisions continues to increase. Gaining and retaining competitive advantage is constantly on the minds of business and government leaders.

At Cisco, we believe that competition is an essential component of innovation and productivity. It is a virtuous cycle, driving innovation, creating new markets for companies, enabling economic growth, and raising the standard of living for countries. Productivity is central to this cycle. The prices of commodities and intermediate products as well as wages for productive workers have risen faster in recent quarters than pricing power. Countries must improve productivity by 2–3 percent a year to maintain their global competitive position, and 3–5 percent a year is possible. By comparison, well-run organizations can improve productivity by as much as 5–10 percent a year.

In recent years, major economic research has revealed a correlation between information and communication technologies (ICT), and gains in productivity. To further explore the nature of this relationship, Cisco is pleased to sponsor the *Global Information Technology Report 2004–2005*, including the presentation of the Networked Readiness Index. This research provides valuable information for countries and companies in areas of opportunity and development to achieve greater participation in the global networked economy.

The Networked Readiness Index has several intriguing aspects. One of the key components of the index is the level of technology usage in a country, measured in terms of individuals, businesses, and governments. As already noted, there is a strong correlation between ICT spending and productivity at the national level, which is demonstrated in this research as a strong correlation between the Networked Readiness Index and global competitiveness.

Perhaps of greater consequence for government and business leaders is the ability or readiness of a country to use technology. While ICT usage is a measure of the present, ICT readiness is perhaps a measure of the future. Education and literacy of citizens, ongoing skills training by businesses, and government education policies are all components of the readiness measurement and play an important role in building the foundations of a country's productivity.

The third component of the Networked Readiness Index consists of environmental factors. These items are usually beyond the control of a single business or organization, such as market conditions, regulatory framework, or available infrastructure. Proactive policies and investments by all levels of government, such as those encouraging broadband network infrastructures, are necessary to clear the way for innovation and productivity gains.

Education and communication are the two great equalizers in this global marketplace. Countries are realizing that their competitive positions are not guaranteed. Tough choices are necessary to decide which investments to make to accelerate innovation, increase productivity, and achieve the resulting economic growth. I hope this and other research Cisco sponsors helps guide the way to those decisions.

John Chambers
President and CEO, Cisco Systems

Information and Communication Technologies (ICT) have clearly had a significant impact on the economies of more developed countries around the world, and are proving to be a powerful tool both for commerce and economic growth and for social and educational advancement.

What we should not miss is the profound impact that ICT is having in emerging markets. Stories from China, India, Russia, the Middle East and eastern Europe, to name a few, have been increasingly heard in the past few years. But there are many other examples of the transformation potential of these technologies.

For example, some exciting programs are happening in Africa. Creative projects there are combining technology with process innovations (such as micro-credit) and helping to close the digital divide and address structural issues. Solar power, mobile and wireless technologies and other components are bringing market information and

communications to more people every day. There is still a long way to go, but the potential is clear.

At the same time, market liberalization is occurring in telecommunications and public sector services throughout eastern Europe and the Middle East, among other parts of the world. These and other environmental changes are encouraging creativity and innovation in these sectors. A latter-stage example can be seen in the 10 new member states of the European Union, as well as those aspiring to join in 2007 or later. These countries have an opportunity to use broadband and other technologies as a path to improve economic competitiveness. Given their ranking in the middle of the index, this may be somewhat surprising. However, looking into the details we can see that the technology readiness of individuals and businesses in these countries is often higher than their overall index ranking. This implies a significant stored potential, waiting to be released by targeted investments and improvements.

When implementing new technologies, compatibility with existing systems will consume less effort, resulting in faster deployments. This presents an opportunity for these countries to catch up with and surpass their peers by more easily adopting advanced technologies that enhance productivity, improving operational efficiency and competitiveness.

It is heartening to see government leaders, educators, business leaders and a variety of technology providers coming together to collaborate on these issues Together, we can realize some tremendous changes in economic competitiveness

Rob Lloyd
President of EMEA Operations,
Senior Vice-President of Cisco Systems

Executive Summary

Augusto Lopez-Claros, World Economic Forum

Soumitra Dutta, INSEAD

The *Global Information Technology Report 2004–2005* makes its appearance at a critical juncture in the recent evolution of the world economy. There is growing evidence that global economic activity may have entered a phase of sustained expansion, with a strong recovery across most markets. This has led to heightened optimism about the near-term outlook, tempered by lingering concerns about the long-term impact of global macroeconomic imbalances, volatility in the oil markets, and an unsettled security situation in the Middle East. The bullish sentiments prevailing in the markets in the late 1990s and the early part of the present decade, which fed the bubble in technology stocks, have largely given way to a more sober mood among investors and a more realistic assessment of the role of information and communication technologies (ICT) in the development process, both in developed markets and among emerging economies.

It is against this background of cautious optimism about global economic prospects, borne of some painful lessons learned in the recent past, that this Report is being published. It builds on the work done in three previous editions and thus may be seen as part of a long-term commitment at both the World Economic Forum and INSEAD to the dissemination of business-relevant research on information technology issues with a strong practical focus.

The Report is divided into three parts. Part 1 contains a series of essays written by knowledgeable practitioners, scholars and experts with an enviable body of relevant experience in the ICT area. Ranging in coverage from an update of the by now well-established Networked Readiness Index rankings, to an examination of the role of the Internet in the provision of government services, to the challenges posed by rapid technological change for the formulation of credible regulatory frameworks, these studies—briefly summarized below—provide fresh insights into some of the most topical issues affecting the ICT industry. The country profiles presented in Part 2 provide valuable background information on the components of each country's networked readiness rankings. As such, they are a useful synthesis—allowing the benefit of international comparisons—of the work underlying the index rankings. The data tables in Part 3 provide additional rankings for all the variables used in the Report.

The Networked Readiness Index

The chapter "An Analysis of the Diffusion and Usage of Information and Communications Technologies of Nations" by Soumitra Dutta and Amit Jain (both at INSEAD) presents the results of research evaluating the relative level of ICT development in 104 nations across the world. Based on empirical analysis of data collected from leading internationally recognized sources such as the World Bank, the International Telecommunications

Union, Pyramid, and the World Economic Forum's Executive Opinion Survey, the level of ICT development of nations is assessed via the estimation of a Networked Readiness Index (NRI). As in previous editions of the *Global Information Technology Report*, the framework developed to estimate networked readiness rests on three fundamental pillars. The first pillar captures aspects of the environment of a given nation for ICT development, such as the regulatory regime and the legal framework for ICT, the available infrastructure, and other factors capturing elements of the market for technological development. The second pillar looks at actual levels of networked readiness of the three main stakeholders in the economy: individuals, businesses, and governments. Finally, the actual levels of usage of ICT by these three groups are also brought in as a third pillar.

The NRI rankings for 2004 are broadly consistent with those published in the 2003–2004 Report. Eight of the top 10 countries listed in that Report still occupy top-ten positions in 2004–2005. Singapore emerges as the global leader on the Networked Readiness Index rankings, with the United States (in first place in 2003–2004) now moving to fifth position and Hong Kong and Japan moving into top-ten ranks. The Nordic countries continue to occupy privileged positions, as in the past. Among the larger emerging markets China and India both moved up the ranks, to 41st and 39th place respectively, significant improvements on 2003–2004. In Latin America, Brazil and Mexico registered drops in rankings, while Chile continues to lead the region by a significant margin.

Dutta and Jain also explore the relationship between the Networked Readiness Index and one of the indices of competitiveness estimated by the World Economic Forum. The link between investment in ICT and a nation's productivity has long been a subject of significant interest for decision makers and economists. The analysis carried out by the authors establishes a strong link between the two, with the three stages of a nation's development identified by the Forum's competitiveness research corresponding to three distinct groups of Networked Readiness scores.

The Emergence of e-Government

Many public sector organizations are trying to broaden delivery of citizen services while improving efficiency. In their chapter "Net Impact: European e-Government" Douglas Frosst, Scott Brown and Andrew Elder from Cisco Systems look at the interactions between people, process and technology in efforts under way to improve productivity in the European public sector. These efforts are part of various initiatives in Europe aimed at strengthening the role of ICT in society. Many of these reflect government programs, such as the European

Commission's e-Europe 2005 Action Plan and Connected Health initiatives. They are part of a broader attempt on the part of EU members to move forward with broad-ranging economic and structural reforms in the context of the so-called Lisbon Strategy, a key component of which seeks to boost aspects of the information society across Europe at the levels of households, businesses and government services. Achieving these goals will require investments in people, process and technology appropriate to the objectives and abilities of the organizations involved.

To help identify the correlation between the aims pursued by these projects and the desired outcomes, Cisco Systems sponsored Net Impact 2004: Europe e-Government, a research project attempting to better understand the impact of Internet technologies on organizations. Under the project's auspices more than 1,400 ICT and business decision makers from government and healthcare organizations in France, Germany, Italy, the Netherlands, Poland, Spain, Sweden and the United Kingdom were interviewed to assess the state of networking infrastructure, business processes, applications, and other organizational behaviors.

Many studies of productivity use a single macro-level measurement to determine productivity gains. Since each organization faces different environmental conditions and objectives, these high-level measures have limited usefulness. Net Impact 2004 identified best practices that led to positive outcomes for business efficiency, service effectiveness, financial optimization and citizen satisfaction. Specific areas of improvement are described within each of these four themes.

Net Impact 2004 concluded that "Connected Organizations" that increased their productivity the most had higher levels of process automation and integration distributed to employees and other key constituents with a sophisticated network. These groups employed formal measurement systems to monitor improvements and provide the necessary feedback. They also encouraged an organizational culture of open communication about these initiatives and a positive focus on process improvement and end-user services. The study by Frosst, Brown and Elder is thus a particularly welcome contribution to a better understanding of the impact on productivity of efforts aimed at boosting the role of Internet technologies in the provision of more efficient services to consumers of government services, such as in the area of healthcare. The Lisbon Strategy seeks to make the European Union (EU) "the most competitive and dynamic knowledge-based economy in the world" by 2010. This study is a fine example of what it will take to achieve the goals of Lisbon pertaining to its information society component.

The Economic Gains of ICT

In a chapter titled "The ICT Sector and the Global Economy: Counting the Gains," Markus Haacker argues that technological advances in information and communication technology over recent decades have transformed the way businesses are operating, and have resulted in changes in the patterns of global trade in goods and services. The prospects for sustained economic growth fuelled by the "new economy" have attracted much attention, although the crash in IT equity prices in 2000 brought home the realization that the economic "laws" of competition (and creative destruction) apply as fully to the ICT sector as to the rest of the economy. Rather than adding to the existing literature on the ongoing ICT-driven transformation of the global economy, Haacker takes a "low-tech" approach focusing on the production and use of ICT equipment. In particular, he argues that, as the use of ICT requires the use of some equipment, data on sales of the latter provide valuable information on the role of ICT across countries, in particular for lower-income countries where detailed national account data on ICT-producing or ICT-using sectors are not available.

Haacker finds that, while productivity gains in the production of ICT goods have been impressive, most of the gains dissipate as these ICT goods are exported. While one would expect that high-income countries, with a larger share of services and more technologically advanced economies, are the primary beneficiaries of productivity gains in the IT sector, the author finds that this is not necessarily the case. While usage of IT products is positively correlated with GDP per capita for higher-income countries, this relationship breaks down for countries with an annual GDP per capita of less than $5,000. Some of the countries posting the highest gains are in this lower-income category, and his findings suggest that trade barriers (or the lack thereof) are an important determinant of usage of ICT goods in lower-income countries.

Regulating the ICT Industry

In their chapter "Is a New Regulatory Framework for Telecommunications Needed for the Twenty-first Century?" McKinsey's Scott Beardsley, Luis Enriquez, Victoria Gerus and Andreas Marschner argue that the liberalization of telecommunication markets during the past two decades has brought substantial benefits to consumers and businesses. In opening previously reserved services to competition, regulators encouraged efficiency improvements in incumbent fixed-line businesses and transferred industry value to consumers through dramatic price reductions and, along the way, sparked significant service innovation. Thus, quite deliberately, the traditional regulatory framework focused on promoting price

competition in formerly monopolistic markets in fixed telephony and data. This approach was not designed, however, to address the recent competitive and technological developments in the industry. These developments, such as fixed-mobile substitution, competition in fixed access, the growth of broadband access and the resulting take-up of voice over Internet Protocol (VoIP) are fundamentally changing the structure of the industry.

The existing "regulatory tool-kit" appears increasingly unsuited to address these types of competitive dynamics. As a result, the application of the classic regulatory approach, the authors argue, is now distorting economic incentives for market players, and thus is putting at risk long-term customer benefits from technological improvements and infrastructure investments. The adequacy of this model to support the industry's future is called into question by the authors, who make a case that a radical reassessment is needed in light of the industry's fundamental economic characteristics and several recent developments, such as its capital-intensive infrastructure, the importance of long-term innovation and the critical effect of scale and network effects on its competitive dynamics, all of which features it shares with many other network industries.

This chapter makes the case for a change in approach and puts forth several thoughts on the direction in which the traditional regulatory framework could be adapted for the next "growth horizon" in the industry. It reminds the reader of the rationale of the traditional regulatory framework and points out its strengths and weaknesses. The authors' approach is broad-based, bringing in, where necessary, useful examples from different countries, contrasting interesting differences in regulatory environments and highlighting the international nature of the problem they have set out to analyse. They then argue that a required return on investment approach to the industry and ongoing processes of technological progress and transformation call for a different line of attack. The chapter's final section explores the implications of these trends and changes for the future of telecommunication regulation. Many of these implications may also be relevant to other sectors in the ICT field that exhibit similar economic characteristics.

The Role of Government Policy

Government policies can have a huge impact, both positive and negative, on how quickly the information society and the infrastructure that supports it develop. While this is no doubt true everywhere, it is particularly so for the developed markets, given the greater penetration of Internet technologies and, at least in the case of the EU, the explicit incorporation of "information society" elements in the formulation of economic and structural reforms, such as in the context of the Lisbon Strategy referred to above. The Internet Society has long stressed the need for governments to address a wide range of challenges, including promoting competition, supporting investment in telecommunications, encouraging Internet deployment and use, reducing the cost of ICT equipment and software, increasing the availability of affordable content, and investing in training and education. The chapter "Building a Sound Foundation for the Information Society" by Elliot E. Maxwell and Michael R. Nelson focuses on one key aspect: the policies that governments should adopt to promote the creation and deployment of advanced telecommunication infrastructures and the applications and content that are made available over them.

Too often have government policies, based on older technologies and assumptions about telecommunications, prevented competition, slowed its growth, or tried to manage it in some fashion. The effect has been to reduce investment and delay the availability of new technologies and services. Maxwell and Nelson use the 1996 Telecommunications Act in the United States as a case study of the need to fully update policy paradigms in order to take advantage of technologies such as cable television, licensed and unlicensed wireless, power lines and satellites to stimulate competition, increase investment, better regulate market power, and make new services, applications, and content available to all.

Rather than continue with regulations that reflect their origins in a world of monopoly telecommunication providers and that treat various distribution technologies differently in a "vertical" or "stove-pipe" fashion, the authors advocate a new "horizontal" model which better reflects the increasing digitization of information, the convergence of technology platforms and services, and the Internet's architecture. It sets forth minimal regulatory requirements for Internet-like "openness", including interconnection of networks, broadband transport available to all on a non-discriminatory basis without government-determined pricing, and rules against "unreasonable" discrimination against applications and content.

These limited requirements reflect the fundamental change resulting from broadband deployment, which is to separate the provision of services from the provision of applications. This new model of regulation would create, in the authors' view, an environment that fosters competition in all the layers (including transport, applications and content), reduce the need for detailed economic regulation, better control market power, and encourage entry and investment by any player using any distribution technology. The chapter also suggests mechanisms to support the availability of advanced services

for the under-served, an elimination of regulations as competition increases, greater flexibility in meeting "provider of last resort requirements" and encouragement of an "open" spectrum policy to facilitate the development of "inter-platform" competition from licensed and unlicensed wireless broadband providers.

Outsourcing Opportunities and Challenges

Outsourcing is now an established fixture in the organizational models of the twenty-first century. At first confined to peripheral business activities such as cleaning, transport, or legal services, outsourcing now encompasses business functions that are closer to the "core", such as manufacturing, customer management, or information technology. As an area of economic activitity, outsourcing has become a multibillion dollar industry, and there is every expectation not only that it is here to stay but that it will expand quickly in coming years. Driven by a rapid and sustained reduction in the costs of cross-border communication, outsourcing will affect in quite tangible ways the business environment of the corporation. The chapter "Next Generation IT Outsourcing: Profits or Perils?" by Mark Melford, Miles Wright and Suvojoy Sengupta of Booz Allen Hamilton is thus a welcome and timely contribution to this volume.

Outsourcing offers strategic and economic benefits that are too compelling to ignore. When it works, the authors point out, outsourcing decreases costs, increases flexibility, enhances expertise, boosts discipline, and provides the freedom to focus on core business capabilities. However, the worlds of business and government are replete with examples of outsourcing decisions gone wrong. And, as suppliers become more tightly integrated into the fabric of a company's basic business operations, the risks attending the failure of these relationships escalate.

IT outsourcing is at the sharp end of this movement. Many companies' IT activities are so close to "core"—that is, so closely intertwined with basic business processes—that an IT outsourcing decision becomes a strategic choice. IT outsourcing is now a developed market where global suppliers are offering differentiated and more flexible service propositions, resulting in increased complexity of the procurement process and relationship models.

Outsourcing in general and IT outsourcing in particular are thus increasingly not just make-or-buy decisions; they are make-or-break decisions. This chapter is aimed at executives facing an imminent or potential IT outsourcing decision. It addresses the following key questions:

- What is the track record of outsourcing so far and what lessons can be learned from it?

- What are the common pitfalls and how can they be avoided?

- How to do it right: key success factors in making your IT outsourcing decision.

Working with clients across a range of industries, the authors identify five common myths about good outsourcing practice, perceptions which can lead to failure of the outsourcing deal. Tapping into a rich trove of experience at Booz Allen Hamilton, they also observe the common characteristics of successful outsourcing decisions, and crystalize six steps which successful companies tend to follow as they identify what should and should not be outsourced and manage new supply relationships. This chapter is, therefore, a highly practical, hands-on primer for managers and decision makers for whom outsourcing, and the choices that stem from its increasingly ubiquitous character in the global economy, are likely to matter a great deal in coming years.

Taiwan and the Emergence of an ICT Giant

For several decades now Taiwan has been one of the fastest-growing economies in the world—a senior member of the "Asian tigers" club, with an impressive track record of sustained increases in per capita income, reflecting the gradual transformation of the economy into a powerhouse of high-technology manufacturing. Indeed, during the past 20 years Taiwan has emerged as a leading producer of information and communication technology products. From motherboards to LCD monitors, from personal computers to wireless local area networks, Taiwanese companies produce a very substantial share of the devices that now find their way into workplaces and homes all over the world. Indeed, technology-intensive industries, mostly in ICT, now make up over half of Taiwan's economy, compared with less than a quarter in the late 1980s. Taiwanese manufacturers collectively produce well over half the global supply of the devices that make up the core of the worldwide ICT industry and infrastructure.

While this would not surprise were it to apply to the United States or some other suitably large economy, it is an extraordinary fact, worth examining, that a small island of some 22 million inhabitants should have attained the status of main ICT supplier to the global marketplace. This edition of the *Global Information Technology Report* presents a case study on the development of the ICT industry in Taiwan. What have been the factors that have propelled it? What has been the role of government policy in nurturing the development of ICT, particularly in the areas of higher education and manpower development, innovation and R&D, and the setting of non-distortionary incentives mechanisms? How has the industry managed to maintain its competitive edge in the

face of growing competition from other producers? What are its long-term prospects against the background of swift changes in the industry brought about by technological change and the natural desire by other countries to want to gain a foothold in the industry? Are there lessons from Taiwan's experience with ICT that have relevance for other countries in search of a niche in the global economy, either in some specialty area of ICT itself or in some other area of particular comparative advantage? The study by F. C. Lin analyses these issues in a way that is at once concise and informative, providing fascinating insights into what is surely one of the more compelling development stories of the past half-century.

The Networked Readiness Index Rankings

The Networked Readiness Index Rankings 2004

NRI RANK	COUNTRY	SCORE
1	Singapore	1.73
2	Iceland	1.66
3	Finland	1.62
4	Denmark	1.60
5	United States	1.58
6	Sweden	1.53
7	Hong Kong	1.39
8	Japan	1.35
9	Switzerland	1.30
10	Canada	1.27
11	Australia	1.23
12	United Kingdom	1.21
13	Norway	1.19
14	Germany	1.16
15	Taiwan	1.12
16	Netherlands	1.08
17	Luxembourg	1.04
18	Israel	1.02
19	Austria	1.01
20	France	0.96
21	New Zealand	0.95
22	Ireland	0.89
23	United Arab Emirates	0.84
24	Korea	0.81
25	Estonia	0.80
26	Belgium	0.74
27	Malaysia	0.69
28	Malta	0.50
29	Spain	0.43
30	Portugal	0.39
31	Tunisia	0.39
32	Slovenia	0.37
33	Bahrain	0.37
34	South Africa	0.33
35	Chile	0.29
36	Thailand	0.27
37	Cyprus	0.25
38	Hungary	0.24
39	India	0.23
40	Czech Republic	0.21
41	China	0.17
42	Greece	0.17
43	Lithuania	0.13
44	Jordan	0.10
45	Italy	0.10
46	Brazil	0.08
47	Mauritius	0.08
48	Slovak Republic	0.03
49	Jamaica	−0.03
50	Botswana	−0.10
51	Indonesia	−0.13
52	Turkey	−0.14

NRI RANK	COUNTRY	SCORE
53	Romania	−0.15
54	Morocco	−0.17
55	Namibia	−0.21
56	Latvia	−0.23
57	Egypt	−0.24
58	Croatia	−0.25
59	Trinidad and Tobago	−0.28
60	Mexico	−0.28
61	Costa Rica	−0.29
62	Russian Federation	−0.36
63	Pakistan	−0.38
64	Uruguay	−0.39
65	Ghana	−0.41
66	Colombia	−0.42
67	Philippines	−0.43
68	Vietnam	−0.46
69	Panama	−0.47
70	El Salvador	−0.49
71	Sri Lanka	−0.49
72	Poland	−0.50
73	Bulgaria	−0.51
74	Gambia	−0.52
75	Kenya	−0.62
76	Argentina	−0.62
77	Uganda	−0.63
78	Dominican Republic	−0.65
79	Serbia and Montenegro	−0.65
80	Algeria	−0.66
81	Zambia	−0.68
82	Ukraine	−0.68
83	Tanzania	−0.71
84	Venezuela	−0.72
85	Macedonia	−0.73
86	Nigeria	−0.73
87	Madagascar	−0.77
88	Guatemala	−0.78
89	Bosnia and Herzegovina	−0.86
90	Peru	−0.91
91	Georgia	−0.94
92	Mali	−0.96
93	Malawi	−0.98
94	Zimbabwe	−1.02
95	Ecuador	−1.08
96	Mozambique	−1.11
97	Honduras	−1.19
98	Paraguay	−1.20
99	Bolivia	−1.25
100	Bangladesh	−1.30
101	Angola	−1.36
102	Ethiopia	−1.52
103	Nicaragua	−1.61
104	Chad	−1.69

Part 1
Chapters

An Analysis of the Diffusion and Usage of Information and Communication Technologies of Nations

Soumitra Dutta, INSEAD

Amit Jain, INSEAD

Overview

The Networked Readiness Index (NRI) is defined as a nation's or a community's degree of preparation to participate in and benefit from information and communication technology (ICT) developments. This chapter presents the results of the computation of the NRI across 104 countries around the globe. The immediate objective is to facilitate the task of policy and decision makers in their endeavor to understand the complexity and the diversity of factors underlying a nation's ICT development, and thereby to assist them in working toward their development objectives. In this way, one hopes that this report would contribute to the development of public policy and to the diffusion and usage of ICT.

The present chapter contains the fourth successive annual NRI report.[1] It represents a continuation of efforts at INSEAD and the World Economic Forum to better understand the impact of ICT on the competitiveness of nations. Building upon the research at INSEAD in the domain and on the three previous reports, the current research provides a continuity of data and analysis for the evaluation of prior decisions and actions, and for the enhancement of planning for the future.

ICT remains critical for the development of nations. Not only does ICT form the basis of rapid and effective communication at all levels—individual, business and government—but it also serves as the infrastructure for commercial transactions. ICT is also playing an increasing role as an enabling mechanism for the delivery of efficient and effective government services. Governments and regulators continue to see progress in ICT as fundamental to national progress. Policies are being put in place to increase ICT penetration and to reduce the digital divide. Tariffs continue to fall and levels of competition to increase to provide incentives for businesses to invest effectively in ICT.

This chapter presents the Networked Readiness Framework that has been used to assess the relative degree of networked readiness and to compute the NRI of 104 countries. The discussion is divided into five main sections. First, there is a brief recapitulation of the Networked Readiness Framework. Second, the results of the research and analysis are presented in the form of a relative ranking of nations based on their degrees of networked readiness. Third, we take a closer look at the three component indexes (and their constituent subindexes) composing the NRI, and how various countries have fared on each of these dimensions. In the fourth section, we examine the relation between the NRI and the Global Competitiveness Index of the World Economic Forum, and the relation between NRI and GDP per capita. Finally, in the fifth section, we look at the trends in regional development with respect to ICT, and it is demonstrated how the NRI could be used to focus on factors that make a given country under- or over-perform. In the conclusion, some of the key challenges faced while we conducted the study are presented.

The Networked Readiness Framework 2004–2005

The Networked Readiness Index is defined as "the degree of preparation of a nation or community to participate in and benefit from ICT developments". The Networked Readiness Framework used to compute the NRI rankings this year (2004–2005) remains identical to that used to compute the NRI rankings for 2002–2003[2] and 2003–2004. The Networked Readiness Framework and its components not only provide a model for evaluating a country's relative development and use of ICT, but also allow for a better understanding of a nation's strengths and weaknesses with respect to ICT.

Figure 1 depicts the structure of the Networked Readiness Framework used in this research. The Networked Readiness Framework is based upon the following premises:

- there are three important stakeholders to consider in the development and use of ICT: individuals, businesses, and governments;

- there is a general macroeconomic and regulatory environment for ICT in which the stakeholders play out their respective roles; and

- the degree of usage of ICT by (and hence the impact of ICT on) the three stakeholders is linked to their degrees of readiness (or capability) to use and benefit from ICT.

Figure 1. **The Networked Readiness Index Framework 2004–2005**

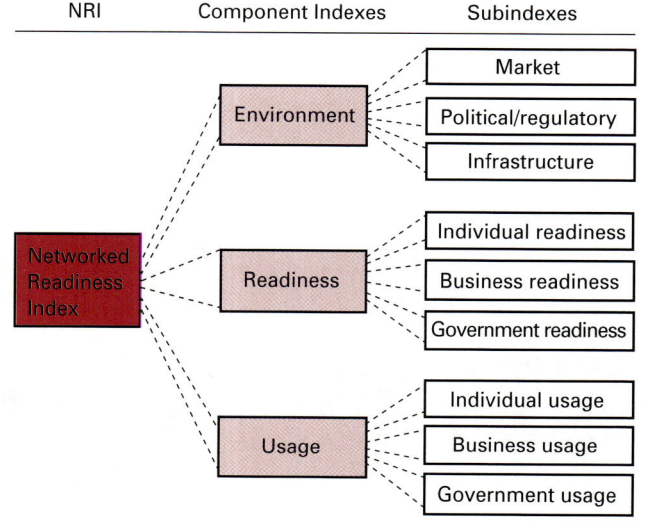

Source: INSEAD

NRI Results for 2004–2005

The overall results for the Networked Readiness Index 2004–2005 are presented in Table 1.[3] Singapore comes out with the top rank, followed by Iceland. Singapore has consistently been ranked in the top three places (third place in the 2002–2003 study and second place in the 2003–2004). This consistent performance of Singapore is a consequence of its government's proactive efforts to promote ICT penetration and usage. Finland, Denmark and the United States occupy the third, fourth and fifth places, respectively. Sweden gets the sixth position, followed by Hong Kong, Japan, and Switzerland. Canada comes in tenth. The following results are worthy of note:

- In the top five places, three positions go to Nordic countries: Iceland (2), Finland (3) and Denmark (4). In addition, Sweden (6) is in the top ten.

- Hong Kong (7) and Japan (8) enter the top ten rankings for the first time, whereas Taiwan is ranked 15th.

- Israel continues to lead the Middle-East group of countries at 18th place, and the United Arab Emirates follows closely at 23rd place. Bahrain also performs well with a rank of 33.

- Estonia (25) leads the new member states of the European Union and other east/central European countries. Malta (28) and Slovenia (32) closely follow.

Furthermore:

- The top-ranked South American countries are Chile (35), Brazil (46) and Mexico (60).

- In Asia, Malaysia is ranked 27th and Thailand 36th. India (39) and China (41) improve their ranking positions by 6 and 10 places respectively.

- Russia is ranked 62nd overall, compared with 63rd in the previous years ranking results.

- South Africa, ranked 34th, leads in Africa, followed by Botswana (50) and Namibia (55).

Evolution of the NRI over time

The networked readiness of a nation is a dynamic measure that evolves over time as a result of policy measures taken by government and business leaders and of changes in the global environment. Looking at the changes in NRI rankings over time (see Table 2), one observes that the United States and Finland have consistently been in the top five positions, whereas Singapore has rapidly progressed from eighth place in 2001–2002 to its current leading position.

Another constant is the performance of the Nordic countries, with Finland, Sweden, Denmark and Iceland present in the top ten places in each of the four years.

Table 1. Networked Readiness Index Rankings 2004–2005

COUNTRY	SCORE	RANK
Singapore	1.73	1
Iceland	1.66	2
Finland	1.62	3
Denmark	1.60	4
United States	1.58	5
Sweden	1.53	6
Hong Kong	1.39	7
Japan	1.35	8
Switzerland	1.30	9
Canada	1.27	10
Australia	1.23	11
United Kingdom	1.21	12
Norway	1.19	13
Germany	1.16	14
Taiwan	1.12	15
Netherlands	1.08	16
Luxembourg	1.04	17
Israel	1.02	18
Austria	1.01	19
France	0.96	20
New Zealand	0.95	21
Ireland	0.89	22
United Arab Emirates	0.84	23
Korea	0.81	24
Estonia	0.80	25
Belgium	0.74	26
Malaysia	0.69	27
Malta	0.50	28
Spain	0.43	29
Portugal	0.39	30
Tunisia	0.39	31
Slovenia	0.37	32
Bahrain	0.37	33
South Africa	0.33	34
Chile	0.29	35

COUNTRY	SCORE	RANK
Thailand	0.27	36
Cyprus	0.25	37
Hungary	0.24	38
India	0.23	39
Czech Republic	0.21	40
China	0.17	41
Greece	0.17	42
Lithuania	0.13	43
Jordan	0.10	44
Italy	0.10	45
Brazil	0.08	46
Mauritius	0.08	47
Slovak Republic	0.03	48
Jamaica	−0.03	49
Botswana	−0.10	50
Indonesia	−0.13	51
Turkey	−0.14	52
Romania	−0.15	53
Morocco	−0.17	54
Namibia	−0.21	55
Latvia	−0.23	56
Egypt	−0.24	57
Croatia	−0.25	58
Trinidad and Tobago	−0.28	59
Mexico	−0.28	60
Costa Rica	−0.29	61
Russian Federation	−0.36	62
Pakistan	−0.38	63
Uruguay	−0.39	64
Ghana	−0.41	65
Colombia	−0.42	66
Philippines	−0.43	67
Vietnam	−0.46	68
Panama	−0.47	69
El Salvador	−0.49	70

COUNTRY	SCORE	RANK
Sri Lanka	−0.49	71
Poland	−0.50	72
Bulgaria	−0.51	73
Gambia	−0.52	74
Kenya	−0.62	75
Argentina	−0.62	76
Uganda	−0.63	77
Dominican Republic	−0.65	78
Serbia and Montenegro	−0.65	79
Algeria	−0.66	80
Zambia	−0.68	81
Ukraine	−0.68	82
Tanzania	−0.71	83
Venezuela	−0.72	84
Macedonia	−0.73	85
Nigeria	−0.73	86
Madagascar	−0.77	87
Guatemala	−0.78	88
Bosnia and Herzegovina	−0.86	89
Peru	−0.91	90
Georgia	−0.94	91
Mali	−0.96	92
Malawi	−0.98	93
Zimbabwe	−1.02	94
Ecuador	−1.08	95
Mozambique	−1.11	96
Honduras	−1.19	97
Paraguay	−1.20	98
Bolivia	−1.25	99
Bangladesh	−1.30	100
Angola	−1.36	101
Ethiopia	−1.52	102
Nicaragua	−1.61	103
Chad	−1.69	104

One must add a note of caution to this analysis because the results of the three research efforts are not directly comparable. The framework used in the 2001–2002 study is different (see Kirkman et al. 2002). Moreover, while the framework and methodology of analysis of the 2002–2003, 2003–2004, and 2004–2005 studies are identical, the underlying data variables used differ to a certain extent. These differences reflect limits on the availability of reliable and up-to-date information for several countries, against the background of an expansion in the coverage of the NRI in recent years, particularly in the developing world.

Table 2. Evolution of Networked Readiness, 2001–2005

COUNTRY (NUMBER OF NATIONS STUDIED)	2001–2002 (75)	2002–2003 (82)	2003–2004 (102)	2004–2005 (104)
Singapore	8	3	2	1
Iceland	2	5	10	2
Finland	3	1	3	3
Denmark	7	8	5	4

COUNTRY (NUMBER OF NATIONS STUDIED)	2001–2002 (75)	2002–2003 (82)	2003–2004 (102)	2004–2005 (104)
United States	1	2	1	5
Sweden	4	4	4	6
Hong Kong	13	18	18	7

COUNTRY (NUMBER OF NATIONS STUDIED)	2001–2002 (75)	2002–2003 (82)	2003–2004 (102)	2004–2005 (104)
Japan	21	20	12	8
Switzerland	16	13	7	9
Canada	12	6	6	10

Interpreting the results

The NRI captures key factors relating to the environment, the readiness and the usage of the three stakeholders in the Networked Readiness Framework (individuals, businesses and governments), and can be used to understand the performance of a nation or a region in relation to ICT readiness and usage. The component index and subindex rankings serve to identify key areas where a nation is under- or over-performing. One would, for instance, be able to identify relative imbalances in development across the three component indexes of Environment, Readiness and Usage, or even go one level deeper.

The index computed in the last three reports was on a scale of 1–7, increasing with the level of a nation's ICT development. Due to a change in the methodology used in the computation of the index, this seven-point scale is no longer used. Instead, the subindex scores are standardized with a mean of 0. This results in the scores of the countries being distributed above and below the mean score of zero.[4] A positive score merely reflects the fact that the country concerned performs better than the mean performance across the 104 countries studied. Likewise, a negative score implies that the ranking performance is below the mean performance of the 104 countries.

One would like to emphasize that, while rankings are useful as relative indicators of a nation's ICT excellence, there are several limitations to the analytic process. Caution should be exercised when one compares countries that are ranked near one another; for instance, there may be very little difference between their index scores—Portugal (ranked 30) and Tunisia (ranked 31) even have the same overall NRI score of 0.39 (but Portugal's is higher at the third decimal place). Additionally, small differences in the index may be outside the limits of statistical significance because some missing observations had to be estimated using analytic techniques such as regression and clustering.

One should also keep in mind that, while the number of countries included in the current study has increased from 82 in the 2002–2003 report to 104 in the current report, a number of nations could not be included in the research due to the limited availability of reliable data. Only those countries that carried out the Executive Opinion Survey of the World Economic Forum were selected. Ranking an even larger set of nations remains a challenge for the future. An overall global ranking needs to account for these missing countries, and any inferences drawn from the current analysis of 104 nations should be made with this limitation taken into consideration.

Finally, the complexity of ICT can get obscured behind the numerical score of the NRI. Countries like India or China, for instance, show enormous internal geographic and demographic divides in ICT readiness and usage.

The Building Blocks of Networked Readiness Index

The NRI provides a relative benchmark of the overall success of a country in participating in and benefiting from ICT. While this is useful, one may need to gain further insights into areas of over- and under-performance of a nation, and to understand the key drivers determining the results. This can be done by looking at the component indexes: Environment, Readiness and Usage. Tables 3, 4, and 5 present the overall results of each component index. Further insight may be obtained from the subindexes composing each component index. The final level of detail can be obtained by observing the 51 variables comprising the subindexes, which are presented in the data tables section of this report.

Environment

The Environment component index is designed to measure the degree of conduciveness of the environment that a country provides for the development and use of ICT. As can be seen from Table 3, the top countries with regards to the Environment are Iceland, Finland and Denmark. The performance of the Nordic countries with respect to the Environment for ICT is consistent with their overall performance on the NRI.

Tables 6, 7 and 8 present the detailed ranking and scores for each of the three subindexes comprising the Environment component: the Market Environment, the Political/ Regulatory Environment and the Infrastructure Environment.

Market Environment

The Market Environment subindex entails the assessment of the presence of the appropriate human resources and ancillary businesses to support a knowledge-based society. The forces that play an important role in determining the market environment for ICT are varied and include fundamental macroeconomic variables like GDP and trade openness, commercial measures like availability of funding and skilled labor, and the level of development of the corporate environment. The leader for this subindex is Finland, followed by the United States and Singapore. Hong Kong and the United Kingdom, in the fourth and fifth positions, are also notable for their performance on the Market Environment subindex.

Political/Regulatory Environment

The priorities of a nation are reflected in its policies and laws, which in turn influence its rate of growth and direction of development. The Political/Regulatory Environment component of the NRI measures the impact

Table 3. **Environment Component Index**

COUNTRY	SCORE	RANK	COUNTRY	SCORE	RANK	COUNTRY	SCORE	RANK
Iceland	2.75	1	Korea	0.17	36	Sri Lanka	−0.51	71
Finland	2.07	2	Hungary	0.13	37	Colombia	−0.53	72
Denmark	1.93	3	India	0.11	38	Malawi	−0.57	73
United States	1.86	4	Greece	0.11	39	Mexico	−0.58	74
Sweden	1.70	5	Thailand	0.06	40	Nigeria	−0.62	75
Singapore	1.67	6	Czech Republic	0.03	41	Bulgaria	−0.63	76
United Kingdom	1.66	7	Indonesia	0.03	42	El Salvador	−0.63	77
Norway	1.63	8	Jordan	0.02	43	Dominican Republic	−0.63	78
Australia	1.63	9	Namibia	0.01	44	Vietnam	−0.68	79
Canada	1.57	10	Lithuania	−0.06	45	Mali	−0.69	80
Switzerland	1.57	11	China	−0.07	46	Philippines	−0.73	81
New Zealand	1.45	12	Brazil	−0.12	47	Madagascar	−0.78	82
Luxembourg	1.44	13	Botswana	−0.12	48	Serbia and Montenegro	−0.79	83
Netherlands	1.42	14	Slovak Republic	−0.12	49	Ukraine	−0.80	84
Hong Kong	1.36	15	Mauritius	−0.15	50	Pakistan	−0.87	85
Japan	1.11	16	Italy	−0.16	51	Macedonia	−0.88	86
Ireland	1.10	17	Costa Rica	−0.21	52	Algeria	−0.88	87
Germany	1.10	18	Morocco	−0.26	53	Argentina	−0.90	88
Austria	1.04	19	Trinidad and Tobago	−0.28	54	Zimbabwe	−0.90	89
Israel	0.90	20	Latvia	−0.30	55	Guatemala	−0.92	90
Estonia	0.80	21	Ghana	−0.31	56	Georgia	−0.96	91
France	0.79	22	Turkey	−0.35	57	Bosnia and Herzegovina	−0.97	92
Taiwan	0.76	23	Jamaica	−0.36	58	Peru	−0.99	93
Belgium	0.63	24	Uganda	−0.38	59	Honduras	−1.04	94
Malaysia	0.62	25	Tanzania	−0.40	60	Mozambique	−1.05	95
United Arab Emirates	0.62	26	Panama	−0.44	61	Venezuela	−1.09	96
Cyprus	0.41	27	Uruguay	−0.44	62	Bangladesh	−1.09	97
Malta	0.37	28	Gambia	−0.44	63	Ecuador	−1.10	98
Portugal	0.36	29	Croatia	−0.46	64	Paraguay	−1.14	99
Spain	0.35	30	Romania	−0.47	65	Ethiopia	−1.15	100
South Africa	0.30	31	Kenya	−0.48	66	Nicaragua	−1.19	101
Slovenia	0.26	32	Egypt	−0.48	67	Chad	−1.27	102
Tunisia	0.22	33	Zambia	−0.50	68	Bolivia	−1.28	103
Chile	0.21	34	Russian Federation	−0.50	69	Angola	−1.31	104
Bahrain	0.19	35	Poland	−0.51	70			

of a nation's polity, laws, regulations and their implementation on the development and use of ICT. The leaders from the Political/Regulatory perspective are Denmark, the United Kingdom and Singapore. Finland is highly ranked at fourth place, whereas Norway is ranked fifth.

Infrastructure Environment

Infrastructure is defined as the level of availability and quality of the key access infrastructure for ICT within a country. A quality ICT access infrastructure facilitates the adoption, usage and impact of these technologies, which in turn promote investment in ICT infrastructure. Infrastructure thus plays a critical role in influencing the networked readiness of a nation. The top ranks along this component go to Iceland, Denmark and Finland. One

notes that China is ranked 62nd on the Infrastructure subindex, whereas India comes in at the 86th place, a very low rank compared with its overall 38th position in Environment component index—an indication of the challenges related to the development of a universally available quality information and communications technology infrastructure for all the socio-economic segments comprising its population.

Readiness

The Readiness of a nation measures the capability of the principal agents of an economy (citizens, businesses and governments) to leverage the potential of ICT. This capability is imparted to a nation by a combination of factors like the presence of relevant skills for using ICT within individuals, access and affordability of ICT for

Table 4. **Readiness Component Index**

COUNTRY	SCORE	RANK	COUNTRY	SCORE	RANK	COUNTRY	SCORE	RANK
Singapore	1.49	1	Lithuania	0.34	36	Serbia and Montenegro	−0.38	71
Finland	1.43	2	Mauritius	0.31	37	Sri Lanka	−0.39	72
Japan	1.39	3	Botswana	0.30	38	Poland	−0.41	73
Taiwan	1.27	4	China	0.29	39	El Salvador	−0.43	74
The United States	1.26	5	Chile	0.28	40	Macedonia	−0.44	75
Denmark	1.17	6	Czech Republic	0.26	41	Philippines	−0.47	76
Switzerland	1.16	7	Portugal	0.23	42	Ukraine	−0.48	77
Germany	1.06	8	Brazil	0.19	43	Venezuela	−0.55	78
Sweden	1.06	9	Jordan	0.17	44	Kenya	−0.56	79
France	1.04	10	Hungary	0.14	45	Panama	−0.56	80
Australia	1.00	11	Indonesia	0.13	46	Argentina	−0.57	81
Hong Kong	0.97	12	Romania	0.12	47	Zimbabwe	−0.57	82
United Kingdom	0.94	13	Jamaica	0.12	48	Zambia	−0.60	83
Luxembourg	0.94	14	Greece	0.10	49	Madagascar	−0.61	84
Iceland	0.93	15	Slovak Republic	0.07	50	Dominican Republic	−0.65	85
Netherlands	0.91	16	Costa Rica	0.02	51	Bosnia and Herzegovina	−0.65	86
Malaysia	0.88	17	Trinidad and Tobago	0.00	52	Nigeria	−0.70	87
United Arab Emirates	0.88	18	Italy	−0.03	53	Guatemala	−0.75	88
Israel	0.87	19	Morocco	−0.04	54	Georgia	−0.93	89
Norway	0.87	20	Cyprus	−0.06	55	Tanzania	−0.98	90
Austria	0.84	21	Namibia	−0.07	56	Uganda	−0.99	91
Canada	0.83	22	Mexico	−0.09	57	Peru	−1.03	92
Tunisia	0.82	23	Pakistan	−0.09	58	Ecuador	−1.03	93
Ireland	0.80	24	Russian Federation	−0.13	59	Honduras	−1.12	94
Korea	0.74	25	Turkey	−0.16	60	Mali	−1.13	95
Belgium	0.73	26	Latvia	−0.17	61	Malawi	−1.16	96
Estonia	0.71	27	Ghana	−0.17	62	Bolivia	−1.20	97
New Zealand	0.59	28	Egypt	−0.18	63	Paraguay	−1.21	98
Spain	0.54	29	Croatia	−0.21	64	Angola	−1.25	99
South Africa	0.52	30	Vietnam	−0.25	65	Mozambique	−1.35	100
Bahrain	0.49	31	Colombia	−0.26	66	Bangladesh	−1.62	101
Thailand	0.48	32	Uruguay	−0.26	67	Ethiopia	−1.84	102
Slovenia	0.43	33	Algeria	−0.27	68	Chad	−1.95	103
India	0.40	34	Bulgaria	−0.28	69	Nicaragua	−2.24	104
Malta	0.36	35	Gambia	−0.36	70			

corporations, and government use of ICT for its own services and processes. As shown in Table 4, Singapore ranks highest on overall Readiness and shows a consistent performance across all three readiness subindexes. Finland is in second place and is supported by a very strong performance in Individual and Business Readiness. Third-ranked Japan benefits from high scores in Readiness in each of the three subindexes.

Detailed results for each of the subindexes used for measuring Readiness can be found in Tables 9, 10 and 11.

Individual Readiness

Individual Readiness measures the readiness of a nation's citizens to utilize and leverage ICT. It is made up of literacy rates, mode and locus of access to the Internet, and the degree of connectivity of individuals. The top three positions on Individual Readiness go to Finland, Singapore and Iceland. Switzerland is ranked fourth and Denmark ranked fifth.

Business Readiness

Business Readiness measures the readiness of businesses to participate in and benefit from ICT. The aim is to focus not just on the largest corporations, but also on small and medium-sized businesses and their willingness to exploit ICT and invest in the ICT skills of their employees. The United States regains its top position in Business Readiness, and is followed by Switzerland. Germany, Finland and Sweden are ranked at the third, fourth and fifth places respectively.

Government Readiness

Government Readiness measures the readiness of a government to employ ICT. It is reflected in the policy-making machinery and internal processes of the government and in the availability of government services online. If the polity of a nation decides to make ICT a priority, this becomes visible in the short- and long-term policy measures and laws that help encourage ICT deployment and use. It is also reflected in the government's own use of ICT and the extent to which it equips its people to do the same. Singapore leads on Government Readiness, followed by Japan and Taiwan. United Arab Emirates and Finland follow in fourth and fifth places respectively. Of note also are Tunisia in ninth place, and Korea, ranked eleventh. The entry of developing countries such as Tunisia and Korea into the upper ranks on the Government Readiness dimension is a reflection of the policies and actions taken by governments to diffuse ICT in the country and in particular in the government.

Usage

The Usage component aims to measure the degree of usage of ICT by the principal stakeholders of the NRI framework—Individuals, Businesses and Governments. In the absence of reliable data about the specific impacts of ICT on the key stakeholders, the Usage component provides an indication of the changes in behaviors, lifestyles, and other economic and non-economic benefits brought about by the adoption of ICT. Singapore, Hong Kong and Sweden are the top three performers in terms of

Table 5. **Usage Component Index**

COUNTRY	SCORE	RANK	COUNTRY	SCORE	RANK	COUNTRY	SCORE	RANK
Singapore	2.04	1	Chile	0.38	36	Namibia	−0.57	71
Hong Kong	1.85	2	Czech Republic	0.34	37	Poland	−0.57	72
Sweden	1.82	3	China	0.30	38	Sri Lanka	−0.58	73
Denmark	1.71	4	Greece	0.29	39	Bulgaria	−0.61	74
United States	1.61	5	Thailand	0.28	40	Dominican Republic	−0.65	75
Japan	1.55	6	South Africa	0.17	41	Guatemala	−0.67	76
Korea	1.53	7	India	0.17	42	Costa Rica	−0.68	77
Canada	1.42	8	Brazil	0.16	43	Peru	−0.72	78
Finland	1.36	9	Jamaica	0.14	44	Ghana	−0.73	79
Germany	1.32	10	Tunisia	0.14	45	Tanzania	−0.75	80
Taiwan	1.31	11	Slovak Republic	0.14	46	Ukraine	−0.76	81
Iceland	1.31	12	Jordan	0.13	47	Serbia and Montenegro	−0.77	82
Israel	1.29	13	Lithuania	0.11	48	Gambia	−0.77	83
Switzerland	1.17	14	Turkey	0.07	49	Algeria	−0.81	84
Austria	1.16	15	Mauritius	0.06	50	Kenya	−0.82	85
Australia	1.08	16	Egypt	−0.07	51	Macedonia	−0.86	86
France	1.06	17	Croatia	−0.09	52	Nigeria	−0.86	87
Norway	1.06	18	Romania	−0.09	53	Madagascar	−0.93	88
United Kingdom	1.03	19	Philippines	−0.09	54	Zambia	−0.93	89
United Arab Emirates	1.01	20	Mexico	−0.17	55	Mozambique	−0.93	90
Netherlands	0.92	21	Pakistan	−0.18	56	Georgia	−0.93	91
Estonia	0.90	22	Morocco	−0.20	57	Bosnia and Herzegovina	−0.96	92
Belgium	0.85	23	Latvia	−0.21	58	Mali	−1.06	93
New Zealand	0.82	24	Argentina	−0.39	59	Ecuador	−1.12	94
Ireland	0.77	25	El Salvador	−0.40	60	Bangladesh	−1.20	95
Malta	0.77	26	Panama	−0.42	61	Malawi	−1.20	96
Luxembourg	0.75	27	Russian Federation	−0.44	62	Paraguay	−1.23	97
Portugal	0.59	28	Vietnam	−0.44	63	Bolivia	−1.28	98
Malaysia	0.55	29	Uruguay	−0.46	64	Nicaragua	−1.40	99
Italy	0.47	30	Colombia	−0.49	65	Honduras	−1.41	100
Hungary	0.46	31	Botswana	−0.49	66	Angola	−1.52	101
Bahrain	0.42	32	Uganda	−0.51	67	Zimbabwe	−1.57	102
Slovenia	0.42	33	Venezuela	−0.54	68	Ethiopia	−1.58	103
Spain	0.40	34	Indonesia	−0.55	69	Chad	−1.83	104
Cyprus	0.38	35	Trinidad and Tobago	−0.55	70			

Table 6. **Market Environment**

COUNTRY	SCORE	RANK	COUNTRY	SCORE	RANK	COUNTRY	SCORE	RANK
Finland	2.19	1	South Africa	0.26	36	Colombia	−0.56	71
United States	2.07	2	China	0.26	37	Poland	−0.62	72
Singapore	1.95	3	Lithuania	0.23	38	Mexico	−0.64	73
Hong Kong	1.92	4	Korea	0.21	39	Vietnam	−0.64	74
United Kingdom	1.78	5	Slovenia	0.20	40	Croatia	−0.67	75
Sweden	1.70	6	Hungary	0.19	41	Madagascar	−0.68	76
Taiwan	1.69	7	Jordan	0.13	42	Malawi	−0.70	77
Iceland	1.56	8	Botswana	0.07	43	Dominican Republic	−0.70	78
Switzerland	1.49	9	Namibia	0.03	44	Serbia and Montenegro	−0.71	79
Norway	1.37	10	Slovak Republic	0.00	45	Mali	−0.73	80
Denmark	1.37	11	Brazil	−0.01	46	Philippines	−0.78	81
Canada	1.29	12	Czech Republic	−0.03	47	Ukraine	−0.85	82
Netherlands	1.26	13	Greece	−0.05	48	Guatemala	−0.86	83
Ireland	1.25	14	Panama	−0.07	49	Pakistan	−0.86	84
United Arab Emirates	1.20	15	Costa Rica	−0.09	50	Uruguay	−0.91	85
Malaysia	1.19	16	Uganda	−0.12	51	Bulgaria	−0.92	86
Japan	1.17	17	Morocco	−0.15	52	Algeria	−0.97	87
Israel	1.16	18	Tanzania	−0.17	53	Zimbabwe	−0.97	88
Luxembourg	1.14	19	Trinidad and Tobago	−0.21	54	Georgia	−0.98	89
Australia	1.12	20	Kenya	−0.21	55	Argentina	−1.06	90
Germany	1.02	21	Mauritius	−0.27	56	Peru	−1.07	91
Austria	0.96	22	Latvia	−0.27	57	Macedonia	−1.15	92
New Zealand	0.91	23	Malta	−0.28	58	Bangladesh	−1.15	93
Estonia	0.85	24	Gambia	−0.29	59	Paraguay	−1.18	94
Belgium	0.70	25	Italy	−0.32	60	Venezuela	−1.20	95
Indonesia	0.62	26	Russian Federation	−0.32	61	Honduras	−1.21	96
India	0.60	27	Sri Lanka	−0.32	62	Bosnia and Herzegovina	−1.22	97
France	0.58	28	Turkey	−0.39	63	Ecuador	−1.23	98
Tunisia	0.57	29	Jamaica	−0.44	64	Nicaragua	−1.23	99
Bahrain	0.55	30	Egypt	−0.45	65	Ethiopia	−1.24	100
Thailand	0.52	31	Nigeria	−0.47	66	Chad	−1.33	101
Chile	0.47	32	Zambia	−0.50	67	Mozambique	−1.36	102
Portugal	0.31	33	Romania	−0.53	68	Bolivia	−1.51	103
Cyprus	0.27	34	El Salvador	−0.54	69	Angola	−1.70	104
Spain	0.26	35	Ghana	−0.54	70			

overall Usage, as shown in Table 5. One can observe variances in country performance across the three subindexes reflecting uneven impact across the three principal stakeholders. For example, Singapore ranks high for Government Usage (1) but relatively low for Individual Usage (12) and Business Usage (9). Another notable example is Estonia, with higher Government Usage (9) than Individual (34) or Business (27) Usage.

Tables 12, 13 and 14 give the detailed results and scores for each of the three subindexes used for measuring Usage.

Individual Usage

Individual Usage gives an indication of the level of adoption and usage of ICT technologies by a nation's citizens. This is done by assessing the deployment of connectivity-enhancing technologies like telephones and Internet connections, levels of Internet usage and money spent online. The Individual Usage rankings differ significantly from those of Individual Readiness. The top performers here are Sweden, Korea, Iceland, Denmark and Hong Kong.

Business Usage

Business Usage measures the level of deployment and use of ICT across businesses in a nation. Business Usage is measured by factors such as the level of business-to-business and business-to-consumer e-commerce, the use of ICT for activities like marketing, levels of online transactions, and the availability and usage of new telephone lines and mobile phones by businesses. The top five performers are Japan, Germany, Sweden, Switzerland and Finland.

Government Usage

Government Usage is the level of use of ICT technologies by the government of a given country. The government, besides making ICT a priority, can also benefit from the use of ICT itself. This usage can help the government streamline services to its citizens and improve its overall functioning. Factors used to measure this include the level of government success in the promotion of ICT and the availability and usage of online government services. The top-ranking countries on this measure are Singapore, Hong Kong, the United States, Denmark and Taiwan. Of note is the United Arab Emirates at seventh place and Estonia at ninth place; these country's governments are taking active steps to promote ICT usage in their own functions.

Exploring the Networked Readiness of Nations

Networked Readiness and the Global Competitiveness Index

The impact of ICT on development has been a subject of continuing interest by economists, who have attempted to demonstrate a link between investment in ICT and economic growth. Given this interest, it would be useful to investigate the relationship between the level of networked readiness of a nation and the Global Competitiveness Index (GCI),[5] one of the World Economic Forum's competitiveness indexes.

Table 7. **Political/Regulatory Environment**

COUNTRY	SCORE	RANK	COUNTRY	SCORE	RANK	COUNTRY	SCORE	RANK
Denmark	1.96	1	Ghana	0.39	36	Nigeria	−0.58	71
United Kingdom	1.86	2	Bahrain	0.32	37	Dominican Republic	−0.63	72
Singapore	1.76	3	Thailand	0.28	38	Mexico	−0.64	73
Finland	1.76	4	Slovenia	0.27	39	Vietnam	−0.68	74
Norway	1.73	5	Cyprus	0.25	40	Philippines	−0.69	75
Australia	1.70	6	Korea	0.23	41	Croatia	−0.73	76
Iceland	1.70	7	Indonesia	0.21	42	El Salvador	−0.73	77
Sweden	1.68	8	Greece	0.21	43	Panama	−0.74	78
New Zealand	1.54	9	Botswana	0.20	44	Poland	−0.75	79
United States	1.51	10	Hungary	0.13	45	Russian Federation	−0.81	80
Germany	1.45	11	Mauritius	0.12	46	Bulgaria	−0.82	81
Netherlands	1.40	12	Morocco	0.11	47	Madagascar	−0.85	82
Luxembourg	1.40	13	Czech Republic	0.08	48	Algeria	−0.97	83
Switzerland	1.36	14	China	0.06	49	Mozambique	−0.98	84
Canada	1.30	15	Brazil	−0.03	50	Pakistan	−0.99	85
Estonia	1.24	16	Jamaica	−0.11	51	Ukraine	−1.10	86
France	1.23	17	Slovak Republic	−0.15	52	Zimbabwe	−1.13	87
Austria	1.19	18	Lithuania	−0.19	53	Macedonia	−1.14	88
South Africa	1.17	19	Uganda	−0.22	54	Honduras	−1.19	89
Malaysia	1.12	20	Malawi	−0.23	55	Peru	−1.21	90
Hong Kong	1.12	21	Tanzania	−0.23	56	Guatemala	−1.22	91
Ireland	1.08	22	Uruguay	−0.26	57	Serbia and Montenegro	−1.26	92
Japan	1.07	23	Costa Rica	−0.28	58	Bosnia and Herzegovina	−1.28	93
Belgium	0.85	24	Gambia	−0.28	59	Bangladesh	−1.32	94
Israel	0.82	25	Trinidad and Tobago	−0.29	60	Georgia	−1.32	95
United Arab Emirates	0.74	26	Italy	−0.30	61	Argentina	−1.35	96
Taiwan	0.73	27	Turkey	−0.35	62	Ethiopia	−1.42	97
Tunisia	0.71	28	Zambia	−0.38	63	Angola	−1.43	98
Malta	0.69	29	Egypt	−0.38	64	Ecuador	−1.44	99
Namibia	0.66	30	Romania	−0.40	65	Venezuela	−1.46	100
Portugal	0.65	31	Kenya	−0.44	66	Paraguay	−1.54	101
Jordan	0.51	32	Sri Lanka	−0.48	67	Nicaragua	−1.61	102
India	0.48	33	Latvia	−0.52	68	Bolivia	−1.64	103
Spain	0.45	34	Colombia	−0.54	69	Chad	−1.68	104
Chile	0.44	35	Mali	−0.55	70			

Table 8. Infrastructure Environment

COUNTRY	SCORE	RANK	COUNTRY	SCORE	RANK	COUNTRY	SCORE	RANK
Iceland	4.98	1	Latvia	−0.12	36	Thailand	−0.62	71
Denmark	2.46	2	Taiwan	−0.13	37	Ecuador	−0.62	72
Finland	2.25	3	Uruguay	−0.15	38	El Salvador	−0.62	73
Canada	2.13	4	Bulgaria	−0.16	39	Botswana	−0.62	74
Australia	2.06	5	Poland	−0.16	40	Egypt	−0.63	75
United States	2.01	6	Slovak Republic	−0.20	41	Guatemala	−0.67	76
New Zealand	1.89	7	Lithuania	−0.21	42	Peru	−0.67	77
Switzerland	1.87	8	Costa Rica	−0.26	43	Namibia	−0.68	78
Norway	1.79	9	Turkey	−0.29	44	Bolivia	−0.69	79
Luxembourg	1.78	10	Mauritius	−0.29	45	Algeria	−0.71	80
Sweden	1.72	11	Bahrain	−0.30	46	Paraguay	−0.72	81
Netherlands	1.60	12	Chile	−0.30	47	Honduras	−0.72	82
United Kingdom	1.33	13	Brazil	−0.30	48	Sri Lanka	−0.73	83
Singapore	1.29	14	Argentina	−0.30	49	Vietnam	−0.73	84
Japan	1.09	15	Trinidad and Tobago	−0.32	50	Philippines	−0.73	85
Hong Kong	1.05	16	Macedonia	−0.35	51	India	−0.74	86
Ireland	0.98	17	Russian Federation	−0.38	52	Nicaragua	−0.74	87
Austria	0.96	18	Bosnia and Herzegovina	−0.41	53	Morocco	−0.74	88
Germany	0.83	19	Serbia and Montenegro	−0.41	54	Indonesia	−0.75	89
Cyprus	0.72	20	Ukraine	−0.44	55	Gambia	−0.76	90
Malta	0.71	21	Malaysia	−0.44	56	Pakistan	−0.77	91
Israel	0.71	22	Mexico	−0.47	57	Ghana	−0.79	92
France	0.55	23	Romania	−0.47	58	Kenya	−0.79	93
Spain	0.33	24	Colombia	−0.49	59	Malawi	−0.80	94
Belgium	0.33	25	Panama	−0.49	60	Angola	−0.80	95
Slovenia	0.33	26	Jamaica	−0.51	61	Nigeria	−0.80	96
Estonia	0.33	27	China	−0.53	62	Mali	−0.80	97
Greece	0.17	28	South Africa	−0.54	63	Mozambique	−0.80	98
Portugal	0.14	29	Dominican Republic	−0.57	64	Ethiopia	−0.80	99
Italy	0.14	30	Georgia	−0.58	65	Tanzania	−0.80	100
Korea	0.08	31	Jordan	−0.59	66	Bangladesh	−0.80	101
Hungary	0.07	32	Venezuela	−0.60	67	Madagascar	−0.80	102
Czech Republic	0.04	33	Zimbabwe	−0.61	68	Uganda	−0.80	103
Croatia	0.00	34	Tunisia	−0.61	69	Chad	−0.81	104
United Arab Emirates	−0.08	35	Zambia	−0.61	70			

The GCI constructed by Sala-i-Martin and Artadi (2004) is a useful benchmark of a nation's competitiveness. It provides a measure of the competitiveness of 104 economies across the globe, and aims to achieve a balance between the microeconomic and macroeconomic foundations of competitiveness as well as its static and dynamic consequences. In addition, the authors explicitly take into account the fact that different nations have different development priorities depending on their level of development. They divide the 104 nations studied into three broad categories or stages based on the nations' per capita GDP[6] and share of exports that take the form of primary goods.

Figure 2 presents a plot between the networked readiness of nations and the GCI (Sala-i-Martin and Artadi, 2004). The three stages of development of nations are clearly

identified by ellipses. One could conclude the following:

1. There is a very tight relationship between the networked readiness of nations and the GCI (coefficient of correlation = 0.9259). As the networked readiness of nations increases, so does the GCI. It is important to caution the reader at this stage that the close correlation does not enable us to infer any causal relationship or to establish the contribution of networked readiness to global competitiveness.

2. The three stages of development of the GCI represent relatively non-overlapping areas of the plot, and hence represent not only stages of distinct GCI but also stages of the networked readiness of nations.

3. The intermediate countries (stage 1–2 and stage 2–3) are relatively evenly dispersed between their borderline

stages. They indicate a transition in networked readiness as well as global competitiveness.

Thus, a nation's networked readiness is inextricably tied up with its global competitiveness. Since the GCI stages are tightly wrapped around the income per capita of a nation, one would be led to conclude that the NRI is also strongly related to GDP per capita.

GDP Per Capita and Networked Readiness

GDP per capita is one data variable that has a strong attraction both as an indicator of global competitiveness (in the previous subsection nations were divided into stages of development based on per capita income) and of networked readiness. However, our opinion is that any attempt to use a single measure to approximate the networked readiness of a nation would be a simplification at best.

Israel, with an annual GDP per capita of US$19,678, has an NRI score of 1.02 and is ranked 18th overall. Greece, with a very similar GDP per capita of US$19,973, has, on the other hand, a score of 0.17 and an overall ranking of 42. One thus sees a wide spread in the NRI score for a given GDP per capita. This is only one of many examples that could be cited.

Nevertheless, a closer look at the relationship between the NRI of nations and GDP per capita is merited. Figure 3 plots GDP per capita against the NRI. The partial log regression plot projects a trend line. One can note immediately the following points:

Table 9. **Individual Readiness**

COUNTRY	SCORE	RANK	COUNTRY	SCORE	RANK	COUNTRY	SCORE	RANK
Finland	1.26	1	Lithuania	0.42	36	Colombia	−0.18	71
Singapore	1.19	2	Italy	0.37	37	Algeria	−0.22	72
Iceland	1.17	3	Greece	0.37	38	Zimbabwe	−0.24	73
Switzerland	1.14	4	Romania	0.37	39	Venezuela	−0.27	74
Denmark	1.09	5	Korea	0.36	40	El Salvador	−0.31	75
Belgium	1.07	6	Latvia	0.30	41	Pakistan	−0.33	76
Sweden	1.06	7	Costa Rica	0.29	42	Bosnia and Herzegovina	−0.35	77
Australia	1.05	8	Portugal	0.27	43	Egypt	−0.37	78
France	1.03	9	India	0.26	44	Dominican Republic	−0.40	79
Hong Kong	1.03	10	Bulgaria	0.22	45	Zambia	−0.44	80
Taiwan	1.00	11	Thailand	0.18	46	Vietnam	−0.56	81
Canada	1.00	12	Jordan	0.17	47	Ecuador	−0.58	82
Netherlands	1.00	13	Russian Federation	0.15	48	Peru	−0.59	83
Austria	0.96	14	Poland	0.15	49	Kenya	−0.61	84
Japan	0.96	15	Croatia	0.14	50	Sri Lanka	−0.62	85
New Zealand	0.94	16	Chile	0.13	51	Guatemala	−0.68	86
United States	0.91	17	Mauritius	0.13	52	Georgia	−0.69	87
Norway	0.90	18	Indonesia	0.13	53	Paraguay	−0.76	88
Ireland	0.85	19	Botswana	0.12	54	Bolivia	−0.78	89
Luxembourg	0.85	20	Morocco	0.11	55	Honduras	−0.78	90
United Kingdom	0.83	21	Trinidad and Tobago	0.11	56	Gambia	−0.83	91
Malaysia	0.76	22	South Africa	0.08	57	Ghana	−0.91	92
Estonia	0.76	23	Turkey	0.07	58	Angola	−1.23	93
Germany	0.72	24	Ukraine	0.02	59	Madagascar	−1.44	94
Slovenia	0.71	25	Namibia	0.02	60	Mozambique	−1.49	95
Israel	0.70	26	China	0.00	61	Nigeria	−1.51	96
Tunisia	0.66	27	Jamaica	−0.01	62	Malawi	−1.89	97
United Arab Emirates	0.63	28	Brazil	−0.02	63	Mali	−1.92	98
Cyprus	0.57	29	Uruguay	−0.03	64	Tanzania	−2.01	99
Czech Republic	0.54	30	Macedonia	−0.06	65	Uganda	−2.07	100
Malta	0.54	31	Panama	−0.07	66	Nicaragua	−2.09	101
Spain	0.50	32	Serbia and Montenegro	−0.08	67	Bangladesh	−2.10	102
Hungary	0.48	33	Mexico	−0.12	68	Chad	−2.31	103
Bahrain	0.45	34	Argentina	−0.13	69	Ethiopia	−2.50	104
Slovak Republic	0.43	35	Philippines	−0.14	70			

- For a given GDP per capita, there is a spread in the NRI scores around the regression plot as presented in Figure 3.

- The impact of GDP seems to be very high at low GDP values, and the NRI score increases rapidly with small increases in GDP.

- Around a GDP per capita of US$6,000–10,000 the curve tapers off and the effect of increasing GDP is much less pronounced. Other factors become more relevant to the NRI score at higher values of GDP per capita.

- The coefficient of correlation is relatively high (0.6998).

Countries widely distanced from the regression plot could be examples of under-performing or over-performing countries. Thus one sees that Singapore leads the NRI ranking and outperforms Norway, which has a significantly higher GDP per capita. Similarly, Taiwan and Estonia would be over-performing on their NRI scores with respect to their GDP per capita.

While this analysis sheds light on the link between GDP per capita and the networked readiness of a country, one might feel the need to examine more closely countries with GDP per capita lower than US$10,000, where the NRI score seems to increase very quickly with small changes in GDP per capita.

Figure 4 plots the Networked Readiness Index of countries with GDP per capita below US$10,000—which still retains a relatively high range of GDP per capita scores from US$611 for Tanzania to a high end of $9,981 for

Table 10. **Business Readiness**

COUNTRY	SCORE	RANK	COUNTRY	SCORE	RANK	COUNTRY	SCORE	RANK
United States	1.52	1	India	0.32	36	Guatemala	−0.30	71
Switzerland	1.42	2	Thailand	0.27	37	Macedonia	−0.32	72
Germany	1.40	3	Italy	0.26	38	Algeria	−0.34	73
Finland	1.33	4	Portugal	0.25	39	Ukraine	−0.36	74
Sweden	1.25	5	Greece	0.23	40	Jordan	−0.36	75
Japan	1.22	6	Slovak Republic	0.19	41	Ghana	−0.39	76
United Kingdom	1.17	7	United Arab Emirates	0.18	42	Peru	−0.39	77
France	1.13	8	Turkey	0.17	43	Dominican Republic	−0.39	78
Denmark	1.13	9	Latvia	0.16	44	Zambia	−0.41	79
Netherlands	1.10	10	Mexico	0.15	45	Kenya	−0.42	80
Belgium	1.08	11	Argentina	0.12	46	Egypt	−0.44	81
Canada	1.04	12	Jamaica	0.12	47	Nigeria	−0.49	82
Australia	1.04	13	Indonesia	0.10	48	Bosnia and Herzegovina	−0.53	83
Israel	1.03	14	Cyprus	0.08	49	Ecuador	−0.64	84
Singapore	1.00	15	Poland	0.08	50	Sri Lanka	−0.65	85
Austria	0.98	16	Romania	0.08	51	Georgia	−0.69	86
Norway	0.92	17	Trinidad and Tobago	0.06	52	Vietnam	−0.69	87
Taiwan	0.91	18	Mauritius	0.05	53	Madagascar	−0.73	88
Iceland	0.88	19	Hungary	0.04	54	Honduras	−0.74	89
Ireland	0.82	20	Botswana	0.02	55	Malawi	−0.75	90
New Zealand	0.76	21	Zimbabwe	0.01	56	Gambia	−0.75	91
Hong Kong	0.75	22	Bahrain	0.01	57	Philippines	−0.75	92
South Africa	0.73	23	Croatia	−0.01	58	Paraguay	−0.83	93
Korea	0.72	24	Malta	−0.02	59	Bolivia	−0.87	94
Spain	0.60	25	China	−0.05	60	Pakistan	−0.87	95
Slovenia	0.58	26	Panama	−0.07	61	Tanzania	−1.04	96
Luxembourg	0.54	27	Colombia	−0.13	62	Uganda	−1.41	97
Malaysia	0.52	28	Morocco	−0.16	63	Angola	−1.41	98
Estonia	0.51	29	Bulgaria	−0.20	64	Mali	−1.80	99
Tunisia	0.51	30	Uruguay	−0.20	65	Chad	−1.91	100
Costa Rica	0.50	31	Russian Federation	−0.23	66	Bangladesh	−2.07	101
Brazil	0.47	32	Venezuela	−0.23	67	Mozambique	−2.11	102
Chile	0.41	33	Namibia	−0.25	68	Ethiopia	−2.31	103
Lithuania	0.37	34	Serbia and Montenegro	−0.25	69	Nicaragua	−3.45	104
Czech Republic	0.35	35	El Salvador	−0.27	70			

Figure 2. **Global Competitiveness Index versus Networked Readiness Index**

Global Competitiveness Index

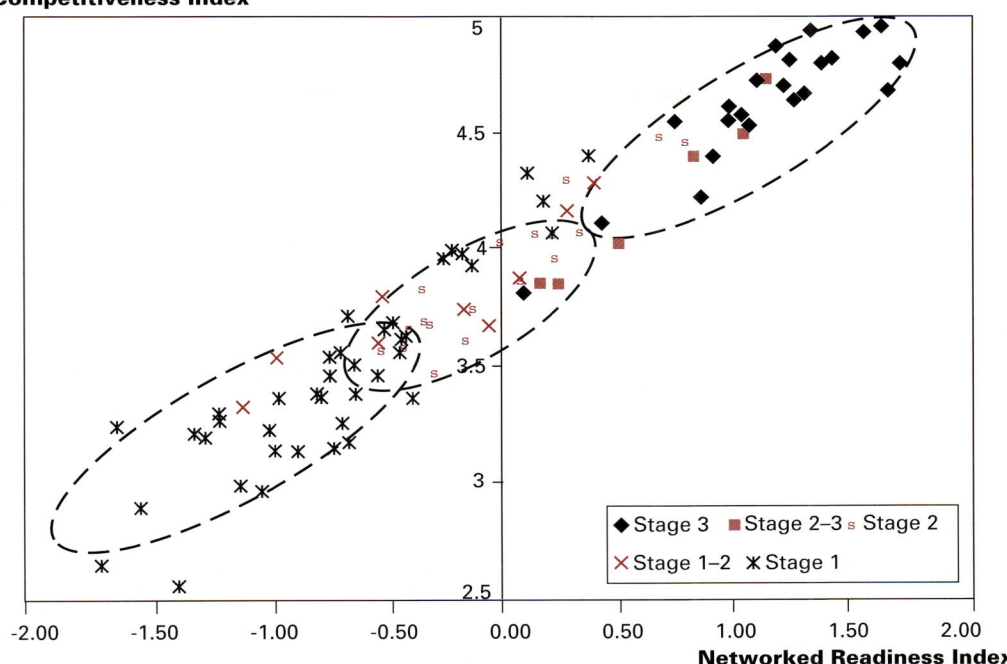

Source: Authors' analysis of data from the World Economic Forum

Latvia. One notes once again the important spread in NRI for a given value of GDP per capita. What is interesting is that the value of the coefficient of correlation decreases from 0.6998 for Figure 3, where all 104 countries are plotted, to 0.3058 for Figure 4, where only the 59 countries with GDP per capita lower than $10,000 are plotted. Thus, in this range GDP per capita does not explain the variance in the observed NRI scores as well as might have been expected. Factors other than GDP per capita must have an effect.

Figure 3. **Networked Readiness versus Gross Domestic Product Per Capita, Partial Log Regression**

Networked Readiness

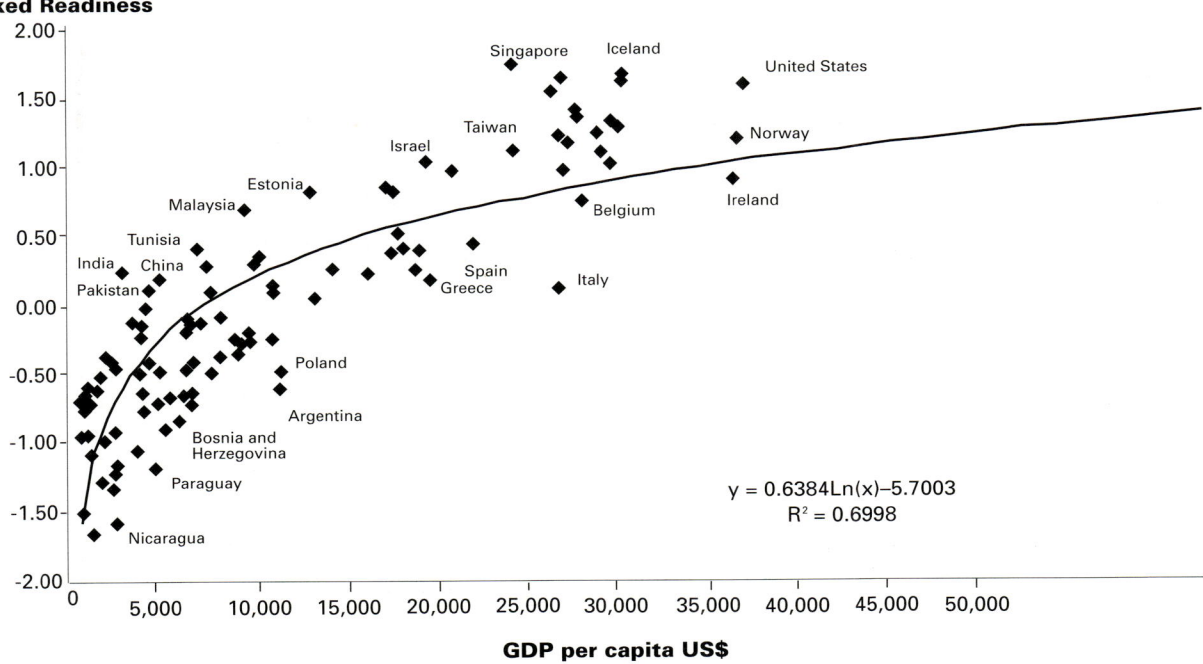

$$y = 0.6384 \text{Ln}(x) - 5.7003$$
$$R^2 = 0.6998$$

GDP per capita US$

Source: Authors' analysis of data from the World Bank

Table 11. **Government Readiness**

COUNTRY	SCORE	RANK	COUNTRY	SCORE	RANK	COUNTRY	SCORE	RANK
Singapore	2.28	1	Malta	0.55	36	Uruguay	−0.54	71
Japan	2.00	2	Uganda	0.53	37	Kenya	−0.65	72
Taiwan	1.89	3	Spain	0.53	38	Bangladesh	−0.69	73
United Arab Emirates	1.83	4	Gambia	0.50	39	Italy	−0.72	74
Finland	1.70	5	Vietnam	0.49	40	Ethiopia	−0.72	75
Luxembourg	1.44	6	Canada	0.45	41	Turkey	−0.72	76
United States	1.37	7	Madagascar	0.35	42	El Salvador	−0.72	76
Malaysia	1.36	8	Mali	0.33	43	Costa Rica	−0.73	78
Tunisia	1.29	9	Chile	0.30	44	Croatia	−0.75	79
Denmark	1.28	10	Egypt	0.27	45	Serbia and Montenegro	−0.81	80
Korea	1.14	11	Lithuania	0.24	46	Cyprus	−0.84	81
Hong Kong	1.13	12	Jamaica	0.24	47	Malawi	−0.84	82
Germany	1.08	13	Indonesia	0.17	48	Bulgaria	−0.87	83
Bahrain	1.00	14	Portugal	0.16	49	Macedonia	−0.93	84
Thailand	0.98	15	Brazil	0.11	50	Zambia	−0.96	85
France	0.96	16	Sri Lanka	0.10	51	Latvia	−0.96	86
China	0.92	17	Tanzania	0.09	52	Bosnia and Herzegovina	−1.08	87
Switzerland	0.92	18	New Zealand	0.08	53	Ukraine	−1.10	88
Pakistan	0.91	19	Belgium	0.05	54	Angola	−1.11	89
Australia	0.90	20	Namibia	0.03	55	Venezuela	−1.15	90
Israel	0.89	21	Slovenia	0.00	56	Dominican Republic	−1.16	91
Sweden	0.88	22	Morocco	−0.06	57	Nicaragua	−1.18	92
Estonia	0.87	23	Romania	−0.09	58	Guatemala	−1.28	93
United Kingdom	0.83	24	Hungary	−0.11	59	Georgia	−1.43	94
Norway	0.79	25	Nigeria	−0.11	60	Poland	−1.46	95
Ghana	0.78	26	Czech Republic	−0.12	61	Zimbabwe	−1.49	96
Mauritius	0.76	27	Trinidad and Tobago	−0.17	62	Panama	−1.53	97
Iceland	0.76	28	Algeria	−0.27	63	Chad	−1.64	98
South Africa	0.75	29	Mexico	−0.29	64	Argentina	−1.69	99
Botswana	0.75	30	Russian Federation	−0.31	65	Honduras	−1.85	100
Ireland	0.73	31	Greece	−0.31	66	Ecuador	−1.87	101
Jordan	0.68	32	Slovak Republic	−0.42	67	Bolivia	−1.96	102
Netherlands	0.64	33	Mozambique	−0.46	68	Paraguay	−2.05	103
India	0.63	34	Colombia	−0.47	69	Peru	−2.10	104
Austria	0.57	35	Philippines	−0.51	70			

Another observation of potential interest is the regional distribution of the NRI scores. Thus, Asian countries such as India, China, Pakistan and Malaysia fall above the trend line, so that the Asian region consistently outperforms the NRI index. Latin America, in contrast, consistently fall short of any potential expectation of networked readiness based on GDP per capita. The difference between levels of ICT development in the Asian pack of countries and those in the Latin American countries cannot be explained to any appreciable extent by GDP per capita.

From the data available, one would be led to conclude that GDP per capita presents a modest explanation of the networked readiness of countries with a GDP per capita less than $10,000. This group of countries represents the countries present in stage 1 and a part of stage 2 of the Sala-i-Martin and Artadi GCI model (Figure 2).

Networked Readiness and Global Competitiveness Index stages of development

Given the strength of the relationships between the Networked Readiness Index and the Global Competitiveness Index, and of the importance of GDP per capita for both benchmarks, one might enquire which key forces determining the NRI are also characteristic of countries present in a given GCI stage of development. In order to answer this question, we conducted a Multiple Discriminant Function Analysis (MDFA) across 86 countries naturally falling into stages 1, 2 and 3.

It was observed from the statistical analysis that three NRI variables—Internet Users per 100 inhabitants, Personal Computers per 100 inhabitants, and Secure Internet Servers per million inhabitants—when used to perform the MDFA minimized the prediction error and automatically classified the countries into three distinct groups. The MDFA was able to explain the classification of 69.8 percent of the countries on the basis of their Networked Readiness Index.

Given this fact, it would be of interest to examine to what extent the classification based on NRI analysis corresponds to that of the GCI analysis. In order to do this, the predictions of the MDFA model were examined with respect to the stages of development of the GCI. Table 15 presents the findings. In all, 89.5 percent of the countries could be correctly classified by the three indicators used in the MDFA model and from the NRI classifications hereby generated. This analysis reinforces the view of the close link between networked readiness and the competitiveness of a nation. Once again, caution must be exercised when one interprets these results for the direction of causality underlying the relationships.

NRI Benchmarking and Country Analysis

The degree of networked readiness of a nation or economy is the result of a multitude of effects. The current research project started with a set of over 78 different variables or indicators for evaluating networked readiness. The 78 variables were narrowed down by statistical analysis to a set of 51. These 51 variables were grouped among the nine subindexes of the NRI framework.

Given the complexity of networked readiness and of ICT, it would be useful to explore in detail each component constituting networked readiness. This section is dedicated

Table 12. **Individual Usage**

COUNTRY	SCORE	RANK	COUNTRY	SCORE	RANK	COUNTRY	SCORE	RANK
Sweden	2.76	1	Croatia	0.34	36	Panama	−0.58	71
Korea	2.56	2	Bahrain	0.30	37	Guatemala	−0.61	72
Iceland	1.94	3	Chile	0.22	38	Morocco	−0.64	73
Denmark	1.77	4	Malaysia	0.16	39	Ecuador	−0.66	74
Hong Kong	1.76	5	Poland	0.14	40	Egypt	−0.70	75
Norway	1.61	6	Slovak Republic	0.13	41	Bolivia	−0.72	76
Switzerland	1.60	7	Lithuania	0.13	42	Georgia	−0.74	77
Canada	1.57	8	Jamaica	0.07	43	Paraguay	−0.74	78
Netherlands	1.43	9	Brazil	0.00	44	Algeria	−0.77	79
Japan	1.41	10	Bulgaria	−0.01	45	Botswana	−0.87	80
United States	1.39	11	Latvia	−0.04	46	Indonesia	−0.90	81
Singapore	1.36	12	China	−0.12	47	Vietnam	−0.91	82
Luxembourg	1.36	13	Mauritius	−0.12	48	Namibia	−0.93	83
Finland	1.27	14	Turkey	−0.13	49	Honduras	−1.03	84
Germany	1.20	15	Costa Rica	−0.18	50	India	−1.04	85
Austria	1.20	16	Argentina	−0.21	51	Pakistan	−1.10	86
Italy	1.17	17	Uruguay	−0.22	52	Sri Lanka	−1.11	87
Belgium	1.17	18	Romania	−0.27	53	Nicaragua	−1.12	88
Israel	1.08	19	Trinidad and Tobago	−0.28	54	Zimbabwe	−1.14	89
Australia	1.03	20	Mexico	−0.32	55	Gambia	−1.15	90
Malta	0.95	21	Serbia and Montenegro	−0.34	56	Kenya	−1.21	91
France	0.94	22	Macedonia	−0.35	57	Ghana	−1.21	92
Cyprus	0.93	23	Thailand	−0.37	58	Zambia	−1.22	93
Spain	0.92	24	Venezuela	−0.37	59	Nigeria	−1.23	94
New Zealand	0.92	25	Bosnia and Herzegovina	−0.37	60	Bangladesh	−1.23	95
Portugal	0.83	26	Colombia	−0.41	61	Mali	−1.23	96
Taiwan	0.83	27	South Africa	−0.44	62	Angola	−1.26	97
Slovenia	0.78	28	Russian Federation	−0.46	63	Tanzania	−1.27	98
United Arab Emirates	0.75	29	El Salvador	−0.48	64	Mozambique	−1.30	99
United Kingdom	0.73	30	Jordan	−0.50	65	Uganda	−1.30	100
Ireland	0.71	31	Peru	−0.51	66	Madagascar	−1.31	101
Greece	0.70	32	Tunisia	−0.52	67	Chad	−1.31	102
Czech Republic	0.70	33	Dominican Republic	−0.54	68	Malawi	−1.33	103
Estonia	0.62	34	Ukraine	−0.55	69	Ethiopia	−1.34	104
Hungary	0.50	35	Philippines	−0.56	70			

Figure 4. **Networked Readiness versus Gross Domestic Product Per Capita, Partial Log Regression for Countries with GDP Per Capita <US$10,000**

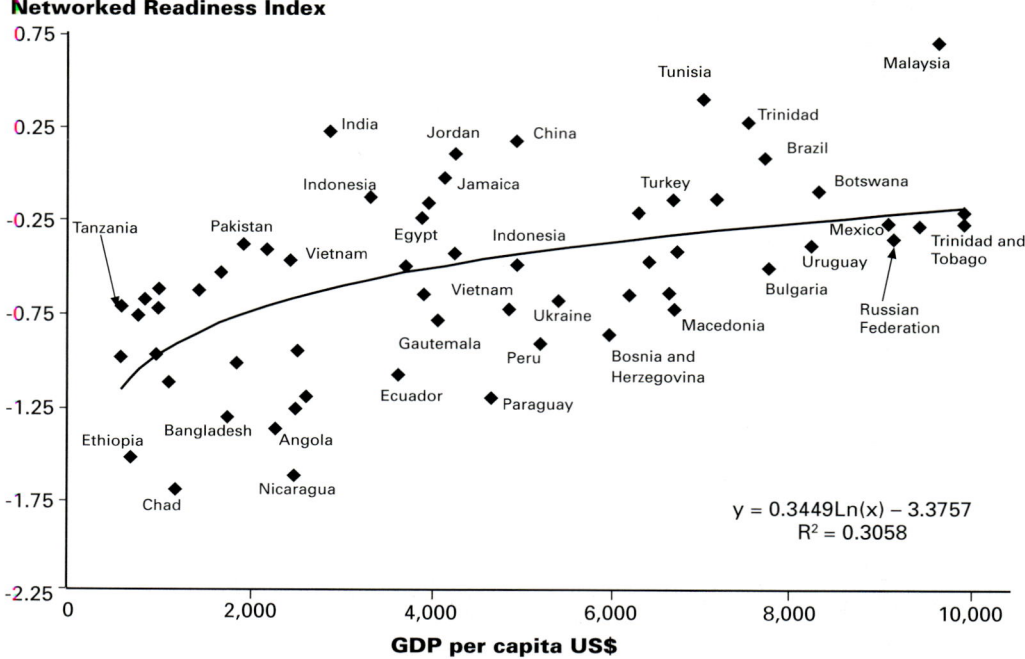

Source: Authors' analysis of data from the World Bank

to furthering such a goal and to the analysis of the performance of the key regions of the world with respect to ICT development. Countries are divided among eight major regions: Africa, Asia, West Europe, Central/East Europe, Latin America, the Middle East, North America, and Oceania.[7] Following this, weighted average regional scores[8] are calculated for each of the eight regions across the NRI index and subindex by weighting each country in a region by its population while computing regional scores.

Regional networked readiness

Countries across the globe show significant variance in their NRI scores. This variance is also present when one compares the key regions; a wide spread in the regional NRI scores can be observed. Table A1 reports the results of the computation of the weighted average NRI scores for the eight key regions analysed in this section. Figure 5 plots these same scores. The figure presents the regions in the order of increasing NRI, from the lowest

Figure 5. **Networked Readiness Index and Subindexes by Region**

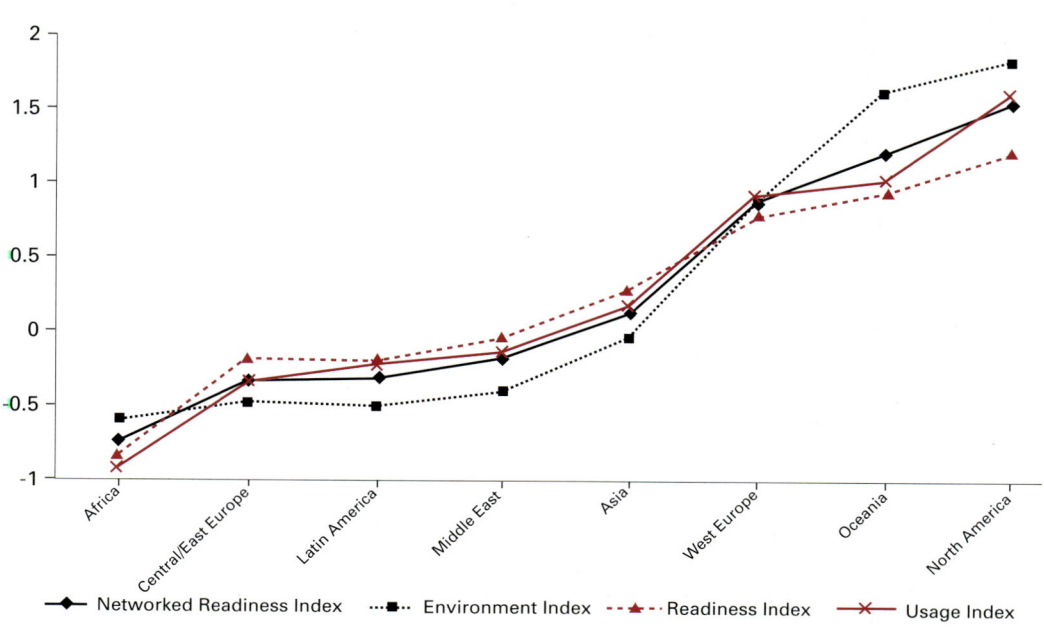

Table 13. **Business Usage**

COUNTRY	SCORE	RANK	COUNTRY	SCORE	RANK	COUNTRY	SCORE	RANK
Japan	1.79	1	Lithuania	0.49	36	Venezuela	−0.38	71
Germany	1.69	2	Spain	0.49	37	Uruguay	−0.41	72
Sweden	1.68	3	Portugal	0.48	38	Colombia	−0.50	73
Switzerland	1.60	4	Bahrain	0.47	39	Poland	−0.62	74
Finland	1.57	5	Tunisia	0.45	40	Zambia	−0.66	75
Iceland	1.49	6	Jordan	0.40	41	Trinidad and Tobago	−0.69	76
Denmark	1.49	7	Turkey	0.38	42	Georgia	−0.73	77
Israel	1.48	8	Panama	0.31	43	Tanzania	−0.73	78
Singapore	1.44	9	Malaysia	0.28	44	Nigeria	−0.80	79
United States	1.41	10	Greece	0.25	45	Kenya	−0.80	80
Taiwan	1.38	11	El Salvador	0.19	46	Ghana	−0.82	81
Hong Kong	1.38	12	Cyprus	0.15	47	Gambia	−0.84	82
United Kingdom	1.26	13	Italy	0.13	48	Ukraine	−0.87	83
Netherlands	1.21	14	Latvia	0.13	49	Indonesia	−0.89	84
New Zealand	1.17	15	Malta	0.11	50	Costa Rica	−0.94	85
Norway	1.14	16	Dominican Republic	0.04	51	Macedonia	−0.97	86
France	1.12	17	China	0.03	52	Bulgaria	−0.97	87
Belgium	1.11	18	Namibia	−0.05	53	Madagascar	−1.10	88
Australia	1.08	19	Jamaica	−0.06	54	Serbia and Montenegro	−1.19	89
Austria	1.05	20	Morocco	−0.07	55	Mozambique	−1.22	90
Canada	0.98	21	Egypt	−0.08	56	Malawi	−1.28	91
Korea	0.96	22	Mauritius	−0.10	57	Algeria	−1.28	92
India	0.83	23	Philippines	−0.10	58	Bolivia	−1.38	93
United Arab Emirates	0.81	24	Croatia	−0.13	59	Paraguay	−1.40	94
Luxembourg	0.80	25	Pakistan	−0.15	60	Ecuador	−1.41	95
Czech Republic	0.76	26	Mexico	−0.20	61	Bangladesh	−1.42	96
Estonia	0.68	27	Sri Lanka	−0.26	62	Bosnia and Herzegovina	−1.44	97
Brazil	0.66	28	Vietnam	−0.27	63	Mali	−1.53	98
Slovak Republic	0.63	29	Romania	−0.28	64	Nicaragua	−1.57	99
South Africa	0.62	30	Botswana	−0.33	65	Honduras	−1.58	100
Ireland	0.60	31	Peru	−0.34	66	Zimbabwe	−1.76	101
Chile	0.59	32	Uganda	−0.34	67	Angola	−2.27	102
Slovenia	0.59	33	Guatemala	−0.34	68	Ethiopia	−2.31	103
Hungary	0.58	34	Russian Federation	−0.35	69	Chad	−2.32	104
Thailand	0.49	35	Argentina	−0.36	70			

weighted average NRI in Africa, to the highest scores for North America.

One observes that the scores of each component index are consistent with that of the overall NRI index on a regional basis. While individual country scores show significant variance across the components of the NRI framework, this variance becomes averaged out on a regional basis.

Regional benchmarking

The regional scores offer the possibility of providing a benchmark against which countries could evaluate their performance. For instance, a closer look at the performance of some of the countries in Latin America reveals that Chile (0.29), Brazil (0.08), and Mexico with a

score of (−0.28) outperform the Latin American average, which has a score of (−0.33). In contrast, Colombia (−0.42) and Peru (−0.91) under-perform the Latin American average.

Chile and Brazil consistently outperform the Latin American average on all the component indexes (Figure 6), whereas Peru strongly underperforms. Focusing on Colombia, one notes that, like Mexico, the country is fairly representative of the Latin American average, except for the Usage Component Index, on which it underperforms.

The NRI permits the reader to delve more deeply and identify a country's key strengths and weaknesses. In the previous paragraph, it was noted that Colombia lagged behind the average Latin American score in terms of the usage of information and communications technologies. In

Table 14. **Government Usage**

COUNTRY	SCORE	RANK	COUNTRY	SCORE	RANK	COUNTRY	SCORE	RANK
Singapore	3.30	1	New Zealand	0.37	36	Russian Federation	−0.50	71
Hong Kong	2.40	2	South Africa	0.35	37	Colombia	−0.56	72
United States	2.04	3	Chile	0.32	38	Nigeria	−0.57	73
Denmark	1.88	4	Switzerland	0.32	39	Argentina	−0.60	74
Taiwan	1.73	5	Hungary	0.29	40	Trinidad and Tobago	−0.68	75
Canada	1.71	6	Romania	0.28	41	Namibia	−0.73	76
United Arab Emirates	1.46	7	Belgium	0.26	42	Latvia	−0.73	77
Japan	1.44	8	Indonesia	0.14	43	Uruguay	−0.76	78
Estonia	1.40	9	Netherlands	0.12	44	Serbia and Montenegro	−0.77	79
Israel	1.31	10	Italy	0.11	45	Venezuela	−0.85	80
Malta	1.26	11	Morocco	0.11	46	Bulgaria	−0.85	81
Finland	1.25	12	Uganda	0.10	47	Ukraine	−0.86	82
Austria	1.24	13	Luxembourg	0.09	48	Zambia	−0.90	83
Malaysia	1.21	14	Cyprus	0.07	49	El Salvador	−0.92	84
France	1.12	15	Mexico	0.02	50	Costa Rica	−0.92	85
Australia	1.12	16	Turkey	−0.02	51	Bangladesh	−0.96	86
United Kingdom	1.11	17	Greece	−0.08	52	Panama	−0.98	87
Germany	1.07	18	Slovenia	−0.12	53	Malawi	−1.00	88
Korea	1.06	19	Vietnam	−0.14	54	Angola	−1.03	89
Sweden	1.01	20	Ghana	−0.16	55	Guatemala	−1.07	90
Ireland	1.00	21	Brazil	−0.17	56	Bosnia and Herzegovina	−1.07	91
China	0.98	22	Spain	−0.23	57	Ethiopia	−1.08	92
India	0.73	23	Tanzania	−0.23	58	Poland	−1.24	93
Thailand	0.71	24	Mozambique	−0.27	59	Macedonia	−1.26	94
Pakistan	0.71	25	Botswana	−0.27	60	Ecuador	−1.28	95
Egypt	0.57	26	Lithuania	−0.28	61	Georgia	−1.32	96
Iceland	0.50	27	Gambia	−0.32	62	Peru	−1.32	97
Tunisia	0.49	28	Slovak Republic	−0.35	63	Dominican Republic	−1.44	98
Jordan	0.49	29	Madagascar	−0.36	64	Nicaragua	−1.50	99
Bahrain	0.49	30	Sri Lanka	−0.36	65	Paraguay	−1.55	100
Portugal	0.45	31	Algeria	−0.37	66	Honduras	−1.61	101
Norway	0.42	32	Mali	−0.42	67	Bolivia	−1.73	102
Jamaica	0.42	32	Kenya	−0.43	68	Zimbabwe	−1.81	103
Mauritius	0.41	34	Czech Republic	−0.44	69	Chad	−1.87	104
Philippines	0.38	35	Croatia	−0.46	70			

order to identify the key areas of weakness, one would need to consider the weighted average regional scores across each subindex, and across the variables comprising the subindexes. These scores are provided in Tables A1–10.

Figure 7 presents the plot of the usage index and subindexes for the five Latin American countries in question, along

Table 15. **Prediction of GCI Stages of Development by Key NRI Variables**

GCI STAGE OF DEVELOPMENT	COUNTRIES	INCORRECT CLASSIFICATIONS	PREDICTION ERROR (%)
Stage 1	41	2	4.9
Stage 2	21	3	14.3
Stage 3	24	4	16.7
	86	9	10.5

with the weighted average usage scores for the region. One notes that Colombia lags behind the region in each area of usage, but the greatest lag occurs in ICT usage by businesses. Peru, by contrast, shows a disproportionately lower performance in government usage.

Figure 8 plots the scores of Colombia in comparison with the Latin American regional scores for the Business Usage Index and its variables. One notes that, in four out of the five dimensions taken into consideration for Business Usage, Colombia lags behind the Latin American weighted average NRI scores. In particular it is found that, according to the Executive Opinion Survey responses, Colombian companies are not as aggressive as the rest of the region in striving to absorb new technologies and to obtain new technologies through foreign technology licensing.

Figure 6. **NRI in Latin America**

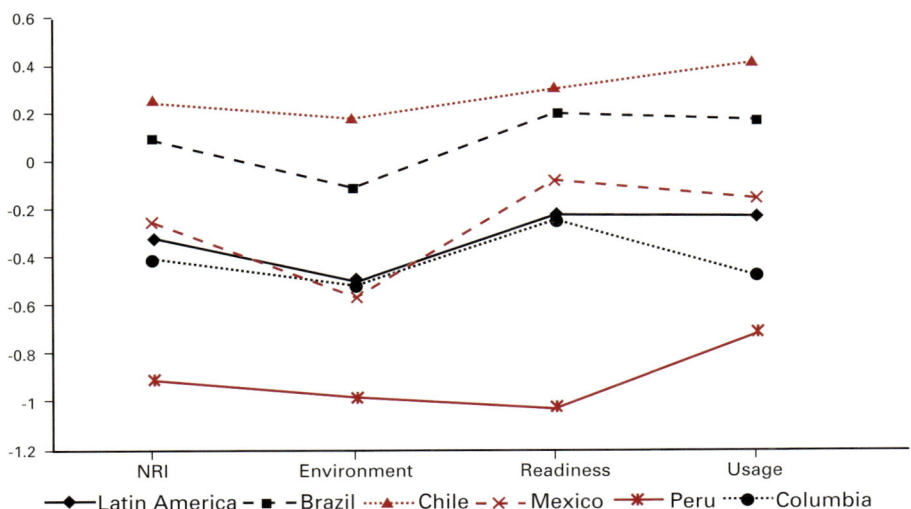

Legend: Latin America — Brazil — Chile — Mexico — Peru — Columbia

Figure 7. **ICT Usage in Latin America**

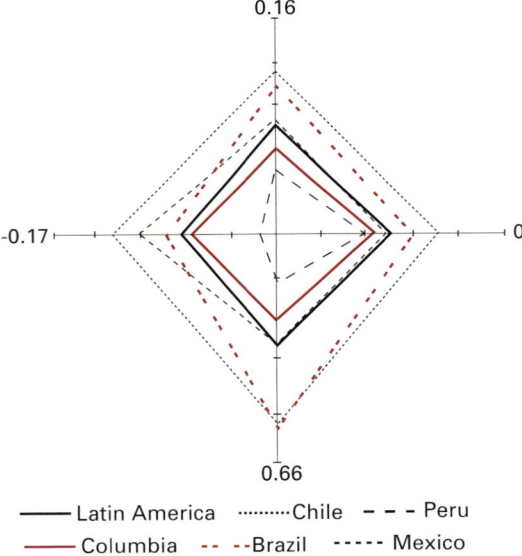

Legend: Latin America — Chile — Peru — Columbia — Brazil — Mexico

In this way, one can identify factors contributing to the under- and over-performance of a country relative to another country or to a region. The NRI provides the tools to disaggregate networked readiness or ICT competitiveness into its constituent elements, and to provide data in a useful and easily interpretable form.

Similar analysis could be done for any country or region of the globe. The NRI is thus useful at several levels: it presents country ICT performance as a single aggregated measure, and it enables the explication of this aggregated performance in terms of the different components of the index.

Research Challenges

Finding the facts

Lack of accurate and reliable data can prove a roadblock to the implementation of even the best laid out frameworks. The goal of our research and analysis has been to provide a scientific and credible interpretation of reality. Thus, an important step has been to collect a complete and high-quality set of data relating to ICT. We used two types of data in our research: soft data, which is subjective data gathered from questionnaires (managed by the World Economic Forum as part of its research for the Global Competitiveness Report), and hard data, which is driven by statistics collected by international multilateral agencies

Figure 8. **Latin America: Business Usage Variables**

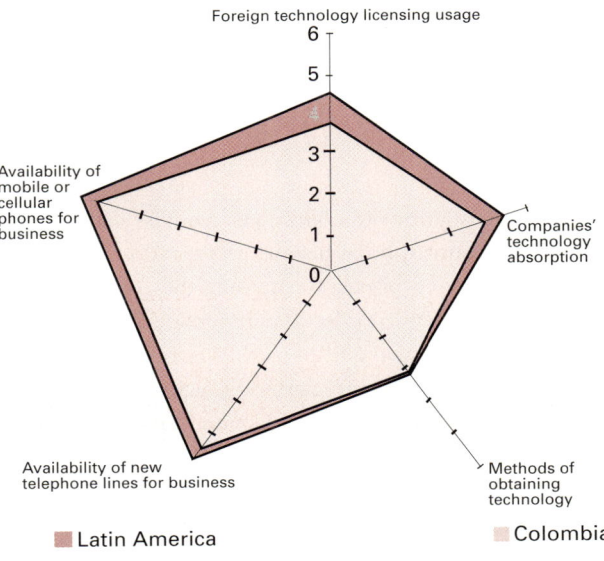

Legend: Latin America — Colombia

(such as the World Bank and the International Telecommunications Union, ITU). Both these sets of data play a crucial role in the overall analysis. The soft data is critical in determining the opinion of the decision makers and influencers who are intimately familiar with a nation's economy and ICT usage; the hard data captures fundamental elements related to the development of infrastructure, human capital and ICT.

Availability of key usage metrics

Key ICT areas such as mobile telephony and the Internet are still undergoing rapid development. As a result, accurate and up-to-date usage metrics are difficult to obtain. For example, data on the cost of access, penetration and usage of mobile phones, the Internet and other ICT indicators is available with a lag of 12–36 months, resulting in the current analysis relying on data as old as that of 2002. Thus, while 36 (70.6 percent) of the variables considered in the study are from 2004, when the research was conducted, there are still 11 (21.6 percent) that are from 2002 and 4 (7.8 percent) from 2003. In the study of rapidly evolving phenomena such as ICT, it would be preferable to analyse contemporary data only.

Table 16. **Age of Data Variables Used**

S. NO	YEAR	NUMBER OF VARIABLES	%
1	2002	11	21.6
2	2003	4	7.8
3	2004	36	70.6

Selection of countries

The use of objective and reliable data is critical in preparing a report of this type. Availability of data has in fact been a key factor in selecting the 104 countries that form part of this study. As a consequence, this research project limits its analysis to the set of 104 countries studied by the Executive Opinion Survey of the World Economic Forum.

Data estimation

Despite our best efforts to collect data from all major international sources, it has been necessary at times to cope with incomplete sets of data for some of the countries under consideration. In order to compensate for this, statistical procedures have been used to estimate missing data,[9] mainly regression and clustering techniques. Control procedures and checks have been devised to ensure that estimations were reasonable and not biased in their representation of the countries in question.

Summary

Networked readiness is a complex phenomenon and the sum of diverse and interrelated forces. Measuring a country's networked readiness remains a significant challenge, and any framework or model representing networked readiness is a simplified representation at best. Further, limitations on the availability of reliable and current data restrict the measurement of the phenomenon to a subset of countries and also to a small number of the underlying forces.

Nevertheless, as has been seen in this chapter, the Networked Readiness Framework and Index can be useful tools for key policy decision makers charting a country's strategic direction in order to enhance national competitiveness. The NRI framework attempts to interpret the underlying complexity of the development and use of ICT in an intuitive and easy-to-comprehend model. The overall NRI is a summary measure of a nation's ability to participate in and benefit from ICT developments. The NRI provides guidance to business leaders and public policy makers for enhancing the impact of ICT on important stakeholders—individuals, businesses and governments.

Governments and policy makers can have a significant impact on the adoption and usage of ICT. For example, our research (Dutta, Lanvin and Paua 2004) has demonstrated that promoting competition and deregulation in the ICT sector leads to decreasing service costs, and that lowered costs result in an increase in consumption of services.

The NRI allows a nation to benchmark its ICT performance and to determine the effectiveness of policy. It also permits a country to learn from the policy and performance of other countries with similar profiles and to identify best practice. The NRI serves to highlight the areas of over- and under-performance of a given country relative to a similar set of countries, and to provide best-practice examples.

Over-performing countries have put ICT on the national agenda and have striven to make it an area of excellence. The former countries have succeeded in going beyond individual measures of national income or national ICT spending in an effort to provide an optimal environment for ICT development, thus promoting high levels of readiness and usage within all three key stakeholders. Singapore, Korea and Estonia are examples of such leaders, and can serve as role models for other nations in their quest for ICT excellence.

Technical Appendix

The Networked Readiness Index 2004–2005 separates environmental factors from ICT readiness and usage, and thus is composed of three indexes. Each component index is further subdivided into three subindexes. Starting with a set of over 78 ICT-related variables, we divided them among the nine subindexes. We then eliminated variables on the basis of the number of countries for which data was available and used analytical procedures such as factor analysis and Cronbach's alpha. Our final index computation is based on a set of 51 variables. [10]

The Networked Readiness Index is defined as follows:

Networked Readiness Index = 1/3 Environment + 1/3 Readiness + 1/3 Usage

I. The Environment component index is defined as follows:

Environment = 1/3 Market + 1/3 Political/Regulatory + 1/3 Infrastructure

I.1. Market Environment is defined by the following variables:

1.01 Availability of scientists and engineers, 2004

1.02 Venture capital availability, 2004

1.03 Sophistication of financial markets, 2004

1.04 Technological sophistication, 2004

1.05 State of cluster development, 2004

1.06 Collaboration in clusters, 2004

1.07 University-industry collaboration, 2004

1.08 Quality of scientific research institutions, 2004

1.09 Subsidies for firm-level research and development, 2004

1.10 Brain drain, 2004

1.11 Ease of access to loans, 2004

1.12 Administrative burden, 2004

1.13 Ease to start a new business, 2004

I.2. Political/Regulatory Environment is defined by the following variables:

2.01 Effectiveness of lawmaking, 2004

2.02 Laws relating to ICT, 2004

2.03 Effectiveness of judiciary, 2004

2.04 Intellectual property protection, 2004

I.3. Infrastructure Environment is defined by the following variables:

3.01 Telephone mainlines, 2002

3.02 Secure Internet servers, 2003

3.03 Internet hosts, 2003

II. The Readiness component index is defined as follows:

Readiness = 1/3 Individual Readiness + 1/3 Business Readiness + 1/3 Government Readiness

II.1. Individual Readiness is defined by the following variables:

4.01 Quality of math and science education, 2004

4.02 Quality of educational system, 2004

4.03 Quality of public schools, 2004

4.04 Internet access in schools, 2004

4.05 Buyer sophistication, 2004

4.06 Buyer dynamism, 2004

4.07 Residential telephone connection charge, 2002

4.08 Affordability of Internet access, 2003

II.2. Business Readiness is defined by the following variables:

5.01 Investment in training, 2004

5.02 Availability of training services, 2004

5.03 Quality of business schools, 2004

5.04 Business investment in R&D, 2004

5.05 Business monthly telephone subscription, 2002–3

5.06 Business telephone connection charge, 2002–3

II.3. Government Readiness is defined by the following variables:

6.01 Government prioritization of ICT, 2004

6.02 Government procurement of ICT, 2004

III. The Usage component index is defined as follows:

Usage = 1/3 Individual Usage + 1/3 Business Usage + 1/3 Government Usage

III.1. Individual Usage is defined by the following variables:

7.01 Cellular mobile subscribers, 2003

7.02 Telephone subscribers, 2002

7.03 Public payphones, 2002

7.04 Telephone lines, 2002

7.05 Television sets, 2002

7.06 Broadband- DSL Internet subscribers, 2002–3

7.07 Broadband-cable modem, 2002–3

7.08 Internet users per 100 inhabitants, 2002

III.2. Business Usage is defined by the following variables:

8.01 Prevalence of foreign technology licensing, 2004

8.02 Firm-level technology absorption, 2004

8.03 Capacity for innovation, 2004

8.04 Availability of new telephone lines, 2004

8.05 Availability of cellular phones, 2004

III.3. Government Usage is defined by the following variables:

9.01 Government success in ICT promotion, 2004

9.02 Government online services, 2003

Regional weighted average Networked Readiness Index and data variables

Table A1. **Networked Readiness Index and Subindexes**

	Africa	Central/ East Europe	Latin America	Middle East	Asia	West Europe	Oceania	North America
NETWORKED READINESS INDEX	−0.74	−0.34	−0.33	−0.19	0.118	0.86	1.19	1.53
Environment Subindex	−0.59	−0.48	−0.51	−0.41	−0.05	0.88	1.6	1.81
Readiness Subindex	−0.78	−0.19	−0.23	−0.03	0.25	0.76	0.93	1.21
Usage Subindex	−0.86	−0.33	−0.24	−0.13	0.16	0.94	1.03	1.57

Table A2. **Market Environment**

	Africa	Central/ East Europe	Latin America	Middle East	Asia	West Europe	Oceania	North America
MARKET ENVIRONMENT SUBINDEX	−0.55	−0.45	−0.51	−0.39	0.23	0.79	1.25	1.96
1.02 Venture capital availability, 2004	2.75	2.97	2.72	2.86	3.35	4.11	4.69	5.60
1.03 Sophistication of financial markets, 2004	3.17	3.42	4.39	3.02	3.84	5.53	6.03	6.59
1.04 Technological sophistication, 2004	2.77	3.31	3.95	3.89	4.04	4.92	5.35	6.34
1.05 State of cluster development, 2004	3.40	3.11	3.38	3.91	4.24	4.33	3.57	5.10
1.06 Collaboration in clusters, 2004	3.53	3.77	3.61	3.88	4.43	4.78	4.20	5.59
1.07 University-industry collaboration, 2004	2.91	3.10	3.14	2.95	3.77	4.32	4.29	5.34
1.08 Quality of scientific research institutions, 2004	3.72	4.22	3.63	3.67	4.28	4.88	5.40	6.17
1.09 Subsidies for firm-level research and development, 2004	2.78	3.11	2.60	3.46	3.92	4.17	4.25	4.53
1.10 Brain drain, 2004	2.64	2.97	3.47	2.84	3.57	4.24	4.09	6.03
1.11 Ease of access to loans, 2004	2.63	2.75	2.88	2.58	3.08	4.03	4.51	4.77
1.12 Administrative burden, 2004	3.06	2.48	2.41	2.71	3.34	3.04	3.30	3.22
1.13 Ease to start a new business, 2004	3.63	3.42	2.93	3.76	4.20	4.17	5.14	5.49

Table A3. **Political/ Regulatory Environment**

	Africa	Central/ East Europe	Latin America	Middle East	Asia	West Europe	Oceania	North America
POLITICAL/REGULATORY ENVIRONMENT SUBINDEX	−0.46	−0.68	−0.59	−0.34	0.08	1.05	1.13	1.47
2.01 Effectiveness of lawmaking, 2004	3.33	2.99	2.43	3.49	4.06	4.22	5.18	4.93
2.02 Laws relating to ICT, 2004	3.23	3.31	3.58	3.39	3.95	4.73	5.23	5.10
2.03 Effectiveness of judiciary, 2004	3.46	2.92	3.28	2.80	4.07	5.34	6.05	5.27
2.04 Intellectual property protection, 2004	3.11	2.85	3.24	3.96	3.48	5.44	5.96	6.14

Table A4. **Infrastructure Environment**

	Africa	Central/ East Europe	Latin America	Middle East	Asia	West Europe	Oceania	North America
INFRASTRUCTURE ENVIRONMENT SUBINDEX	−0.76	−0.32	−0.43	−0.57	−0.71	0.79	1.06	1.99
3.01 Telephone mainlines, 2002	23.55	260.45	170.80	122.69	118.66	572.20	523.64	639.36
3.02 Secure Internet servers, 2003	1.71	5.26	6.25	5.44	4.67	96.50	297.67	453.18
3.03 Internet hosts, 2003	6.61	76.31	114.95	38.78	44.99	492.11	1407.69	151.05

Table A5. **Individual Readiness**

	Africa	Central/ East Europe	Latin America	Middle East	Asia	West Europe	Oceania	North America
INDIVIDUAL READINESS SUBINDEX	−1.34	0.14	−0.19	−0.17	−0.03	0.75	0.98	0.91
4.01 Quality of math and science education, 2004	3.31	4.86	2.98	3.73	4.45	4.80	5.20	4.49
4.02 Quality of educational system, 2004	2.96	3.49	2.55	3.04	3.50	4.35	5.21	4.46
4.03 Quality of public schools, 2004	2.54	4.03	2.56	3.27	3.19	5.16	5.69	5.04
4.04 Frequency of Internet access in schools	2.31	3.52	3.44	3.33	3.88	5.05	5.96	5.91
4.05 Buyer sophistication, 2004	3.26	3.61	3.54	3.19	4.26	5.25	5.67	5.88
4.06 Buyers dynamism, 2004	3.97	4.38	4.35	4.71	4.71	5.35	5.89	6.03
4.07 Residential telephone connection charge, 2004	153.43	33.68	28.35	66.05	64.40	2.67	3.91	1.31
4.08 Affordability of Internet access, 2003	272.69	10.14	13.04	7.26	23.65	1.14	1.10	0.67

Table A6. **Business Readiness**

	Africa	Central/ East Europe	Latin America	Middle East	Asia	West Europe	Oceania	North America
BUSINESS READINESS SUBINDEX	−0.84	−0.09	0.07	−0.26	−0.07	0.96	1.14	1.45
5.01 Investment in training, 2004	3.33	3.30	3.63	3.63	3.91	4.95	5.21	5.57
5.02 Availability of training services, 2004	3.77	4.07	4.20	3.95	4.35	5.47	5.53	6.25
5.03 Quality of business schools, 2004	3.76	3.83	4.55	3.74	4.40	5.57	5.67	6.59
5.04 Business investment in R&D, 2004	2.94	3.16	3.15	2.91	3.78	4.41	4.18	5.59
5.05 Business monthly telephone subscription, 2002–3	12.83	2.41	4.17	1.61	6.24	1.03	0.83	1.15
5.06 Business telephone connection charge, 2002–3	159.02	65.14	47.69	123.56	82.61	2.96	4.00	2.07

Table A7. **Government Readiness**

	Africa	Central/ East Europe	Latin America	Middle East	Asia	West Europe	Oceania	North America
GOVERNMENT READINESS SUBINDEX	−0.17	−0.62	−0.57	0.27	0.47	0.58	1.83	1.27
6.01 Government prioritization of ICT, 2004	4.19	3.83	3.90	5.04	5.12	4.70	4.79	5.16
6.02 Government procurement of ICT, 2004	3.69	3.41	3.43	3.62	4.05	4.19	4.36	4.66

Table A8. **Individual Usage**

	Africa	Central/ East Europe	Latin America	Middle East	Asia	West Europe	Oceania	North America
INDIVIDUAL USAGE SUBINDEX	−1.02	−0.31	−0.23	−0.56	−0.63	0.94	1.23	1.32
7.01 Cellular mobile subscribers, 2003	6.90	28.94	23.40	15.11	15.13	84.05	70.77	52.98
7.02 Telephone subscribers, 2002	7.25	51.03	36.62	24.41	24.44	135.87	117.35	111.72
7.03 Public payphones, 2002	0.66	1.62	4.55	1.45	4.25	3.10	4.18	5.07
7.04 Telephone lines, 2002	1.93	26.17	18.15	13.22	14.01	56.34	52.67	61.68
7.05 Television sets, 2002	22.51	97.90	87.71	92.53	63.75	95.75	96.48	97.34
7.06 Broadband-DSL Internet subscribers, 2002–3	0.06	1.65	2.03	1.80	5.75	25.16	6.23	25.01
7.07 Broadband-cable modem, 2002–3	0.20	2.53	1.91	7.47	3.41	21.75	9.32	56.10
7.08 Internet users per 100 inhabitants, 2002	1.23	8.28	8.30	4.86	5.90	36.52	48.21	54.33

Table A9. **Business Usage**

	Africa	Central/ East Europe	Latin America	Middle East	Asia	West Europe	Oceania	North America
BUSINESS USAGE SUBINDEX	−0.63	−0.26	−0.19	0.01	0.46	1.04	1.92	1.66
8.01 Prevalence of foreign technology licensing, 2004	4.07	4.01	4.45	4.43	4.79	5.01	5.69	5.05
8.02 Firm-level technology absorption, 2004	4.02	4.41	4.39	4.97	5.11	5.09	5.59	6.19
8.03 Capacity for innovation, 2004	2.83	3.54	3.17	3.10	3.86	5.26	4.27	5.64
8.04 Availability of new telephone lines, 2004	4.25	5.10	5.72	5.30	5.51	6.44	6.37	6.48
8.05 Availability of mobile or cellular phones, 2004	5.20	6.03	6.27	5.52	5.88	6.70	6.55	6.41

Table A10. **Government Usage**

	Africa	Central/ East Europe	Latin America	Middle East	Asia	West Europe	Oceania	North America
GOVERNMENT USAGE SUBINDEX	**−0.49**	**−0.49**	**−0.45**	**0.36**	**0.49**	**0.69**	**1.46**	**1.98**
9.01 Government success in ICT promotion, 2004	3.85	3.30	3.40	4.61	4.61	4.07	4.25	4.31
9.02 Government online services, 2003	1.91	2.77	2.68	2.59	3.32	4.06	4.42	6.40

The Regional Weighted Average Networked Readiness Scores were computed by classifying countries into different geographical regions as follows:

Table A11. **Country Classification by Region**

I. Africa

1. Angola
2. Botswana
3. Chad
4. Ethiopia
5. Gambia
6. Ghana
7. Kenya
8. Madagascar
9. Malawi
10. Mali
11. Mauritius
12. Morocco
13. Mozambique
14. Namibia
15. Nigeria
16. South Africa
17. Tanzania
18. Uganda
19. Zambia
20. Zimbabwe

II. Asia

1. Bangladesh
2. China
3. Hong Kong
4. India
5. Indonesia
6. Japan
7. Korea
8. Malaysia
9. Pakistan
10. Philippines
11. Singapore
12. Sri Lanka
13. Taiwan
14. Thailand
15. Vietnam

III. Central/East Europe

1. Bosnia and Herzegovina
2. Bulgaria
3. Croatia
4. Czech Republic
5. Estonia
6. Georgia
7. Hungary
8. Latvia
9. Lithuania
10. Macedonia
11. Malta
12. Poland
13. Romania
14. Russian Federation
15. Serbia
16. Slovak Republic
17. Slovenia
18. Turkey
19. Ukraine

IV. Latin America

1. Argentina
1. Bolivia
2. Brazil
3. Chile
4. Colombia
5. Costa Rica
6. Dominican Republic
7. Ecuador
8. El Salvador
9. Guatemala
10. Honduras
11. Mexico
12. Nicaragua
13. Panama
14. Paraguay
15. Peru
16. Uruguay
17. Venezuela

V. Middle East

1. Algeria
2. Bahrain
3. Egypt
4. Israel
5. Jordan
6. Tunisia
7. United Arab Emirates

VI. North America

1. Canada
2. Jamaica
3. Trinidad and Tobago
4. United States of America

VII. Oceania

1. Australia
2. New Zealand

VIII. West Europe

1. Austria
2. Belgium
3. Cyprus
4. Denmark
5. Finland
6. France
7. Germany
8. Greece
9. Iceland
10. Ireland
11. Italy
12. Luxembourg
13. Netherlands
14. Norway
15. Portugal
16. Spain
17. Sweden
18. Switzerland

Notes

1 The first *Global Information Technology Report* (2001–2002) was edited by Kirkman et al. (2002) at the Centre for International Development at Harvard University, in association with the World Economic Forum. The reports for 2002–2003 (S. Dutta, B. Lanvin and F. Paua 2003) and 2003–2004 (Dutta S., B. Lanvin and F. Paua 2004) were the result of a joint effort between INSEAD, Infodev at the World Bank, and the World Economic Forum. The current report is thus the fourth report in this series.

2 For more information on the development of the Networked Readiness framework and other efforts in the domain, see Dutta and Jain (2003).

3 The Networked Readiness Index 2004–2005 uses a different methodology for data analysis from that used for the 2002–2003 and 2003–2004 GITR reports. In previous reports, the NRI was reported on a scale of 1–7, with an increasing score representing a higher level of networked readiness. In the present report, the scores are the result of subindex computations using factor analysis techniques. As a result, the factor and subindex scores

have a mean of zero. These subindex scores are averaged to compute the component indexes and finally the NRI. The NRI and component indexes thus also have a mean score of zero. Countries having negative scores simply have scores below the mean score for the 104 countries in this research project, and countries having positive scores have scores above the mean for the 104 countries studies.

4   For more information on the computation of the Networked Readiness Index and the methodology used in data analysis, the interested reader is referred to a note titled "Data analysis and index computation: Methodology" available from the authors upon request.

5   The World Economic Forum (WEF) has published two competitiveness indexes in the past: the Growth Competitiveness Index, which refers to the macroeconomic, institutional and technological determinants of productivity, and the Business Competitiveness Index (BCI), which captures the microeconomic components of productivity. In 2004 the WEF publishes the Growth Competitiveness Index and a new index, the Global Competitiveness Index (GCI), which integrates both macroeconomic and microeconomic factors in one framework. The GCI will eventually replace the Growth Competitiveness Index as the leading WEF indicator of a nation's competitiveness.

6   In the words of Sala-I-Martin and Artadi (2004):

"A country belongs to the factor driven stage (stage 1) if its GDP per capita is below US$2,000 or the fraction of its exports in the form of primary goods is above 70 percent.

A country with a per capita income between US$3,000 and $9,000 that does not export more than 70 percent in primary goods belongs to the second stage.

A country with more than US$17,000 per capita income and less than 70 percent of the exports in primary goods belongs to the third stage.

Countries with a per capita income between US$2,000 and $3,000 are said to be in transition from stage 1 to stage 2. Countries with per capita income between US$9,000 and $17,000 are said to be in transition between stages 2 and 3."

7   Two countries analysed in this research—Australia and New Zealand—are classified under Oceania.

8   The regional scores were computed as a weighted average of the scores of the countries comprising the region. The weighted average regional score was calculated using the following formula:

Weighted average variable score of region = ($\sum$(Variable score of Country*Population of country))/ Population of the region) over all the countries in the given region.

9   A certain number of countries with incomplete data-sets were included in the analysis. Our decision not to drop these countries from the Networked Readiness Index was based on the conviction that it was better to tap into the rich information provided by the Executive Opinion Survey and estimate some of the missing hard data than to ignore the survey data altogether because of limitations in the hard data. Nevertheless, some caution is warranted in the interpretation of the results for the countries near the bottom of the rankings. As the hard data sources improve in coming years, this problem should be gradually lessened.

10 Our research used the most recent data available from the concerned sources, namely, World Economic Forum (2004) and data from World Bank (2004) and ITU (2004).

References

Dutta, S. and A. Jain. 2003. "The Networked Readiness of Nations", in S. Dutta, B. Lanvin, and F. Paua, eds, *The Global Information Technology Report 2002–2003: Readiness for the Networked World*. New York: Oxford University Press. Online. http://unpan1.un.org/intradoc/groups/public/documents/APCITY/UNPAN014654.pdf

Dutta, S., B. Lanvin, and F. Paua, eds. 2003. *The Global Information Technology Report 2002–2003: Readiness for the Networked World*. New York: Oxford University Press.

——— ——— ———. 2004. *The Global Information Technology Report 2003–2004: Towards an Equitable Information Society*. New York: Oxford University Press.

ITU (International Telecommunications Union). 2004. *World Telecommunications Indicators Database*. Geneva: ITU.

Kirkman, G., P. Cornelius, J. Sachs, and K. Schwab, eds. 2002. *The Global Information Technology Report 2001–2002: Readiness for the Networked World*. New York: Oxford University Press.

Sala-i-Martin, X. and E. Artadi. 2004. "The Global Competitiveness Index", in M. Porter, K. Schwab, X. Sala-i-Martin, and A. Lopez-Claros, eds., *The Global Competitiveness Report 2004–2005*. Basingstoke: Palgrave Macmillan.

World Bank. 2004. *World Development Indicators 2004*. Washington, DC: World Bank.

World Economic Forum. 2004. *Executive Opinion Survey*. Geneva: World Economic Forum.

Chapter 2

Net Impact: European e-Government:

How People, Process and Technology Achieve Greater Productivity in the European Public Sector

Douglas Frosst, Cisco Systems

Scott Brown, Incepta Marketing Intelligence

Andrew Elder, Incepta Marketing Intelligence

The European public sector is faced with significant pressure to deliver citizen services more efficiently and simultaneously work to broaden their reach. This pressure comes from citizens and elected officials, as well as from the increasing importance of public sector services in this age of global economies. A country's public service infrastructure not only serves the needs of its constituents, it is also a major component in the cycle of productivity that generates economic growth and higher standards of living. Elected officials have embraced a spectrum of pending public sector initiatives, including the European Commission's *e-Europe 2005 Action Plan* and *Connected Health* initiatives, to promote long-term change. As a result, European organizations must choose the technology, staff and process investments they believe will help them most effectively meet pending mandates and increase overall organizational productivity.

This chapter discusses *Net Impact 2004: Europe e-Government* (Momentum Research Group 2004), a research project aimed at identifying the investments in information and communication technologies (ICT) and business practices that are leading to significant improvements in operating productivity of e-services across European public sector organizations. Participation in the study was limited to "Connected Organizations", that is, organizations with one or more active enterprise business applications distributed throughout their network (see section on methodology). In all, more than 1,400 ICT and business decision makers from the following eight countries participated in the study: France, Germany, Italy, the Netherlands, Poland, Spain, Sweden and the United Kingdom.

The chapter concentrates on the opportunities and challenges in healthcare, which is high on the list of priorities in most countries and one of the largest public services.

Key Findings

The primary aim of *Net Impact* was to understand the interrelationships between people, processes and technology, and the resulting productivity gains. Do Connected Organizations derive higher levels of productivity from their greater investment in sophisticated network infrastructures, Internet business applications and process re-engineering?

Net Impact identified several common themes. Generally, Connected Organizations that increased their productivity the most:

- invested in sophisticated network infrastructure beyond the minimum necessary to support their applications;

- re-engineered business processes for efficiency and effectiveness prior to application deployment;

- automated individual business processes with Internet applications and integrated them with other service functions;

- oriented the organizational culture toward process improvement and service delivery; and

- deployed formal measurement systems to track operational performance.

Although the resulting formula of network + application + process + culture seems simple, the implementation has many variations, as does the typical cake recipe (butter + sugar + eggs + flour). As with baking a cake, simply combining ingredients does not guarantee a positive outcome. The sequence and timing of the mix is essential.

The Net Impact Series

Net Impact 2004 is the fourth in a series of Net Impact research projects evaluating the effect of Internet technologies on organizations.

The first two studies quantified the macroeconomic impact of ICT on the national economies of the United States, United Kingdom, France, Germany (Momentum Research Group, Litan and Varian 2001), Italy (Momentum Research Group et al. 2002) and Canada (CEBI et al. 2002). One significant finding from these early studies was that companies implementing Internet business applications experienced reduced operating costs and increased revenues on a scale that affected the current and projected productivity growth rates in each country.[1]

In 2003, Cisco Systems® sponsored Net Impact 2003: United States Private Sector (Momentum Research Group 2003), the first study of its kind to identify technology and business practices leading to significant increases in the productivity of individual organizations. It focused on the practices that enhanced productivity (in the form of operational outcomes) within the Customer Service & Support and Sales functions of more than 300 U.S. companies in several vertical markets.

Net Impact 2003 built on past findings to demonstrate that U.S. enterprises that aligned their Internet business applications, networking technologies and business processes to leverage new, technology-enabled capabilities experienced greater operating improvements and cost savings than companies that did not.

The objectives of Net Impact 2004 included testing past research findings and identifying business and technology practices that are currently accelerating citizen service productivity in Europe.

Methodology

Net Impact 2004 targeted the European public sector, specifically government and healthcare organizations, to evaluate the productivity relationship between applications, networking infrastructure, business processes and other organizational behaviors within these organizations.

Net Impact 2004 sought to identify:

- the extent to which organizations have adopted or implemented networked applications;

- organizational attitudes or cultural traits that encourage adoption of ICT; and

- a benchmark of current business and technology best practices for achieving enhanced productivity in the delivery of public sector e-services.

The European organizations targeted in Net Impact 2004 are not necessarily "average" public sector organizations. In order to isolate technology-driven best practices, the study focused on organizations that were most likely to be experiencing some sort of impact—positive or negative—from technology investments.

Earlier research demonstrated that the use of Internet business applications is a driver of increased organizational productivity. With this in mind, organizations targeted by Net Impact 2004 were required to have at least one networked business application supporting the delivery of citizen services. In addition, because many aspects of productivity are based on the efficiency of people-based processes, at least one networked application had to be accessible to more than 20 percent of the relevant employees.

To differentiate these more sophisticated organizations from the "average" European organization (which may or may not be supporting enterprise-wide business applications), the organizations that participated in Net Impact 2004 are referred to throughout this chapter as "Connected Organizations." This filter was put in place to increase the chances of identifying leading-edge practices that are disproportionately improving citizen service productivity in Europe. However, it does mean that the study sample is not necessarily representative of the entire range of public sector organizations.

The pool of Net Impact 2004 respondents was managed to provide a distribution of public sector organizations across multiple variables including:

- Country of origin

- Organization focus—healthcare or government

- Organization type and level—local, regional or national

- Size of organization

More than 30,000 organizations across eight countries (France, Germany, Italy, the Netherlands, Poland, Spain, Sweden and the United Kingdom) were contacted for this study, but only 1,112 qualified for and ultimately completed the study. Of these, 58 percent were various types of government organizations (local, regional and national). The remaining 42 percent were from the healthcare segment, including primary care organizations such as clinics and hospitals, as well as national healthcare agencies, regional or local health councils and health–insurance organizations. Healthcare organizations in Poland were not targeted for this study.

Figure 1. **Net Impact 2004 Respondent Organizations by Segment**

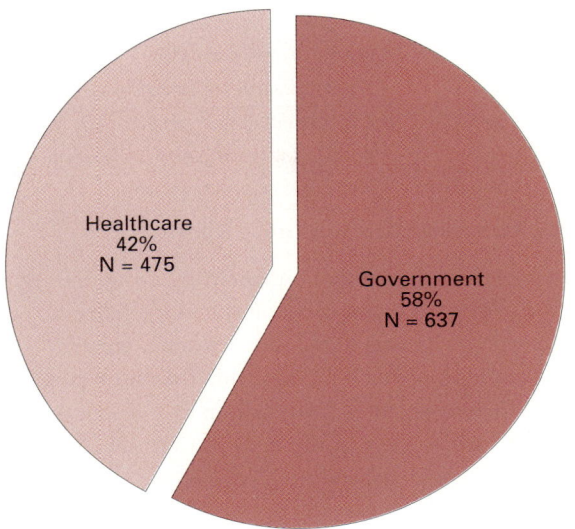

Source: Momentum Research Group (2004)

Figure 2. **Net Impact 2004 Healthcare Respondents**

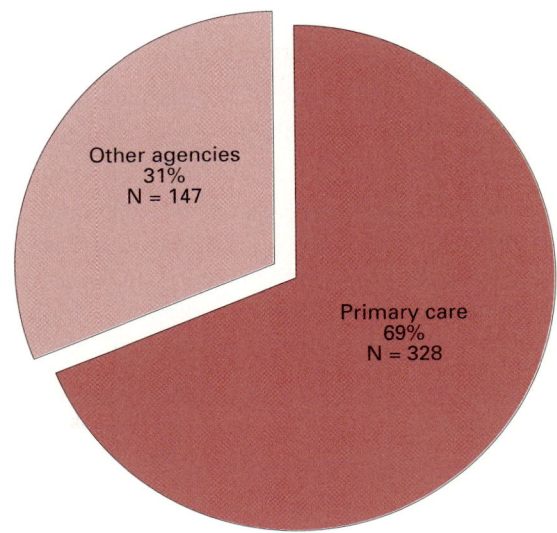

Source: Momentum Research Group (2004)

The interviews were conducted between 17 November 2003 and 29 January 2004 by phone by interviewers fluent in the local language.

Business plus technology perspective: a two-stage design

Net Impact 2004 used a two-stage design to incorporate responses from both technical and business experts within the same organization in order to obtain a more holistic view of an organization's practices and results.

Technical decision makers

The first stage of the survey process qualified companies through an interview with a senior technical manager responsible for organization-wide technology investments and/or implementation. Typical technical decision makers interviewed included Chief Information Officers (CIOs), ICT department heads and ICT directors and managers. This interview focused on:

- Internet business application adoption and current level of deployment

- Degree of network sophistication

- Broadband access and usage

- Perceived obstacles to future technology implementations

- Business conditions influencing adoption of new technology and business processes

Business decision makers

Improvements in operating outcomes are more readily identified for discrete business processes rather than at an aggregate level. As a result, past *Net Impact* studies focused on specific business functions such as Sales, Marketing and Finance with easily identified business processes. Due to the lack of common "business functions" across all public sector organizations, *Net Impact 2004* focused on areas that serve as an interface for the delivery of citizen services. *Net Impact 2004* also focused on citizen services because of the current number of high-profile initiatives to improve public sector services across Europe.

The focus on citizen services resulted in the second set of *Net Impact 2004* interviews being conducted with senior business decision makers overseeing areas such as records management, healthcare admissions, and permitting and licensing departments. The similarity of goals of this range of business decision makers allowed *Net Impact 2004* to focus on specific operating outcomes for citizen service interactions such as time–to–service resolution, average cost per service transaction and average cases resolved per employee. Typical business decision makers interviewed included CEOs, department heads, directors and program or project managers.

Business decision maker interviews focused on:

- Levels of process automation

- Measurement systems used to track the productivity of delivering services

- Estimated improvements in operating outcomes

- Organizational attitudes toward technology

The combination of technology and business data from a single organization allowed the identification of productivity best practices—organizational and technological activities common among organizations with the greatest improvements in their citizen-service operating outcomes.

Productivity

The relationship between productivity and technology is frequently discussed by corporate and political leaders because of its impact on economic growth and standards of living. In recent years, productivity has become a bell-wether economic statistic and figures prominently in discussions about interest rates, wages and fiscal policy.

Although the value of productivity is discussed at a macroeconomic level, European public sector organizations are focusing on how they can improve the quality, scope and cost of delivering citizen services—all as budgets decrease or are held constant. But which investments yield the greatest improvements? Which investments will help an organization reach its specific goal? Where should public sector organizations focus their finite budgets and resources?

Significant productivity gains possible in citizen services

Many studies of productivity use a single macro-level measurement—often financial or labor-based—to determine productivity gains. Since each organization faces different environmental conditions and has unique goals, these abstract measures rarely help decision makers understand what actions they can take to more effectively meet their organization's operating goals.

The *Net Impact* studies have found that productivity gains result from inputs working together, including business processes, applications and network infrastructure as well as factors such as worker skills, choice of tools and the general business environment. Each type of operating outcome had a slightly different set of best practices. For example, the best practices for containing costs are different from those for increasing the use of e-services.

A corporate action was identified as a best practice if it had a statistically significant relationship (correlation) with one of the 12 operating metrics tracked in *Net Impact 2004*. To aid in identifying top-line trends, the 12 operating metrics were aggregated into four productivity themes, discussed below:

Table 1. **Net Impact 2004 Productivity Metrics**

Productivity theme	e-Services metrics included in this theme
Efficiency	• Average time to resolution • Average cases resolved per employee • Total cases resolved per day/week/month • Citizen satisfaction • Average cost per case resolution
Services volume	• Number of citizens using the service • Number of website visitors • Number of cases filed online • Percentage of relevant services available online
Cost containment	• Citizen services operating costs • Percentage of cases resolved through self-service
Revenue or fee growth	• Annual fees or revenue collected

Source: Momentum Research Group (2004)

Net Impact 2004: found that organizations using the best practices identified across these productivity themes experienced on average three to seven times greater improvement in operational outcomes than their peers not using every best practice. This adds up to a significant operational advantage that can be applied to achieving organizational goals and meeting citizen demands more quickly, efficiently or effectively.

Overall, findings from *Net Impact 2004* supported the primary hypothesis: organizations that invest in (and align) their process re-engineering, networked applications and network infrastructure achieve significantly greater productivity enhancements. Although the exact combination of practices correlating with the greatest improvements varies between the operating outcomes tracked, the following best practice themes carry across the citizen-service activities. *Net Impact 2004* found that organizations with the greatest improvements in their citizen service productivity:

1. *Invest in sophisticated network infrastructure beyond the minimum necessary to implement business applications across the organization.* Organizations with the most significant improvements are not only deploying networked business applications, they have a network infrastructure that:

- allows greater access to internal employees and relevant external users;

- are more reliable than three 9's (99.9 percent or less than nine hours of unplanned network downtime annually);

Figure 3. **Net Impact 2004: Key Findings**

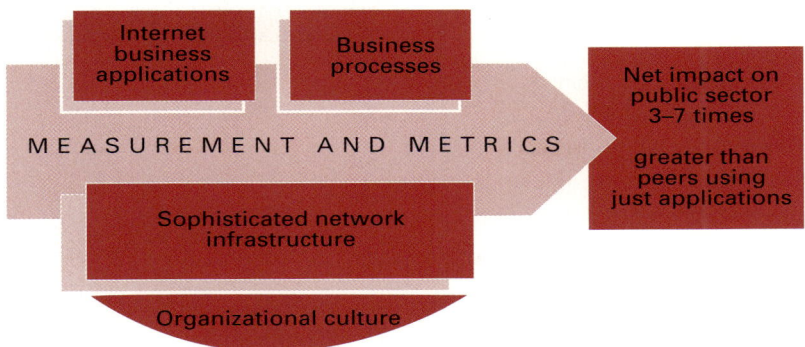

Source: Momentum Research Group (2004)

- uses more sophisticated network-based technologies such as storage area networks (SAN), virtual private networks (VPN) and remote disaster recovery; and

- provides adequate bandwidth to all organizational sites to optimize the use of networked applications and data.

The network plays a critical role in the *Net Impact 2004* equation—it provides a foundation for deploying applications, and distributes capabilities and data across the organization. Because of the foundational nature of the network, its importance as an enabler of organizational productivity can be overshadowed by higher-profile investments such as new enterprise applications, or major innovations in organizational processes.

2. *Re-engineer business processes to leverage new technology capabilities, especially prior to the deployment of networked applications.* The most effective organizations changed their business processes to take advantage of technology's capacity to make their workflow more efficient. This re-engineering manifests itself through such means as the automation of repetitive tasks or placing information at the fingertips of front-line workers to improve the quality and timeliness of their decisions.

The act of process re-engineering, either before or after a networked application is deployed, was shown to be a significant driver of productivity in many of the operating outcomes tracked by *Net Impact 2004*. Sixty-eight percent of respondents currently re-engineer their citizen service business processes at some point during the implementation of new technology.

What is surprising is the effect the *timing* of process re-engineering can have on the final outcome: on average, organizations that re-engineer *after* the deployment of a network application can experience almost a 50 percent smaller improvement in cost containment outcomes than organizations that re-engineer prior to application deployment. This finding could lead to substantial improvements in the European public sector—almost half of all the respondent organizations currently start their process re-engineering after the deployment of a new networked application.

3. *Automate citizen service processes through network applications and integrating the resulting data and processes with other service functions.* In past *Net Impact* studies the mere presence of a network application, such as a Finance and Accounting or Customer Relationship Management (CRM) system, was identified as a predictor of increased organizational productivity. *Net Impact 2004* found that network applications play more of a foundational role in the European public sector by enabling the automation and integration of repetitive citizen service processes.

4. As part of their productivity efforts, European Connected Organizations have been able to automate 30–60 percent of the business processes involved in their delivery of citizen services. The automated processes most frequently associated with improvements across citizen service outcomes include:

- Case management

- Billing and collection

- Information and service delivery

The automation of such processes not only increases the speed of the service delivery but also reduces data-re-entry errors and facilitates the transmission of data to other, complementary processes or service areas. In fact, some initiatives being undertaken in the European public sector, such as integrated patient records, are based on integrating processes and information to increase the efficiency and effectiveness of the overall service delivery experience.

5. *Orient their organizational culture toward service delivery.* The internal culture and orientation of the organization plays an important role in realizing the greatest productivity improvements. Organizational behaviors that help improve citizen service outcomes include:

- regular communication throughout the organization of the strategy for developing and delivering citizen services—44 percent of organizations currently do this;

- benchmarking of service performance against peers or published standards—48 percent of organizations currently do this; and

- the consistent alignment of business processes, network applications and network infrastructure—55 percent of organizations do this.

Making a single individual responsible for aligning technology and organizational behavior does not necessarily lead to greater organizational productivity. The identification of a dedicated "technology champion" did not increase Connected Organizations' performance on any citizen service metrics except for average cost per case resolution. This suggests that the responsibility for leveraging technology in the pursuit of productivity should be shared and distributed throughout the organization.

6. *Measure citizen service efforts*. Supporting the adage that "you can't manage what you don't measure," the deployment of formal measurement systems to track operational performance can of themselves lead to operational improvements. For example, organizations that track the following metrics tend to experience greater improvements than their peers not tracking these metrics:

- Citizen satisfaction—13 percent greater improvement

- Average cost per case resolution—12 percent greater improvement

- Average cases resolved per employee—7 percent greater improvement

Developing measurement systems should be a quick means of increasing productivity for the 10 percent of Connected Organizations reporting that they currently do not track any metrics for their citizen service activities.

Healthcare

Healthcare services are high on the priority list of most nations. Citizens are pressuring their governments to improve efficiency and accessibility to these services. Healthcare is also an economic issue. A secure and healthy workforce is an important contributing factor to a country's growth potential. Healthcare spending targets, service delivery options and alternative forms of public and private care are frequent topics of debate.

When one looks at broad statistical indicators (from the *World Development Indicators 2004*—World Bank 2004), it seems clear that spending alone is not a primary indicator of a successful healthcare system. For the countries included in this study, spending on healthcare occupies a fairly narrow range when measured as percentage of a nation's GDP. This varies from a low of 7.5 percent (Spain) to a high of 10.8 percent (Germany), with most of the countries hovering around 8 percent of GDP. When measured as average spending per capita, the range is somewhat greater, from US$1,088 (Spain) to $2,412 (Germany).

Table 2. **Healthcare Spending 2001***

	% of GDP	Per capita (US$)
France	9.6	2,109
Germany	10.8	2,412
Italy	8.4	1,582
Netherlands	8.9	2,138
Spain	7.5	1,088
Sweden	8.7	2,150
UK	7.6	1,835

* Data are for most recent year available
Source: World Bank (2004)

However, other measures paint a different picture and reveal an even greater disparity. The number of physicians per 1,000 people ranges from 2.0 (United Kingdom) to 4.3 (Italy) and the number of hospital beds per 1,000 from 3.6 (Sweden) to 10.8 (Netherlands). Infant mortality rates, average lifespan and other measures of a country's health show similar discrepancies. None of these measures are highly correlated with healthcare spending, indicating that money alone will not necessarily improve services.

Table 3. **Healthcare Indicators 2001***

	Physicians/ 1,000 people	Hospital beds 1,000 people
France	3.3	8.2
Germany	3.3	9.1
Italy	4.3	4.9
Netherlands	3.3	10.8
Spain	3.3	4.1
Sweden	3.0	3.6
UK	2.0	4.1

* Data are for most recent year available
Source: World Bank (2004)

The e-Health component of the European Commission's e-Europe *Action Plan* (European Commission URL) is aimed at improving access to, and the quality and effectiveness of, healthcare for European citizens. Exploiting new ICT to reach these goals is a strategic component of the plan. Some specific technology-related components include a European electronic health card interconnecting organizations through health information networks, and putting health services online.

What does productivity mean in healthcare?

One of the first questions that arises when studying productivity in the public sector is how to define it. Typical private-sector measurements such as revenue or profits are not always meaningful in the public sector.

For government respondents to this study, the primary driver of their technology initiatives was increasing service efficiency and reaching more people. Expanding services with existing resources, accelerating the speed at which the organization operates and improving citizen satisfaction were the goals cited by 80 percent or more of public sector respondents.

Healthcare, however, had somewhat different goals. At the top level, quality of care and access to care are typically the primary objectives of healthcare organizations. What can serve as a proxy for these objectives from an e-services perspective?

In healthcare Connected Organizations, the desire for managing costs was tied with the desire to accelerate the speed at which the organization operates, with 79 percent of organizations citing this goal. Citizen satisfaction was close behind at 75 percent.

Citizen satisfaction and increasing speed would seem to be a direct surrogate for quality of care. Costs would seem to be a reasonable surrogate in this context for access to care, because money saved with these initiatives could be applied to other areas.

Figure 4. **Net impact 2004 Key Productivity Drivers by Segment**

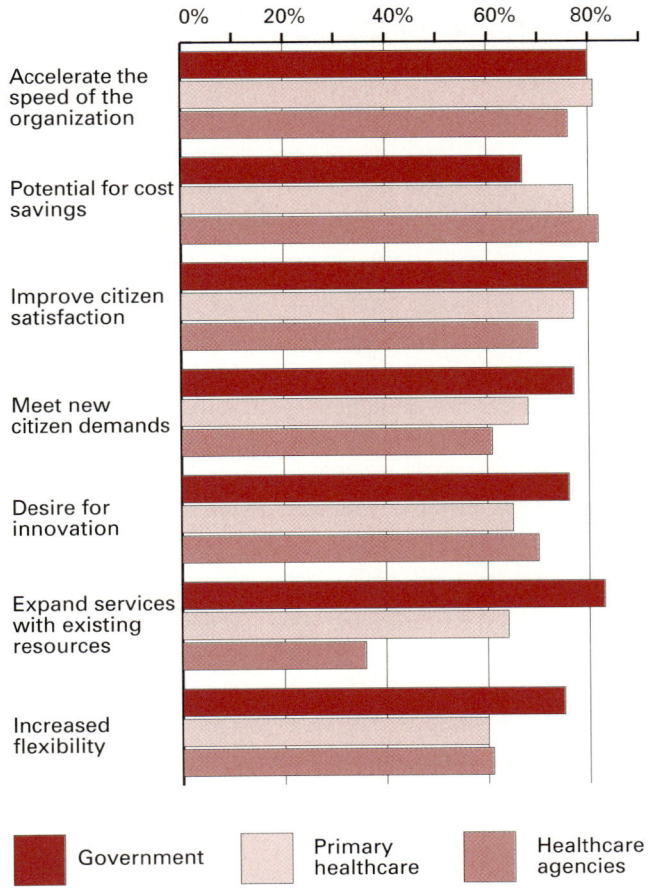

Source: Momentum Research Group (2004)

Most notable is the fact that the healthcare segment had overall lower scores than the government segment on almost all of the goals. Only the potential for cost savings had higher results for healthcare than for government. Citizens may welcome this focus on financial improvements, but *Net Impact 2004* found that it carried some risk. Organizations that focus on cost savings through formal measurement systems may be saving money at the expense of citizen satisfaction.

Managing costs

If healthcare decision makers place such a high priority on using technology to managing costs and reimbursements, are they achieving their goals? Unfortunately, only 20 percent of healthcare respondents reported a decrease in e-services operating costs over the past 12 months, while 35 percent reported an increase. These results are not as good as those of their government counterparts. This situation is undoubtedly aggravated by the constant upward pressure on healthcare budgets. Many European Union (EU) countries are experiencing annual healthcare budget increases as well as demand growth due to aging populations and the rise of chronic health conditions.

The potential for savings is significant. *Net Impact 2004* showed that organizations using the best practices of appropriate investments in network technology, business processes and network applications realized on average a 10–30 percent reduction in annual operating costs (the healthcare average was a one percent *increase*). Prudent network engineering practices, such as providing sufficient network capacity for application upgrades planned in the next 12 months and deploying sophisticated traffic management tools, contributed significantly to e-services cost-containment efforts.

If a greater percentage of healthcare organizations are trying to reduce costs through technology investments, why have they not experienced greater savings? The adoption of cost-containment best practices by healthcare Connected Organizations provides some important clues to answering this question. The study indicates that the leading operational indicators for e-services cost reduction are the percentage of cases resolved through self-service and the number of citizens using the service. Doubling either of these metrics translates to an estimated 10 percent or greater reduction in annual operating costs, as well as a greater than 20 percent increase in citizen satisfaction. However, instead of focusing on these end-user outcomes, healthcare Connected Organizations are more likely to be developing e-services for internal use by their organizations. With only half of them providing an external web portal, these organizations have yet to focus their efforts on areas that have the potential to yield the greatest cost reductions. This issue is covered in more detail in the Online Services section below.

Box 1. Financial Best Practices

20–30 percent improvement
Applications

- Web interface: workforce training and collaboration tools, resource allocation and management

Networking technologies

- Network uses load balancing, content distribution and caching to optimize network applications

- Network will support any applications that may be deployed in the next 12 months

- Network is accessible by remote employees

Business process

- Functions: integrated record systems

Organizational culture and behavior

- Business processes are re-engineered (regardless of timing) to use new technology capabilities

- Data standards are uniform throughout the organization

Ironically, a strong focus on cost reduction may have a negative effect on citizen satisfaction. *Net Impact 2004* found that organizations formally tracking general annual operating costs tend to experience citizen satisfaction improvements 6–8 percent lower than their peers.

Reimbursements and billing

Healthcare business models vary across the surveyed countries, ranging from budget-based systems to reimbursement for services. Given the complex nature of healthcare services, it is not surprising that billing and budget collection processes have the highest levels of automation in healthcare Connected Organizations and are also among the most integrated with other tasks. Making the organization efficient at collecting the necessary budget or fees is understandably important.

Using the Web for these applications provides an additional boost to productivity, and yet billing and fee collection applications are the least likely to use a Web interface (only 36 percent). This may be the result of existing government systems or the difficulty of modifying systems with multiple constituents, or it could be due to a lack of communication between business and technical decision makers. Only 20 percent of healthcare technical decision makers think they have been asked to achieve improvements in billing or reimbursements through the technology infrastructure.

Improving billing or reimbursement efficiency requires a different set of practices than those most effective at managing costs. Implementing a comprehensive disaster recovery solution was a networking best practice used by Connected Organizations with the most significant improvements in this area over the last 12 months. A network that is down cannot collect money, and an organization unprepared for unexpected network events will require more time for recovery, compounding the issue. Fewer than 25 percent of healthcare respondents reported that they had an active disaster recovery solution in place.

Organizational culture also plays a role in productivity gains. A cultural best practice for improving reimbursements was benchmarking service performance against peers or published standards. This competitive outlook on services has been shown to have a positive impact in many industries. However, only 40 percent of primary healthcare Connected Organizations benchmark their service delivery efforts on a regular basis. Development of benchmarking criteria and best practice sharing is a core element of the eHealth action plan. As EU legislation and a common European health card affect the free movement of services and citizens, regionally and then internationally, this comparative benchmarking will increase in importance.

Twenty-first century efficiency

European healthcare respondents' desire to improve their operational efficiency was second only to managing costs, with 79 percent wanting to accelerate the speed of the organization and 75 percent wanting to improve customer satisfaction. *Net Impact 2004* identified several e-services efficiency metrics, including average time to case resolution, average number of cases per employee and average cost per case resolution. From an overall healthcare perspective, reducing medical errors is a critical part of efficiency. Citizen satisfaction is a likely proxy for this, and it was also closely identified with improvements in efficiency.

European public sector organizations are using twenty-first century business processes, the Web and new networking technologies to automate tasks, with impressive results. *Net Impact 2004* found that business process-based practices are the most significant factors for improvements in service efficiency. Automation of repetitive transactions, such as billing and collection or information and service delivery, provides a significant boost in efficiency.

Automation and integration of processes has a considerable impact on reducing errors, by eliminating data re-entry and allowing for validation of rules (such as drug interactions). Citizens who can research information about their medical issues, make appointments online, have a

35–55 percent improvement
Applications

- Finance and accounting application installed

- Web interface: workforce training and collaboration tools

- Application data is integrated with data from international databases

- Greater numbers of processes accessed through a Web-based interface

Networking technologies

- Deployment of storage area network (SAN)

- Deployment of a network-user authentication system such as public key infrastructure (PKI)

- Network uses quality of service policies and tools to prioritize throughput

- Network will support any applications that may deployed in the next 12 months

Business process

- Functions: data mining and analysis

- Automation of billing and collection, information and service delivery, Case management processes

- Integration of case management processes

Organizational culture and behavior

- Strategic plan for developing and delivering services is regularly communicated throughout the organization

- Business processes are re-engineered (regardless of timing) to use new technology

doctor file their prescription with a pharmacy and perhaps even pay for it online consume far less time and resources than those who must queue up at the clinic, phone for information or request forms by mail. Reducing the response time for interactions like these is not only the primary objective; it also has a direct impact on citizen satisfaction.

Healthcare Connected Organizations in this study are successfully implementing several e-services efficiency best practices. Given the mission-critical nature of many of their applications, their networks tend to be more sophisticated, with higher reliability than the average government network. Some of this reliability could be attributed to their greater use of traffic management and scalability tools, such as load balancing and quality of service policies. They are also more likely to use a centralized database or data-warehouse, and have higher levels of integration between their applications.

Healthcare Connected Organizations reported higher levels of process automation than government organizations in most of the key areas. Billing and reimbursement top their list of automated processes, but they also had higher than average levels of automation of case management and knowledge management processes. The one process where healthcare organizations fall behind is automation of information and service delivery. Further automation of business processes like these is important because they were identified as the most significant contributors to efficiency improvements across the European public sector.

After the processes are automated, the next priority is ensuring that employees have access to the necessary applications. This appears to be a crucial focus for primary healthcare respondents. The median rate of access by relevant workgroups is only 50 percent, compared with 70–90 percent for government or healthcare agencies. Some of this may be a result of the mobile, non–office work environment in many primary-care facilities. However, cost-effective technologies such as wireless networks and handheld computers are now available to address this challenge.

In addition to automating service processes, another best practice that broadly drives efficiency outcomes is the deployment of data mining and analysis tools. Healthcare organizations tend to have large amounts of data that must be utilized to respond to requests and to deliver effective treatment. These tools are closely linked to the networking best practice of storage area networks (SANs). The sheer volume of data stored by healthcare organizations means that the higher average speed, network efficiency and availability that SANs offer can better support employees and citizens who need access to this information.

Of course, value can be derived from data analysis only if the data is stored electronically. European healthcare Connected Organizations appear to be making good progress in the difficult area of collecting and organizing data electronically. Centralized databases are used by 58 percent of healthcare Connected Organizations, and 56 percent have added data-mining tools. Although only 28 percent have introduced a storage area network; this is slightly above the government average, supporting the belief that the healthcare field has somewhat more sophisticated networks than the general public sector. Because the surveyed organizations represent the more technically advanced of the general public sector population, this remains an untapped opportunity for most.

Connected Organizations implementing all of the efficiency best practices experience on average between six and eight times greater improvement in efficiency outcomes than their peers implementing just network applications. Overall, *Net Impact 2004* found that Connected Organizations implementing the best practices for efficiency realized an improvement in efficiency of 35–55 percent. The average improvement of efficiency outcomes across the healthcare respondents was slightly lower at 21 percent. It is important to remember that these are relative percentage gains, and depend on the size, complexity and starting point of each organization.

Online services

In view of the pressure to provide more healthcare services with existing resources, it is surprising how low the healthcare business respondents rank extending services as a reason for their technology initiatives. This is at the top of the list for governments, recorded by 83 percent of government Connected Organizations. However, there seems to be a significant gap between primary care organizations and healthcare agencies. Sixty-four percent of the primary care respondents listed this as an organizational goal, ranking it sixth on their list. But the agencies in the study reported the need to extend services with existing resources only 36 percent of the time. Given the significant cooperation required among healthcare organizations, this miscommunication or difference of opinion raises some concern.

Further analysis identifies another potential communication problem between the business decision makers and technical decision makers in the healthcare organizations. Technical decision makers reported most often that they were being asked to extend services through technology, and cost saving was a distant fourth on the list. This is almost an exact inverse of the business decision makers' perceived goals for technology investments.

The business decisions makers' attitudes to technology may provide some clues about this perception difference. Only 49 percent of primary healthcare business decision makers believe that their organization consistently aligns their business processes, network applications and network infrastructure. An even lower 38 percent of them felt that their technology investment was tied directly to the delivery of specific services, compared with 56 percent in the government. On the other hand, 82 percent of technical decision makers felt that the IT department worked closely with organizational leaders to ensure that technology solutions fulfilled organizational needs.

Perhaps business decision makers are just skeptical. The disjunction between technology investments and the perceived service objectives may be a significant contributor to the generally stronger focus on the easier-to-measure financial metrics. ICT decision makers are aware of this communication issue, with 37 percent of

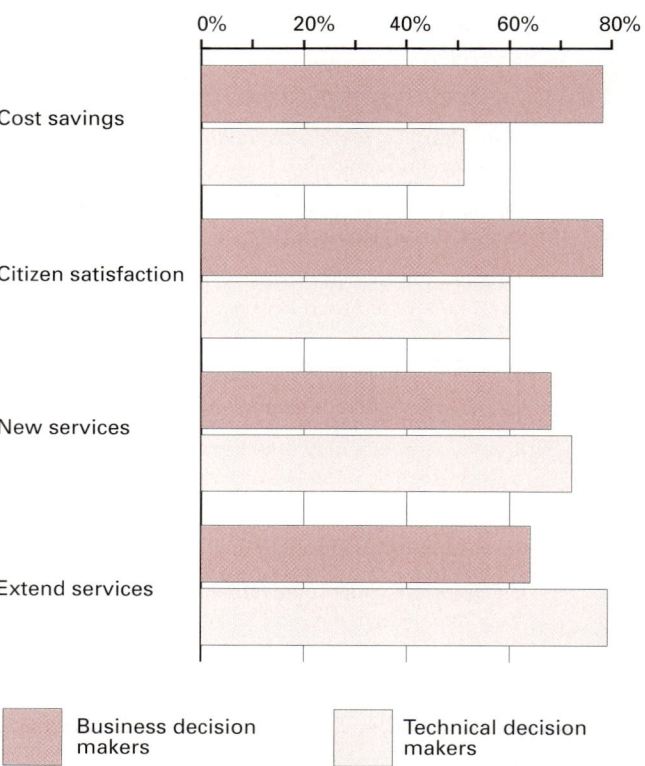

Figure 5. **Net Impact 2004 Primary Healthcare Differing Objectives**

them reporting unclear or ill-defined objectives as a barrier to future productivity gains through technology.

Forty-six percent of healthcare organizations are currently providing some e-services to citizens, but most are not measuring the effects of these services. Fewer than 10 percent of surveyed healthcare Connected Organizations are tracking measures such as percentage of self-service cases, number of requests filed online, or percentage of services available online.

Online healthcare services are obviously different from typical e-government services such as taxes or permits. Citizens are showing an increasing interest in researching healthcare and treatment information, so that they can play a more active role in their own care. With the aging of the European population and corresponding increase in chronic over acute medical conditions, the trend toward greater research and advocacy in healthcare treatment is expected only to increase. But providing healthcare services online does not necessarily mean an immediate jump into Web-based diagnostics or telemedicine solutions.

Government organizations have achieved significant operational savings by automating information queries (on average, it is the second most automated government process). This frees personnel from answering basic questions to work on more value-added tasks. Healthcare

---

Box 3. Service Best Practices

30–50 percent improvement
Applications

- Web interface: workforce training and collaboration tools

- Greater number of data sources (both inside and outside the organization) integrated with deployed enterprise applications

Networking technologies

- Deployment of real-time intrusion detection technology

- Having a layered security system designed to monitor traffic and detect intrusion

- Use of storage area network

Business process

- Functions: e-learning, support of regulatory compliance

- Automation of information service and delivery

- Integration of information service and delivery processes

Organizational culture and behavior

- Focus is on providing e-services directly to citizens, end customers

- Organization works to consistently align applications, network and business processes

- Business processes are re-engineered (regardless of timing) to use new technology capabilities

---

service, the number of website visitors and the percentage of relevant services available online. Public sector organizations that implemented the service volume best practices had an estimated 30–50 percent improvement in their volume metrics compared with the prior year. There is a significant opportunity waiting to be grasped by many healthcare organizations.

Box 4. Satisfaction Best Practices

45–65 percent improvement
Applications

- None identified

Networking technologies

- None identified

Business process

- Functions: data mining and analysis

- Automation of information and service delivery, workforce collaboration and training, problem diagnosis and resolution processes

Organizational culture and behavior

- Organization's strategic plan for developing and delivering services is regularly communicated throughout the organization

- Business processes are re-engineered (regardless of timing) to take advantage of new technology capabilities

- ICT department works closely with organizational leaders to ensure technology fulfils organizational needs

---

organizations have at least the same savings potential, as the public searches online for reputable and reliable sources of health information. Extending a basic health information service with electronic patient records, drug prescriptions and even appointment scheduling can provide significant boosts to productivity, and result in the desired cost savings.

However, privacy regulations and cultural reservations may be barriers to the online extension of personalized healthcare information. Technical decision makers in the healthcare field consider privacy regulations to be one of the greatest obstacles to future productivity growth. One technical solution to this problem is the use of network user authentication systems, such as public key infrastructure (PKI), which already show higher levels of adoption in healthcare than in the government organizations.

Net Impact 2004 measured differences in service volumes through metrics such as the number of citizens using the

Improving citizen satisfaction: a goal for many

Citizen satisfaction is the most commonly tracked productivity metric by the healthcare respondents, supporting the hypothesis that it is a proxy for quality of care in this study. In addition to the 47 percent of primary care respondents who measure it, 34 percent of them rate citizen satisfaction as the one metric they would most like to improve.

Previous *Net Impact* studies found that citizen or customer satisfaction is strongly correlated to the operational efficiency of the organization. Specifically, it has been found that, the more quickly the organization can respond and resolve the requests of its citizens, the higher the satisfaction rating. This trend was further substantiated in the *Net Impact 2004* study, which demonstrated that a 100 percent increase in efficiency metrics such as average time to case resolution and total cases resolved per time period would result on average in 37 percent and 30 percent increases in citizen satisfaction, respectively. The cumulative effect of increasing organizational efficiency is too complex to model, but its effect on citizen satisfaction has been demonstrated at a foundational level and can be intuitively understood.

It takes more than efficiency, however, to dramatically affect citizen satisfaction. *Net Impact 2004* found that focusing on services volume outcomes also has a significant positive impact on citizen satisfaction. The more citizens are able to use public sector services, the higher the citizen satisfaction. In fact, a 100 percent increase in organizational metrics such as number of citizens using services and number of visitors to an online portal created average customer satisfaction increases of 45 percent and 29 percent respectively.

In general, the leading indicators for improving citizen satisfaction with e-services are getting more people to use the services, especially online, or delivering faster turnaround times for requests. Overall, the organizations implementing the identified best practices for citizen satisfaction reported a 45–65 percent improvement in citizen satisfaction over the last 12 months.

People, process and technology

Net Impact 2004 found that process re-engineering to leverage new technology capabilities was a significant contributor to improvements in efficiency and cost reduction. When sequenced appropriately, processes and technology have an even greater impact—organizations that, in addition to other best practices, applied process re-engineering before the deployment of applications realized cost savings of 20–30 percent over 12 months. Those that re-engineered *after* application deployment were more likely to achieve only half that level of savings.

The net effect is that a slight difference in timing imposes as much as a 50 percent penalty on the possible improvement in operational costs. The importance of redesigning processes prior to technology implementation becomes quite clear when one considers examples, such as integrated prescription and drug interaction systems, or using automated mobile-phone text messages to on-call staff for scheduling issues. In these and other examples, the process should drive the technology specifications to meet the desired outcome rather than having the process constrained by the capabilities of an already deployed technology.

Primary healthcare organizations in this study were the most likely to re-engineer their business processes at some point, with 80 percent of them reporting positively compared with 70 percent of government organizations or healthcare agencies. However, they were least likely to do so *before* application deployment. Only 18 percent of them reported following this best practice. Compare this with the financial industry, considered by many studies to be

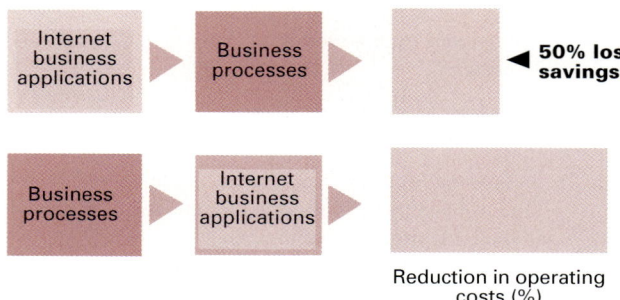

Figure 6. **Timing is Important**

Source: Momentum Research Group (2004)

one of the most productive users of technology. It is no coincidence that 40 percent of the financial services companies studied in *Net Impact 2003* re-engineered their processes *before* application deployment.

Organizations that spend millions of euros implementing and integrating complex ICT projects should not forget the corresponding investments in people. Organizational and employee issues present by far the biggest perceived barrier to future productivity growth. More than 35 percent of healthcare respondents identified the following factors as potentially the most limiting for future productivity gains:

- Internal resistance to changing processes
- Inability to absorb new technologies
- Inability to change staff behavior
- Centralized decision making
- No government standards for data integration

In contrast, 20 percent of respondents identified lack of budget as the single largest organizational obstacle to future productivity improvements. Whereas this should not suggest that the majority of public sector e-service initiatives have sufficient funding, it does indicate that business and technical decision makers are realizing the importance and difficulty they face in integrating technology into the daily practices of the organization.

Overcoming organizational inertia

Net Impact 2004 identified three practices that successful Connected Organizations commonly implemented to help counteract organizational inertia that can result if the people element is overlooked. The two cultural behaviors and one technology practice tended to have a positive effect across citizen service outcomes:

1. *Communicate regularly the strategic plan for services delivery.* It is easy to see how clear and consistent communication of objectives throughout the organization would help lower resistance to process and behavioral change while clarifying leadership support. Engaging the clinicians in

this dialogue is important as it is their working practices that are most likely to be affected. This virtually no-cost best practice is already being implemented by 42 percent of healthcare Connected Organizations.

2. *Maintain focus on providing e-services to citizens or end-customers.* This does not imply that internal or back-office functions should become a lower priority. In fact, many e-Government projects require a certain amount of internal process change and automation before the external aspects can be implemented. However, maintaining a focus on citizens appears to be as important for public sector organizations as a strong customer focus is for corporations. The vast majority (88 percent) of healthcare Connected Organizations is supporting internal e-services, but only 46 percent currently have external *Connected Health* initiatives targeted to citizens or end customers.

3. *Deploy Web-enabled workforce collaboration and training applications.* Web-enabled training applications are extremely time-efficient, allowing people to take courses specifically suited to their needs and schedule. This facilitates the development of individual and customized training programs for clinicians and other healthcare professionals, which can be crucial to successful project implementations. Although only 19 percent of the primary healthcare Connected Organizations indicated support for e-learning applications, 28 percent of the agencies said they did. Perhaps training is being focused at this level, instead of in the clinics and hospitals.

Conclusions

Productive countries are not necessarily those with the cheapest labor costs or lowest tax rates. As the *Net Impact* research shows, service response times, breadth of services, flexibility and innovation carry more weight in the public sector's cost/benefit analysis than just reducing costs. As we move into an increasingly global economy, the importance of public sector services continues to increase. While people have limited mobility between countries, corporations and capital do not. The public service infrastructure of a country is an important part of the cycle of productivity that generates economic growth and increases in living standards, and healthcare is a significant component.

Healthcare is clearly a knowledge-intensive activity. In both *Net Impact 2003* and *Net Impact 2004*, the automation initiatives identified with the largest economic impact in each industry were centered on the largest collections of knowledge. If this holds true for healthcare, it provides some details for a prescription for the industry.

The largest body of knowledge in healthcare is centered on patients. This implies that the greatest potential to improve access to, and the effectiveness and quality of, healthcare is at the clinics, hospitals and medical offices—

where the patients are. Re-engineering patient interaction processes and then automating repetitive tasks is an obvious, if complex, recommendation in this area. Interconnecting and integrating these processes across healthcare providers and their respective agencies would bring even greater benefits. Reducing duplication, cutting the opportunity for errors and ensuring treatment of the whole patient would not only reduce unnecessary medical expenses, it would simultaneously improve citizen satisfaction (and quality of care). This would also build the necessary foundation for supporting the inter-country interconnections outlined in the eHealth plans.

Leveraging this integrated knowledge base further with diagnostic and decision-support tools would bring yet more benefits. Identifying chronic medical issues early can reduce treatment costs and significantly improve quality of life. Patients can often be treated with simpler techniques, allowing them to manage their own care at home or by phone. Primary healthcare providers in this study appear to be making progress in their automation and integration activities, but are not necessarily reaping the full potential of their investments due to low levels of employee access to the applications.

Preventive action to reduce demand on the healthcare system is another potential benefit of this knowledge base. Providing reliable and accessible online information about public health issues, healthy living practices and disease prevention is an important first step. Exploiting this online resource to encourage appropriate use of self-service health management tools would further reduce routine demands on the system, which would be important in an era of aging populations and increasing demand.

Increasing self-service rates was identified in both the latest *Net Impact* studies as a leading indicator of productivity gains. However, half of the healthcare Connected Organizations do not have an external Web portal and even fewer have a focus on external e-services. This is a foundational component of *Connected Health* services, as well as central to the e-Health strategy, and should be given higher priority.

No one can predict the future of healthcare, but ICT and *Connected Health* solutions will likely play an increasing role in helping the industry face its challenges. This does not mean simply spending more on healthcare technology. It means following the model of the most productive organizations in the world, first redesigning business processes, then automating with appropriate investments in networks and applications, and then integrating across departments, organizations and even countries. This is a complex task, but healthcare can learn from other industries and its own leading adopters to leverage ICT as an enabler of change. Greater sharing and coordination is vital to the long-term success of public healthcare.

Note

1 Additional findings from past *Net Impact* studies can be found at Net Impact (URL).

References

CEBI (Canadian e-Business Initiative), Momentum Research Group, R. J. McClean, D. A Johnston, and M. Wade. 2002. "Net Impact Study Canada". Online. http://www.netimpactstudy.com/ca/

European Commission. URL. Online. http://europa.eu.int/information_society/eeurope/2005/all_about/ehealth/index_en.htm

Momentum Research Group. 2003. "Net Impact 2003: United States Private Sector". Online. http://www.netimpactstudy.com/nis_2003.html

—— 2004. "Net Impact 2004: Europe eGovernment". Online. http://www.netimpactstudy.com/nis_2004.html

—— , R. Litan and H. Varian. 2001. "Net Impact: National Economic Benefit". Online. http://netimpactstudy.com/nis_2002.html

—— , F. Pennarola, F. Giavazzi, M. Dallochio and E. Valdani. 2002. "Net Impact Study Italia". Online. http://www.netimpactstudy.com/it/

Net Impact. URL. Online. http://netimpactstudy.com

World Bank. 2004. *World Development Indicators 2004*. Washington, DC: World Bank. Online. http://www.worldbank.org/data/wdi2004/

Chapter 3

The ICT Sector and the Global Economy:

Counting the Gains

Markus Haacker, International Monetary Fund

Introduction

Technological advances in information and communication technology (ICT) over recent decades have transformed the way businesses are operating in many sectors, and have resulted in changes in the pattern of global trade in goods and services. To name just a few examples, complex production processes in the motor industry have become more differentiated and decentralized, and by global sourcing of components companies are able to reduce production costs; labor-intensive services in banking have increasingly been automatized, and through online banking customers can conduct a wide range of transactions that in (most of) the twentieth century would have required the attention of a bank clerk (and a visit to the bank). At the same time, banks can offer increasingly complex products at the retail level; much has been written about new categories of trade in services, such as call centers and back-office accounting functions.

These changes are substantial, and the transformation of the global economy driven by advances in ICT is by no means complete. The prospects for economic growth in and enabled by the "new economy" have attracted much attention, although the crash in IT equity prices in late 2000 has brought home (for some painfully) the realization that the economic "laws" of competition (and creative destruction) fully apply to the ICT sector. Rather than adding to the existing literature on the ongoing transformation of the global economy driven by ICT,[1] we take a low-tech approach and focus on the production and use of ICT equipment. Just as in a criminal investigation—where there is a murder there must be a weapon—for our purposes it generally applies that, where an economic activity is enabled by advances in ICT, some ICT goods must be used in the process. Focusing on the data on sales and purchases of ICT goods that are available for a large number of countries also enables us to cast a much wider net than the existing literature on the economic benefits of technological progress in the IT sector, much of which is confined to the United States or the OECD countries.[2]

Following a discussion of how to measure technological progress in the ICT sector, we will assess the economic role of ICT using two different (but complementary) approaches. First, we focus on the contribution of technological progress to GDP growth and the growth of real demand. Distinguishing between GDP (that is, the value of production) and real demand (consumption and investment) is important because the major producers of ICT goods, while enjoying high productivity gains in this sector, export most of their production, whereas countries that import most or all the ICT goods they require nevertheless benefit from lower prices for these products. Second, we estimate the "social surplus" from declining prices of ICT goods. When prices of ICT (or any other) goods are high, such goods are applied exclusively to their most valuable uses. As the prices decline, ICT goods already in use become cheaper, and at the same time they are also used for a wider range of applications. By estimating a demand curve for ICT products, it is possible to calculate the social surplus in a similar way to the consumer surplus in microeconomic theory. As this approach requires data on sales of ICT products only, the analysis of the "social surplus" covers a larger number of countries, including low-income countries. As many countries import most if not all their ICT goods, we also assess the impact of trade barriers on the dissemination of ICT products.

Measuring the Rate of Technological Progress

Throughout this chapter, we use the decline in relative prices of various IT products to measure the rate of technological progress.[3] However, unlike in other sectors (say, textiles or steel), where the prices of specific products—at least over longer time periods—reflect production costs, and falling prices can be relatively easily interpreted in terms of productivity gains, in the IT sector it is the characteristics of the products that change, and products that have a high market share at some stage may quickly become outdated and economically irrelevant. To measure the technological gains in the production of computers or other IT equipment and components, it is much more important to look inside (at the specifications of their components) than outside (at their price tags).

Measuring the rate of technological progress in the IT sector is, therefore, not a straightforward matter. The target is moving, and it is also not possible to simply break down IT equipment into the sum of its parts on a level where productivity gains could be measured more adequately (there may be many parts, their composition changes over time, and they add to the features of the assembled product in ways that are not directly related to the technical features of any individual part). As a

consequence, the methods of measuring price changes in the IT sector differ substantially across countries (see, for example, Schreyer 2002). Indeed, van Ark et al. (2003), focusing on OECD countries, note that differences in national data on the prices of IT products for these countries primarily reflect differences in methodologies.

Table 1 illustrates the very substantial declines in the prices of IT products. For example, the prices of computers and workstations (in the US producer price indices) have declined by over 95 percent since 1995. The table also provides information about the prices of IT components; in particular, the prices of microprocessors declined very rapidly (to about one-tenth of a percent of the initial price) over the same decade, reflecting the fact that more powerful processors were rapidly driving older ones from the market. Adjusting the price data illustrated in Table 1 for exchange rate changes then yields international price indices such as those used in the cross-country analysis below.

Before we turn to a discussion of the macroeconomic gains arising from the IT revolution from a global perspective, we need to discuss how to estimate price changes in IT products for a large number of countries. Data as detailed as those for the United States are not available for most countries, and where they are, different methodologies frequently mean that the price data are not comparable. Our measures of the decline in prices of IT goods for individual countries are therefore based on price data from the US national accounts and the US producer price indices (Table 1).[4] These price indices were obtained in a way that takes into account the changing characteristics of IT products, based on price differences between products within a certain category (such as computers) but with different technical specifications. By comparing different computers within a given time period, it is thus possible—with a small leap of faith—to compare the prices of "typical" IT products over time. The second reason why we adopt this approach is that IT products are generally highly tradable goods, where transport costs account for only a small proportion of the value of the products.[5] Third, this method requires that product markets are competitive and prices do adequately

Table 1. **Changes in Relative Prices of IT Goods in the United States (Accumulated % Change over the Period Indicated)**

IT GOODS	SOURCE	1985–1995	1995–2004*	1985–2004*
Computers, peripheral equipment, and software	BEA (CPI)	−87.5	−91.7	−99.0
Computers and workstations	BLS (PPI)	…	−95.2	…
MOS memory devices	BLS (PPI)	−58.1	−83.9	−93.2
Digital monolithic integrated circuits	BLS (PPI)	−55.0	−76.1	−89.3
Microprocessors	BLS (PPI)	−53.2	−99.9	−99.9

* Latest available month for 2004

CPI: Consumer price index
PPI: Producer price index

Sources: Own calculations, based on data from US Bureau of Economic Analysis (BEA) (URL) and from Bureau of Labor Statistics (BLS) (US Department of Labor) (URL).

Table 2. IT Products in the US National Accounts Statistics (in units indicated)

	ANNUAL RATE OF GROWTH AT CONSTANT PRICES (IN PERCENTAGE POINTS)		RATE OF CHANGE IN PRICES (IN PERCENTAGE POINTS)		SHARE OF NOMINAL GDP (%)		
	1985–1995	1995–2003	1985–1995	1995–2003	1985	1995	2003
GDP	2.9	3.6	2.8	1.4	100.0	100.0	100.0
Final sales of computers	21.5	32.5	−13.5	−19.1	1.1	1.0	0.9
Personal consumption expenditures: computers, and peripheral equipment	48.2	56.8	−16.5	−21.0	0.07	0.33	0.42
Non-residential investment: computers and peripheral equipment	21.6	22.2	−12.1	−15.6	0.80	0.75	0.76
Non-residential investment: software	15.3	9.1	−2.7	−0.7	0.56	0.57	0.61

Source: US Bureau of Economic Analysis (URL)

reflect costs, rather than changes in mark-ups.[6] On this, the *New York Times* (10 December 2004), in the context of the impending sale of IBM's personal computer division, aptly comments: "The bloom has long been off the PC market, profits have plummeted, and the computer has become utterly generic." Still, on a year-to year basis, the assumption of an immediate pass-through from exchange rate changes to the prices of IT goods appears to be somewhat strong for final products, in light of existing stocks and market power on the retail level. This is not a big concern for our analysis, though, which focuses on longer time periods. At least for standardized components (such as microprocessors), our approach thus can be expected to yield good results.

Measuring the Macroeconomic Impact

As a point of departure for our broader cross-country analysis, we provide a macroeconomic assessment of the role of the IT sector, using national account data for the United States. This makes sense as the United States, which is both an important producer and user of IT products, has relatively good national account's data on this sector, and has inspired much of the research on the macroeconomic aspects of technological progress in the IT sector. After illustrating the key issues in the context of the United States, we broaden the focus to assess the impact of ICT across countries. This involves developing a much simpler approach, built around available databases of IT production and usage across countries.

Several lessons from a macroeconomic perspective about the role of the IT sector can be drawn from Table 2. First, while the share of the IT producing sector in GDP (represented by the line item "final sales of computers") is about 1 percent, its contribution to economic growth (owing to high productivity growth in the IT sector) is much higher. For example, with 1995 used as a base year, if a sector accounting for 1 percent of GDP grows at an annual rate of 32.5 percent, this adds 0.325 percentage points to aggregate growth, a substantial proportion of the

aggregate growth rate of 3.6 percent. Second, over the period covered in the table, computers have become important consumer goods, as consumer expenditure on computers and related goods as a percentage of GDP rose sixfold between 1985 and 2003, and accounted for about 5 percent of spending on durable consumer goods, and amounted to 10 percent of consumer expenditure on motor vehicles (the largest item in the consumer-good category) and parts in 2003. Third, the most important component of IT spending is investment. Over the period 1985–2003, investment in computers and peripherals increased by over 20 percent a year in real terms. However, as the relative price of IT goods declined, the share of investment in computers and peripherals in nominal GDP likewise declined somewhat. It is also noteworthy that a very substantial proportion of IT-related investment is spent on purchasing or developing software. Fourth, as investment is more IT-intensive than GDP overall, rapidly falling prices of IT goods relative to the GDP deflator mean that the relative price of investment goods overall falls. As a consequence, a given investment rate (as a percentage of GDP) translates into a higher rate of capital accumulation.[7] Finally, while the above observations also apply to other IT producers and users, a particular feature of the data for the United States is that the share of IT hardware production in nominal GDP has declined, while the share of IT hardware expenditures in nominal GDP has risen for the period 1985–2003.

These observed trends have inspired work to assess the direct and indirect contributions of the ICT sector to economic growth, mainly through technological progress in the IT sector (represented in the national accounts by falling prices of domestically produced IT goods) and capital deepening (as the relative price of investment goods falls). While it is possible to draw inferences on the magnitude of the first of these effects directly from national accounts data, estimates of the role of capital deepening are usually based on a calibration that uses a simple model of economic growth. Along these lines, van Ark et al. (2003) provide estimates of the impact of the IT

Table 3. **Contributions of the ICT Sector to Labor Productivity Growth, Several Countries, 1995–2000 (%)**

COUNTRY	LABOR PRODUCTIVITY GROWTH	TOTAL ICT CONTRIBITION	CONTRIBUTION OF ICT CAPITAL	CONTRIBUTION OF PRODUCTIVITY GROWTH IN ICT PRODUCTION
France	1.35	0.56	0.32	0.22
Germany	1.76	0.53	0.37	0.16
Ireland	5.88	3.70	0.68	3.02
Italy	1.13	0.53	0.38	0.15
United Kingdom	1.76	0.97	0.65	0.32
United States	2.21	1.15	0.75	0.40
European Union	1.43	0.60	0.40	0.20
Japan	1.9	1.69	1.14	0.55
OECD	2.1	1.36	0.73	0.63

Note: The total contribution of the ICT sector is calculated as the sum of the entries in the two right-hand columns

Sources: van Ark et al. (2003) except Japan and OECD: Davies, Brookes and Potter (2000), observations relate to 1996–99 only

revolution on labor productivity which highlight the role of capital deepening (from the use of IT capital) and productivity gains within the ICT sector (Table 3, which is supplemented by some data from Goldman Sachs— Davies, Brookes and Potter 2000).[8] For a group of advanced industrialized countries, these estimates show that the estimated contribution of ICT to productivity growth is substantial but by no means uniform across countries, accounting for between 30 percent and almost 90 percent of productivity growth in these countries. In terms of the productivity gains within the ITC sector, Ireland—the most significant producer of ICT hardware located in Europe—stands out: there, the gains contribute 3 percent to aggregate productivity growth. Also, the extent of ICT-related capital deepening—reflecting investment rates in ICT equipment—differs very

substantially across countries, from 0.32 for France to 0.75 for the United States (and 1.14 for Japan, but that number relates to a shorter period).

The second major issue that needs to be taken into account when one analyses the benefits of the ICT revolution across countries is that the pattern of IT use and production differs very substantially between countries. The production of ICT goods is concentrated in several countries, where final sales of ICT products account for a very substantial proportion of GDP and thus—with rapid productivity gains in this sector—of GDP growth (Table 4). It is important to note that the share of IT production in GDP reflects gross sales of ICT products, not on the IT sector's value added. This creates two potential data problems related to the treatment of intermediate inputs: double counting of ICT products (which are also used as intermediate inputs) and, if a country imports (or exports) most of the components of IT products, misattribution of the productivity gains. To mitigate this problem, the measure of IT production includes only final ICT products (such as computers), plus net exports of active components (such as microprocessors), that is, IT intermediate inputs for which the rate of technological progress is most pronounced. This means that the estimated productivity gains of a country that assembles IT products from imported components are lower than in a country that also produces these components (or even exports them). As described above, the productivity gains in the IT sector are measured by the relevant price indices from the US national accounts, adjusted by changes in the respective exchange rates. The contribution of IT-related productivity gains to the growth of GDP (or domestic demand) is then measured according to the change in the price index for IT products (or expenditure) relative to the GDP deflator.

Several lessons can be drawn from the data summarized in Table 4. First, the impact of productivity gains in the

Table 4. **The Impact of IT-Related Productivity Gains on GDP and Domestic Demand Growth, 1996–2000**

COUNTRY	IT PRODUCTION (FINAL SALES, % OF GDP, 2000)	RELATIVE PRICES OF IT PRODUCTS (ANNUAL % RATE OF CHANGE)	CONTRIBUTION OF PRODUCTIVITY GAINS TO GDP GROWTH (%)	IT EXPENDITURE (% OF GDP, 2000)	PRICE INDEX FOR IT EXPENDITURE (ANNUAL % RATE OF CHANGE)	CONTRIBUTION PRODUCTIVITY GAINS TO DOMESTIC DEMAND GROWTH (%)
Australia	0.2	–9.0	0.03	1.8	–15.6	0.25
France	1.5	–4.9	0.06	1.7	–7.7	0.11
Germany	1.1	–4.6	0.04	1.6	–10.3	0.12
Ireland	14.2	–18.1	2.1	1.4	–12.9	0.2
Japan	2.6	–12.8	0.37	1.2	–13.6	1.17
Korea	9.7	–10.4	0.85	3.0	–9.2	0.28
Malaysia	31.5	–15.0	3.31	1.4	–10.7	0.21
Singapore	39.2	–17.6	6.71	2.7	–10.3	0.30
South Africa	0.3	6.4	–0.01	2.4	–9.8	0.19
Sweden	4.3	2.7	–0.10	2.4	–12.6	0.31
UK	2.0	–16.2	0.30	2.1	–16.3	0.31
US	1.8	–15.1	0.28	2.4	–16.7	0.39

Source: Bayoumi and Haacker (2002)

IT-producing sector on GDP growth is substantial in many countries, reflecting primarily the differences in the ratio of IT sales to GDP. Reflecting exchange rate fluctuations and differences in the composition of IT production and the extent of net exports (or imports) of active components, the change in relative prices of IT products differs across countries, so that the relationship between IT sales and GDP growth is not uniform. In South Africa and Sweden the relative price of IT products actually rose, so that the contribution to GDP growth is negative.

Table 4 also shows that the productivity gains in IT production largely dissipate as the major producers export most of their product. While among the countries covered in Table 4 the ratio of IT sales to GDP differs by a factor of almost 200, IT expenditure as a percentage of GDP differs by a factor of only 2 across these countries. As productivity gains translate into falling prices, IT exporters suffer a decline in their terms of trade that eliminates most of the gains. While the major producers of IT goods also have relatively high rates of IT expenditure, and the impact of productivity gains from IT production on domestic demand growth is therefore relatively great, they lag behind countries like the United States or the United Kingdom, where IT production is much less important. At the same time, countries like Australia or South Africa, where the IT producing sector is very small, enjoy benefits in terms of domestic demand growth that are even greater than those of Ireland, one of the most significant producers.

IT, Productivity and the Social Surplus

In order to assess the overall impact of declining prices of IT products on the economy, it is necessary to find a way to "add up" the impacts on a year-to-year basis, discussed above, as they accumulate over time. The method used below is based on the notion of a "social surplus". The social surplus, like the consumer surplus in microeconomic theory, is based on the recognition that initially, when the costs of using IT products were high, they were used only for high-priority applications. As IT products have become cheaper, their range of uses has expanded. For example, not only have computers become omnipresent in offices and common in private homes, their usage has deepened as computing power of the most common configurations has expanded.[11]

Figure 1 illustrates the social surplus that arises as technological progress results in a decline in the price of some IT goods from P_0 to P_1. This results in "windfall" gains (represented by the rectangle containing light shade) for the original users of IT goods, purchasing quantity Q_0. However, as IT goods become cheaper, their usage also expands to Q_1. Accordingly, the increase in the

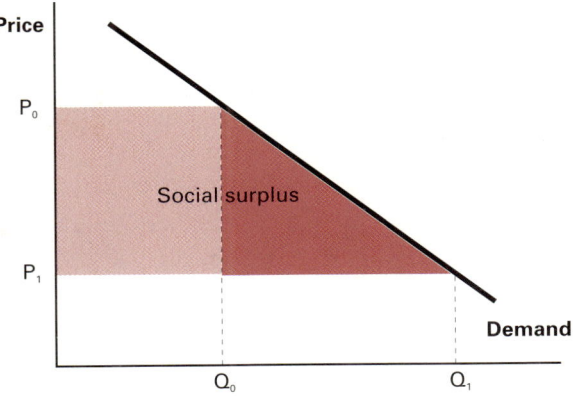

Figure 1. **The Demand for IT Goods and the Social Surplus**

social surplus also includes the gains of these users (represented by the triangle containing dark shade). In order to estimate the social surplus for individual countries, we make use of data on national expenditures on IT hardware and communications equipment from WITSA (2004). The empirical approach, which is described in more detail in Bayoumi and Haacker (2002), involves a panel data estimation which, in addition to the relative price of the respective IT goods, includes the level of GDP per capita and a time trend. Thus, rather than tracing the shape of the demand curve by drawing a line through the observations that have become available over time, this approach allows the demand curve to shift in response to changes in GDP per capita or technological progress in related fields that expands the range of applications of IT products (crudely captured by the time trend).

Digital Planet 2004 (WITSA 2004), a survey of the global information economy, focuses on IT spending across countries, distinguishing between computer hardware, software, services, and communications spending, and covers 69 countries.[12] While spending on software and services accounts for a substantial share in IT spending (31 percent in 2003, and rising), we focus on computer hardware and communications, the areas most directly affected by advances in technology. The countries covered include the major producers and consumers of IT and communication products, but also numerous smaller economies and low-income countries. In terms of the analysis presented below, it is convenient to group the countries by income level, distinguishing between countries with a level of GDP per capita higher than US$5,000[13] and those with a level of GDP per capita below US$5,000.[14] (Data on GDP per capita, in current US$ for the year 2003, were taken from the World Economic Outlook database [IMF URL] of the International Monetary Fund.) The lowest-income countries included in this latter group, with GDP per capita less that US$500, are Pakistan ($493), Kenya ($436), and Bangladesh ($369).

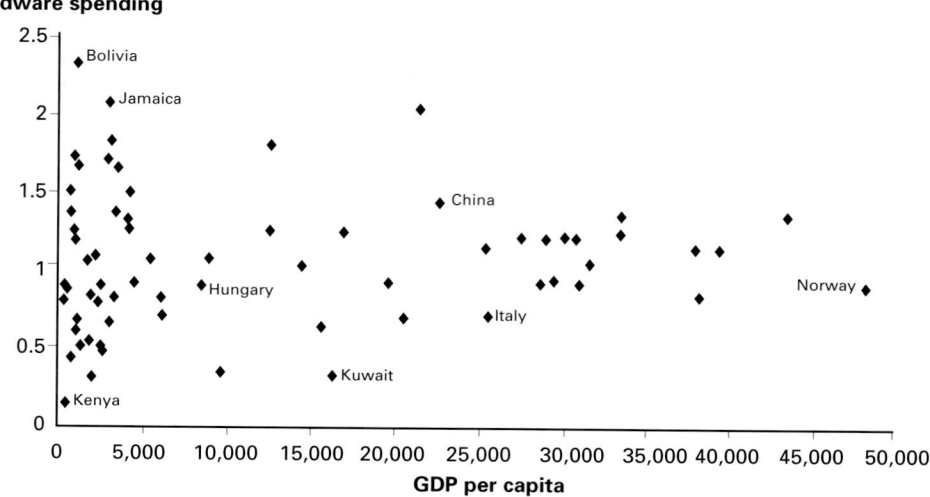

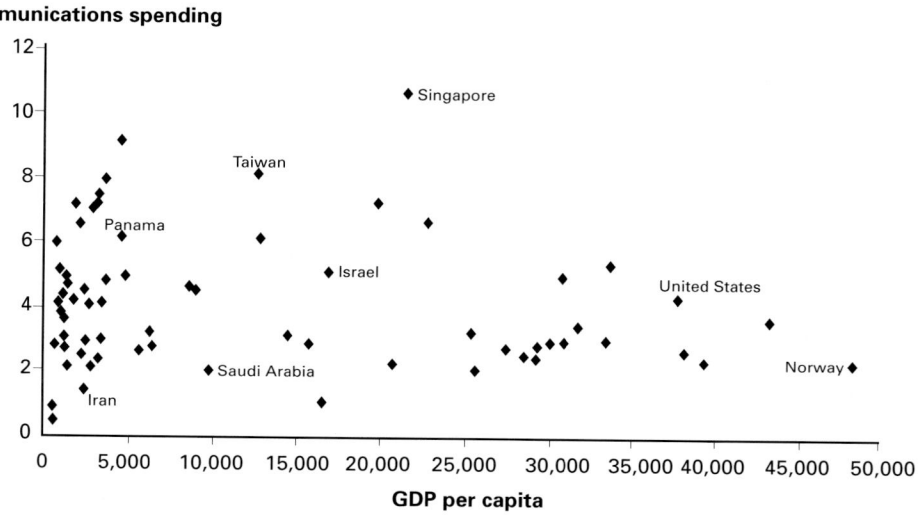

Figures 2A and 2B show the level of IT hardware and communications spending across countries as a percentage of GDP. The most notable feature of the data for both forms of spending is the high dispersion for countries with relatively low income, especially for the 36 countries with level of GDP per capita of less than $5,000. For the second group of countries (16 countries with levels of GDP per capita between US$5,000 and $25,000), the dispersion is still relatively high (the standard deviation is somewhat lower for IT hardware, but higher for communications spending, than in the first group). Finally, for the 17 countries with levels of GDP per capita exceeding US$25,000, the dispersion is much lower both for IT hardware and for communications spending.

For the countries with lower levels of GDP per capita, the quality of statistical data is likely to be lower, contributing to the extent of dispersion across countries. For that reason, we adopt a more robust approach for these countries (using a somewhat arbitrary cut-off point of US$5,000) and look primarily at the ranking of countries

within this group. While the cross-country differences in spending for the second group are similar to the first group, they are less likely to reflect measurement error, combining this group with the higher-income countries also makes sense because the second group also includes the lower-income OECD countries and some of the largest producers of IT products.

To estimate the gains from declining prices of ICT products (as illustrated in Figure 1), it is necessary to estimate the demand functions for IT and communication products. Our data for 68 countries (from WITSA 2004) were available for the years 2000–2003 only. Longer time series (1992–1999) for IT and communications spending were available for 41 of the 68 countries from an earlier version of WITSA's *Digital Planet*, but a change in contractor appears to have introduced some changes to data for individual countries so that it was not possible to merge the datasets. We therefore use price and demand elasticities estimated for the 41 countries for the period 1992-99, but adjust the intercept for these and the other 27 countries in line with 2000-2003 data.[15]

Figure 3. Social Gains from Falling Prices of IT Products (% of GDP) and GDP per capita (in current US$), 2003 (for countries with GDP per capita of less than US$5,000)

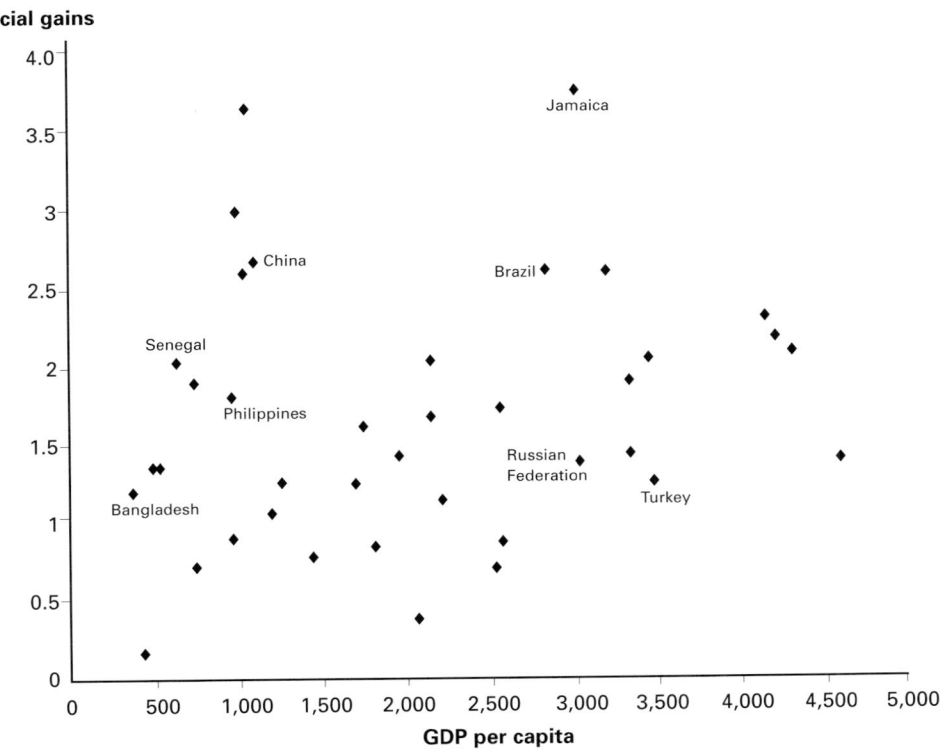

Figure 3 shows social gains from lower prices of IT products for countries with GDP per capita of US$5,000 or less. While there is a mild positive correlation between GDP per capita and the social gains (an increase in GDP per capita is associated with an increase in social gains of 0.1 percent of GDP), this relationship, in light of the high dispersion, is not statistically significant. However, some insights can be gained from looking at which countries enjoy relatively high (or low) gains. Among the 10 countries posting the highest gains (Jamaica, Bolivia, Honduras, China, Uruguay, Brazil, Ukraine, Malaysia, Costa Rica, and Panama) are seven from Latin America and the Caribbean. Next (rank 11–13) are three African economies (South Africa, Senegal, and Cameroon), with gains of around 2 percent of GDP. As the latter two economies are among those with the lowest levels of per capita income in the sample ($634 and $740 respectively), this also indicates that low-income countries can enjoy substantial benefits from cheaper IT products.

Figure 4 illustrates the gains for the higher-income countries (with GDP per capita exceeding US$5,000). The major gains in this group accrue to important IT producers (Singapore, Korea), which also have relatively high IT expenditure. The 10 countries gaining most also include the high-income countries or regions with relatively high levels of IT production (Hong Kong, Israel, Japan, and the United Kingdom). However, a high level of IT production is not a necessary condition for high gains, as the examples of Germany, Sweden, and

Switzerland show. It is noteworthy that, compared with an earlier study (Bayoumi and Haacker 2002), the United States has slipped from rank 1 to rank 13 as a result of the strong US dollar. The five countries posting the lowest gains in this group include three of the southern European Union countries (Greece, Italy, Spain), and two countries that generate a large share of GDP from resource extraction (Kuwait, Saudi Arabia).

To turn to communication equipment, Figure 5 shows the distribution of gains for the countries with GDP per capita lower than US$5,000. In this group, the largest gains are posted by several eastern European countries (Bulgaria, Ukraine, Russia, and Romania, at ranks 1, 2, 4, and 10 respectively), reflecting high investments in their telecommunications sectors. Some of the Latin American and Caribbean countries featuring high IT spending are also included in this group (Jamaica, Costa Rica, Panama). Other countries with relatively high gains are Jordan, Turkey, and Zimbabwe. Interestingly, Jordan and Zimbabwe have very low levels of IT spending, which suggests that the pattern of usage of ICT products is not uniform across countries. However, Iran and Kenya are among the countries realizing the lowest gains on both counts.

As for the higher-income countries (Figure 6), the major gains (as with the lower-income countries) accrue to several eastern European countries (Czech Republic, Hungary, Slovakia, and Poland, ranking 1, 3, 5, and 10

Figure 4. **Social Gains from Falling Prices of IT Products (% of GDP) and GDP per capita (in current US$), 2003 (for countries with GDP per capita of more than US$5,000)**

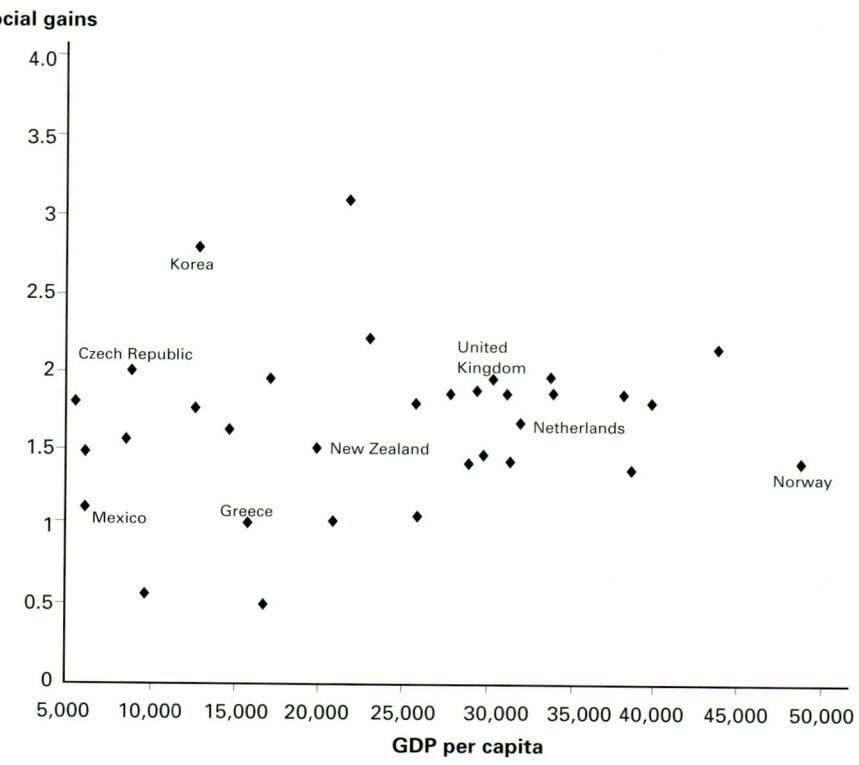

respectively). In addition to the Czech Republic, several other countries or regions that rank high in terms of gains from using IT products also enjoy high gains from using communication equipment: Korea, Singapore (both major producers of IT products), Hong Kong, and Israel. Other countries with relatively high gains are New Zealand and the United States.

Barriers to Trade Impede Access to ICTs

As German poet Heinrich Heine observed in 1844, trade barriers are not effective in preventing information flows across borders.[16] With the breakthroughs in ICTs over the last decades, the dissemination of information has become both cheaper and harder to control. However,

Figure 5. **Social Gains from Falling Prices of Communication Equipment (% of GDP) and GDP per capita (in current US$), 2003 (countries with GDP per capita of less than US$5,000)**

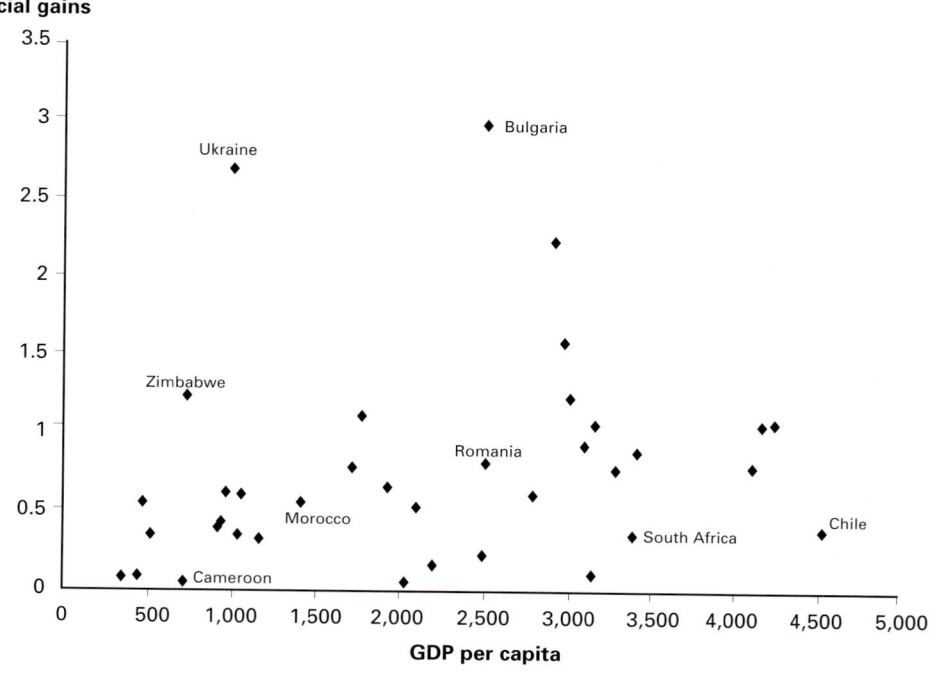

Figure 6. **Social Gains from Falling Prices of Communication Equipment (% of GDP) and GDP per capita (in current US$), 2003 (countries with GDP per capita of more than US$5,000)**

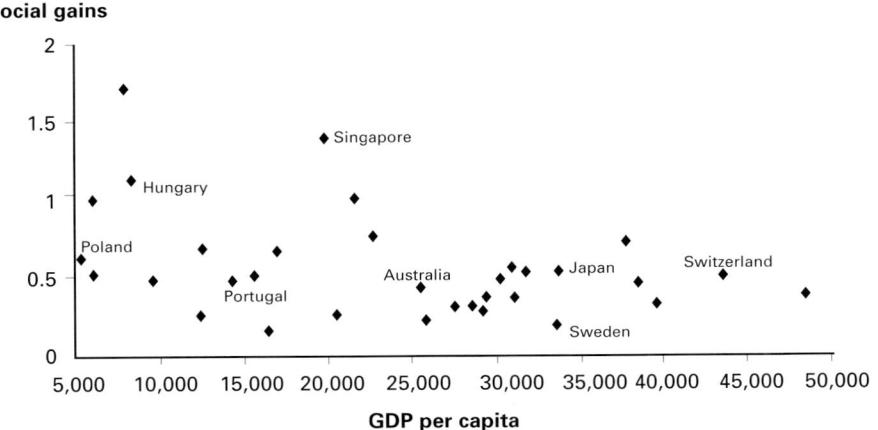

participating in and utilizing the information flows enabled by advances in ICTs requires access to ICT equipment and is much facilitated by a well-developed communication infrastructure. As most countries import this equipment, trade barriers therefore do impede information flows (in the same way as German censors in Heine's day impeded dissemination of his ideas by keeping his books out of circulation).

How effective are trade barriers in impeding access to information? Figures 7A and 7B illustrate the relationship between tariff rates and sales of IT hardware for all countries where data were available, and for the group of countries with income per capita of less than US$5,000.

These figures suggest that trade barriers are effective in reducing access to information technologies, especially for countries with lower GDP per capita.[17] Six of the 10 lower-income countries who gain most are also among the 10 countries with the lowest average tariff rates in this group (Bolivia, Costa Rica, Honduras, Malaysia, Panama, Ukraine), and six of the 10 lower-income countries who gain least are among the countries with the highest average tariff rates (Egypt, Iran, Kenya, Morocco, Tunisia, and Zimbabwe). For communication technologies (not shown), the picture is similar. Among the 10 countries who gain least are 8 of the 12 countries with the highest tariff rates. A notable exception is Zimbabwe (the fourth

Figure 7A. IT Products: Social Gains and Average Tariff Rates (All Countries)

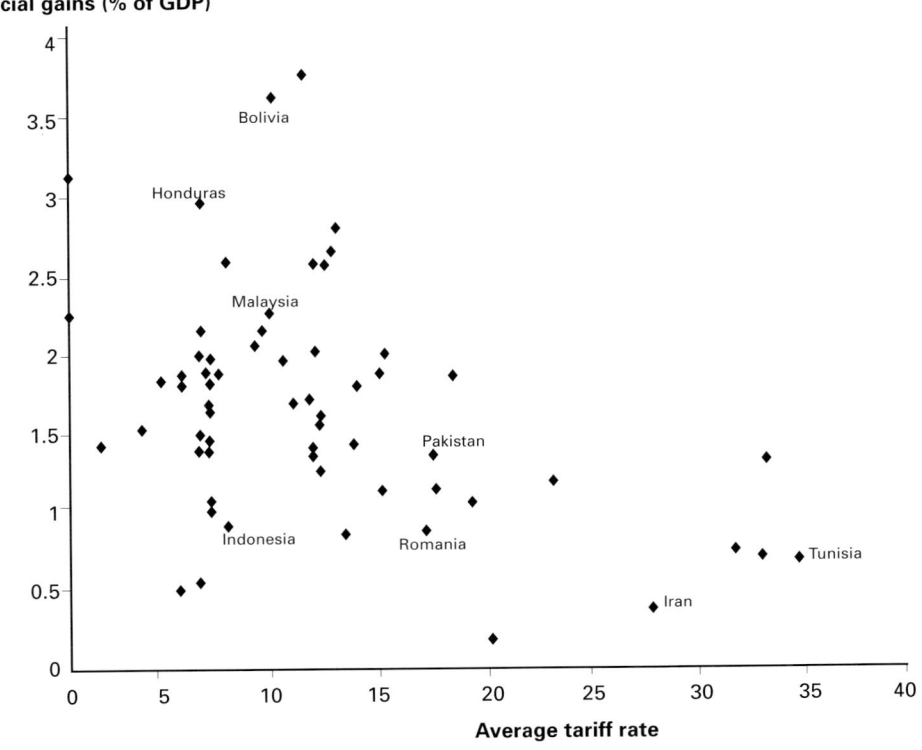

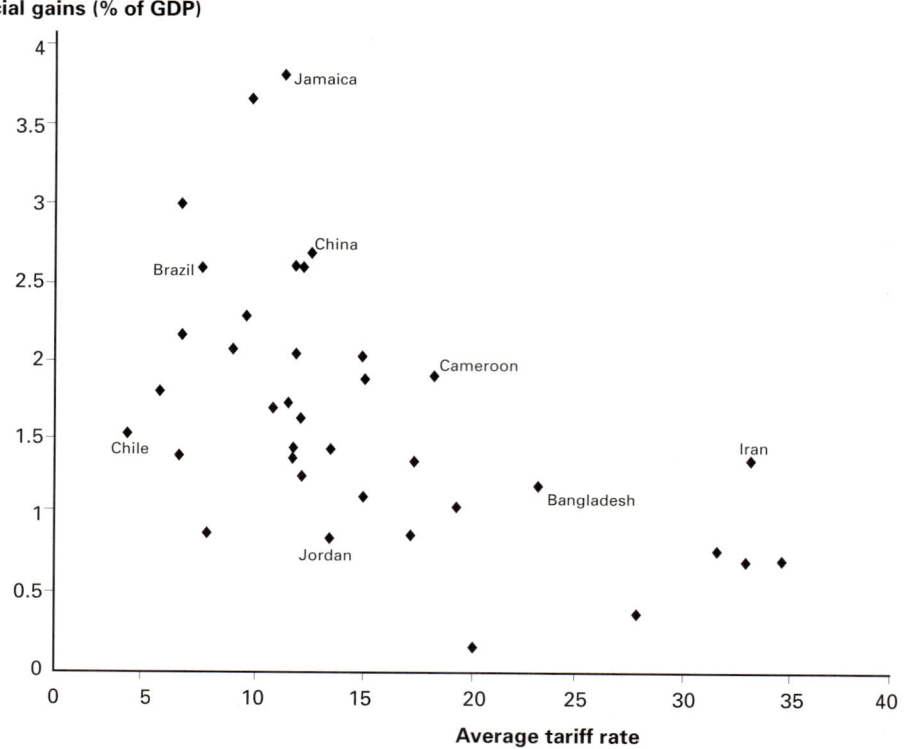

highest gains among lower-income countries, but the third-highest tariff rates), possibly reflecting the fact that the cellular phone sector in this country is relatively competitive.

Conclusions

We take the low road in assessing the contribution of the ICT sector to the global economy. While much of the new economy may take a weightless or invisible form, we postulate that professionals utilizing ICT equipment do not differ so much from people engaged in more mundane work in the sense that in front of each ICT professional there is a computer (and quite possibly a cell-phone pressed to his or her ear). More generally, rather than measuring the global impact of the ICT sector by the output of ICT producing sectors and the productivity gains in IT-using sectors, we focus on the demand for ICT products and their role as inputs to the production process. As our approach needs relatively little data (in its simplest form, it requires data only on sales of the main ICT products across countries), it facilitates comparisons between larger groups of countries, including low-income countries for which detailed national account data are not available.

First, we assess the contribution of technological progress in ICT production to economic growth and growth in real demand (consumption or investment). While the contributions of ICT production to economic growth is

impressive in some countries, we find that most of these gains migrate to the importers of ICT products as increases in productivity result in declining terms of trade. While some of the major producers also enjoy significant gains from using ICT products, the major beneficiaries also include countries like South Africa or Australia, which do not feature any significant production of ICT products.

Second, we estimate the social surplus associated with the use of ICT products. Like the consumer surplus in microeconomic theory, the social surplus measures the gains to consumers and producers alike from purchasing ICT products at declining prices. For spending on IT hardware and on communications, we observe a high dispersion in lower-income countries. For this group (defined here as countries with per-capita income of less than US$5,000), some Latin American and Caribbean countries are among the main users of IT hardware (as a percentage of GDP). As for communications spending, the major beneficiaries among the countries with lower income per capita include several east European countries, presumably because they are spending substantial amounts on upgrading their communication systems. Among the high-income countries, those with a significant ICT production also tend to be among the major beneficiaries in terms of the use of ICT products. If we exclude these countries, spending on IT hardware is positively correlated with GDP per capita, and some of the major beneficiaries (like Germany, Sweden, and Switzerland) have very small IT-producing sectors. For the high-income group, there also is some overlap among the countries that gain most from falling prices in IT hardware and communications, respectively. This group includes Korea and Singapore (also major producers), as well as Hong Kong and Israel.

Finally, we assess the role of trade restrictions in the dissemination of ICTs. While new technologies have reduced the costs of processing and disseminating data, utilizing them requires ICT equipment and some infrastructure. Our findings suggest that trade barriers are effective in reducing access to ICTs as they impede access to ICT equipment.

Notes

1 On that literature, we comment (with W. H. Auden): "Perhaps the roses really want to grow, the vision seriously intends to stay; if I could tell you I would let you know."

2 Many of the most important papers on the economic impacts of technological progress in the ICT sector are included in special issues of the *Journal of Economic Perspectives* (Vol. 14, 2002) and the *Review of Income and Wealth* (Vol. 48, No. 2, 2002). OECD (2003; 2004) provide an exhaustive discussion of the economic impact of ICT, including in-depth studies using firm-level data.

3 We define the change in the "relative price" of an IT product as the change in the price of the respective product relative to the change in the GDP deflator. As the IT products generally also form part of GDP and thus are included in the denominator, treating the GDP deflator as given introduces a bias which, in the above example using US price data, is small as the various products account for a small proportion of GDP. As most other countries do not apply the "hedonic" price indices as used in the US National Accounts, the bias does not arise when the method is applied to these countries.

4 An alternative approach is based on the assumption that the relative price of IT goods (compared with a deflator for investment goods or GDP) within countries is constant across countries. This would mean that the decline in the relative prices of IT goods, as detailed in Table 1, applies to all countries. As we look at a large number of countries (many of which import essentially all their ICT equipment) and need to accommodate potentially large exchange-rate fluctuations between countries, we treat ICT equipment as tradable goods the prices of which fluctuate in line with the exchange rate.

5 As a consequence, IT production is highly concentrated in a small number of countries.

6 These issues are discussed in more detail in van Ark et al. (2003).

7 However, since most IT equipment depreciates faster than other forms of investment, this—other things equal— would result in a lower rate of net capital accumulation.

8 Other important cross-country studies include Daveri (2001; 2003), Colecchia and Schreyer (2002), Davies, Brookes and Potter (2000), and Roeger (2001). See also van Ark (2002) for a summary of the literature.

9 We do not necessarily assume that the pass-through from changes in the exchange rate to the prices of IT goods is immediate, as we are working with annual averages.

10 This, however, is less relevant in our context as we are primarily interested in the overall macroeconomic gains, including any windfall gains or losses to traders.

12 WITSA (2004) also provides a breakdown of spending by industry for selected countries. As the emphasis here is on the role of IT spending across a wide sample of countries, we do not distinguish between industries.

13 This group includes Australia, Austria, Belgium, Canada, the Czech Republic, Denmark, Finland, France, Germany, Greece, Hong Kong, Hungary, Ireland, Israel, Italy, Japan, Korea, Kuwait, Mexico, Netherlands, New Zealand, Norway, Poland, Portugal, Saudi Arabia, Singapore, Slovakia, Spain, Sweden, Switzerland, Taiwan, the United Kingdom, and the United States.

14 The latter group includes Argentina, Bangladesh, Bolivia, Brazil, Bulgaria, Cameroon, Chile, China, Colombia, Costa Rica, Ecuador, Egypt, Honduras, India, Indonesia, Iran, Jamaica, Jordan, Kenya, Malaysia, Morocco, Pakistan, Panama, Peru, Philippines, Romania, Russia, Senegal, South Africa, Thailand, Tunisia, Turkey, Ukraine, Uruguay, Venezuela, and Zimbabwe.

15 The underlying estimates are discussed in more detail in Bayoumi and Haacker (2002). The price elasticities used here are 1.31 for IT hardware, and 0.87 for telecommunications equipment.

16 "Ihr Thoren, die ihr im Koffer sucht! Hier werdet ihr nichts entdecken! Die Kontrebande, die mit mir reist, Die hab' ich im Kopfe stecken" (Heine, 1844).

17 Similar exercises, using the IMF's rankings of non-tariff barriers, tariffs, and the trade regime overall, also suggest that higher trade barriers are associated with lower dissemination of IT equipment.

References

van Ark, B. 2002. "Measuring the New Economy: An International Comparative Perspective," *Review of Income and Wealth* 48, pp. 1–14.

——, J. Melka, N. Mulda, M.Timmer and G. Ypma. 2003. *ICT Investments and Growth Accounts for the European Union 1980-2000: Final Report on "ICT and Growth Accounting" for the DG Economics and Finance of the European Commission*. Groningen: Growth and Development Centre, University of Groningen. Online. http://www.ggdc.net/pub/online/gd56-2(online).pdf

Bayoumi, T. and M. Haacker. 2002. *It's Not What You Make, It's How You Use IT: Measuring the Welfare Benefits of the IT Revolution Across Countries*. IMF Working Paper No. 02/117. Washington, DC: IMF.

Colecchia, A. and P. Schreyer. 2002. *The Contribution of Information and Communication Technologies to Economic Growth in Nine OECD Countries*. Economic Studies No. 34. Paris: OECD. Online. http://www.oecd.org/dataoecd/4/41/2496902.pdf

Daveri, F. 2001. "Information Technology and Growth in Europe." Unpublished manuscript. Milan: Innocenzo Gasparini Institute for Economic Research.

——. 2003. *Information Technology and Productivity Growth Across Countries and Sectors*. Working Paper No. 227. Milan: Bocconi University/Innocenzo Gasparini Institute for Economic Research. Online. http://ideas.repec.org/p/igi/igierp/227.html

Davies, G., M. Brookes and S. Potter. 2000. "The IT Revolution—New Data on the Global Impact," Goldman Sachs, *Global Economics Weekly*, No. 00/37, 18 October, pp. 3–21.

Heine, H. 1844. *Deutschland: Ein Wintermärchen*. Hamburg: Hoffman und Campe.

IMF (International Monetary Fund). URL. *World Economic Outlook Database*. Online. http://www.imf.org/external/ns/cs.aspx?id=28

OECD (Organisation for Economic Cooperation and Development). 2003. *ICT and Economic Growth: Evidence from OECD Countries, Industries and Firms*. Paris: OECD. Online. http://www1.oecd.org/publications/e-book/9203031E.PDF

——. 2004. *The Economic Impact of ICT: Measurement, Evidence and Implications*. Paris: OECD. Online. http://www1.oecd.org/publications/e-book/9204051E.PDF

Roeger, W. 2001. *The Contribution of Information and Communication Technologies to Growth in Europe and the United States: A Macroeconomic Analysis*, Economic Papers No. 147. Brussels: European Commission. Online. http://europa.eu.int/comm/economy_finance/publications/economic_papers/2001/ecp147en.pdf

Schreyer, P. 2002. "Computer Price Indices and International Growth and Productivity Comparisons." *Review of Income and Wealth 48*, no. 1, pp. 15–31.

US Bureau of Economic Analysis. URL. National Economic Accounts. Online. http://www.bea.doc.gov/bea/dn/nipaweb/index.asp

US Department of Labor. URL Bureau of Labor Statistics, Producer Price Indexes. Online. http://www.bls.gov/ppi/home.htm

WITSA (World Information Technology and Services Alliance). 2004. *Digital Planet 2004: The Global Information Economy*. Arlington, VA: WITSA.

Is a New Regulatory Framework for Telecommunica-tions Needed for the Twenty-first Century?

Scott Beardsley, McKinsey & Company Inc. Belgium

Luis Enriquez, McKinsey & Company Inc. Belgium

Victoria Gerus, McKinsey & Company Inc. Belgium

Andreas Marschner, McKinsey & Company Inc.

Austria

Introduction

There can be no doubt that the liberalization of telecommunications and data markets has brought substantial benefits to consumers in many countries over the past two decades. By liberating service provision in many areas, by allowing competitive entry, and by putting in place retail and wholesale price control regimes, regulators encouraged efficiency improvements in incumbent fixed-line businesses and transferred industry value to consumers through dramatic price reductions while sparking significant service innovation in existing and new areas such as broadband and wireless.

As a result of this initial approach, and with little fundamental change since then, the traditional fixed regulatory framework has focused on promoting price competition in formerly monopolistic markets in fixed telephony and data, and treated new areas such as wireless, mobile and broadband as relatively distinct markets. That framework was not designed to address the recent competitive and technological developments in the industry that have led to increasing convergence of services. These developments, such as fixed-mobile substitution of both access and calls, competition in fixed access, the growth of broadband access, and the resulting take-up of Voice over Internet Protocol (VoIP), are fundamentally changing the structure of the telecommunications industry. The existing "regulatory toolkit," however, is not suited to address these types of competitive dynamics. As a result, the application of the classic regulatory approach is now distorting economic incentives for market players and thus is putting at risk long-term customer benefits from technological improvements and infrastructure investments. The adequacy of this model to support the industry's future needs calls for a radical reassessment in the light of the industry's fundamental economic characteristics and these recent developments (which it shares with many other network industries): its capital-intensive infrastructure, the importance of long-term innovation, and the critical effect of scale and network effects on its competitive dynamics.

This chapter sets out the need for change and puts forward several thoughts on the way in which the traditional regulatory framework for telecommunications could be adapted for the next "growth horizon" of the industry. The first section addresses the rationale of the traditional framework and points out its strengths and shortcomings. The second section demonstrates how a required return on investment and technological progress calls for a different approach. The final section defines the implications for future telecommunications regulation. Many of these implications may be relevant also for other sectors in the information and communication technology (ICT) field that exhibit similar economic characteristics.

The Traditional Regulatory Framework: A Messy Compromise?

We have hit a wall in our ability to regulate the industry.

(Michael Powell, Chairman Federal Communication Commission, February 2004)

In most of the post-war period up to the 1980s and 1990s, the telecommunications industry in most countries was a state-owned monopoly. It was the conventional wisdom at the time that the supply of telecommunications services and the underlying publicly switched telephone network (PSTN) was a natural monopoly. Regulation therefore consisted of legislation and/or a license contract that established monopoly rights of supply for the incumbent operators, which were also often state-owned. As telecommunications access and services were considered essential public services, governments normally set prices and performance targets so as to ensure that they were supplied to the public at "affordable" rates, and imposed substantial public service and rollout obligations. Affordability meant subsidized access prices; monthly fees were kept below costs and heavily controlled—especially for the consumer segment. A raft of cross-subsidies funded this arrangement out of usage prices, which were kept far above costs: long-distance and international callers subsidized local ones; urban callers subsidized rural ones; heavy users subsidized light ones; business users subsidized residential users; and so on.

In the 1980s and 1990s, governments around the world began challenging the conventional wisdom and subscribing to the liberalization of the telecommunications sector. They were driven primarily by three forces:

1. a broad political shift in the view of the government's role in the economy;

2. a view that technological progress was not being effectively exploited by monopoly service providers to meet the evolving demands of business and residential users; and

3. multilateral agreements, particularly in recent times, that added external pressure to open up telecommunications markets.

The intellectual foundation of this market liberalization was the economic orthodoxy that argued that businesses operate most efficiently where they are governed by market forces. Regulators then set out to create a regulatory regime that would, in theory, mimic the outcome of competitive markets. These regulatory arrangements would simulate competition through asymmetric regulation of the large fixed incumbents. Regulation would try to offset the substantial scale and scope economies and network effects that favored large players.

The competitive supply of telecommunications would bring lower prices, a wider choice of services, more technological and other innovation and better customer service. At the same time, governments recognized that telecommunications networks were crucial strategic assets at the heart of their national economies. Where these networks were underdeveloped, competitive markets were expected to attract private-sector capital to expand and upgrade them and to introduce new services. The key question was how to bring this about rapidly, and this was a particular problem in countries that had yet to develop basic telecommunications infrastructures.

One option was to create incentives to encourage entrants to build alternative infrastructure and compete directly with the incumbent operator in each market. But that would have taken time and been extremely expensive. Besides, in many cases a natural monopoly in building connections to homes and businesses would still have made entry unprofitable. Initially regulators reasonably chose a different option: to encourage entry by operators that could both piggyback on the incumbent's network (such as leased lines or managed data services) and use the incumbent's network infrastructure facilities where necessary. Although competitors often were allowed to compete across all services, new entrants often would not choose to provide all the services provided by the incumbent. Instead, they would compete in the most profitable services (such as long-distance calls) and customer segments, such as business. Competitors would be allowed access to the incumbent's network at a price that would pay a regulated rate of return sufficient to ensure network investment in some areas, thus creating effective service-based competition. End users, so the new regulatory conventional wisdom went, would be indifferent to networks and technology—they would be buying a service. Note that this opening of facilities did not amount to "unbundling" (that is, where operators can buy elements of the network at a cost-based price and can connect their own networks to them); this was a much simpler network opening.

This process did not result in "deregulation" of the industry. It was instead "regulation" to open up markets. The regulatory regime that exists today is far more complex and extensive than that which prevailed in the era of state-owned monopolies. And such intervention is far more frequent and extensive than in many other industries.

Service-based competition (as opposed to competition based on building alternative networks) required new entrants to gain *indirect* access to end users, so that customers could choose between competitive suppliers of voice telephony. The regulators therefore required incumbent operators to provide interconnection to all who made reasonable requests and, in case incumbents offered network access only at excessive prices and effectively prevented competition, they introduced further regulation to limit the wholesale access prices that could be charged. They often based prices on the costs of a hypothetical, efficient operator and allowed only regulated rates of return at an imposed average cost of capital. This approach created problems: by forcing the incumbent to charge only the lowest allowable prices, explicitly assuming efficiency gains and implicitly assuming scale benefits (since pricing was based on the costs of the largest player in the market), it virtually guaranteed that entry into many markets would not be profitable for smaller, infrastructure-based players who lacked economies of scale. Since it also focused on allocating existing infrastructure at "fair" prices, it also had the perverse effect of reducing the incentives to deploy new infrastructure, which was (and is) greatly needed in many countries.

The traditional regulatory framework concentrated on fixed networks, at least implicitly, because when liberalization began in many countries mobile services were a relatively small market that did not effectively compete with voice services. However, as mobile services have grown in importance and hence in market power, they are increasingly offering alternatives to fixed-line telephones and, in many countries with underdeveloped

fixed infrastructure, providing the primary telecommunications infrastructure. But, beyond the basic licensing conditions, the regime under which they operate is substantially unregulated, with some exceptions such as controls on mobile termination rates in countries where the calling-party-pays (CPP) system operates.

By many measures, the early telecommunications liberalization appears to have been a success (Beardsley et al. 2002). The initial regulatory framework led quite quickly to significant welfare gains for residential and business customers through increased quality of service, lower prices and greater choice of services and suppliers (Figure 1). For example, in the OECD average call prices for residential and business customers have fallen by close to 50 percent in real terms since 1990 (Figure 2).

But these welfare gains mostly were short-term gains that flowed from the elimination of gross cross-subsidies. The broader regulatory toolkit developed to support market opening was designed to reduce prices and to be able to intervene at the slightest sign of deviations, whether in profits or in behavior, from a stylized competitive ideal. In this context, it is possible that some welfare gains were

Figure 1. **Liberalization has Driven Significant Improvements in Quality of Service**

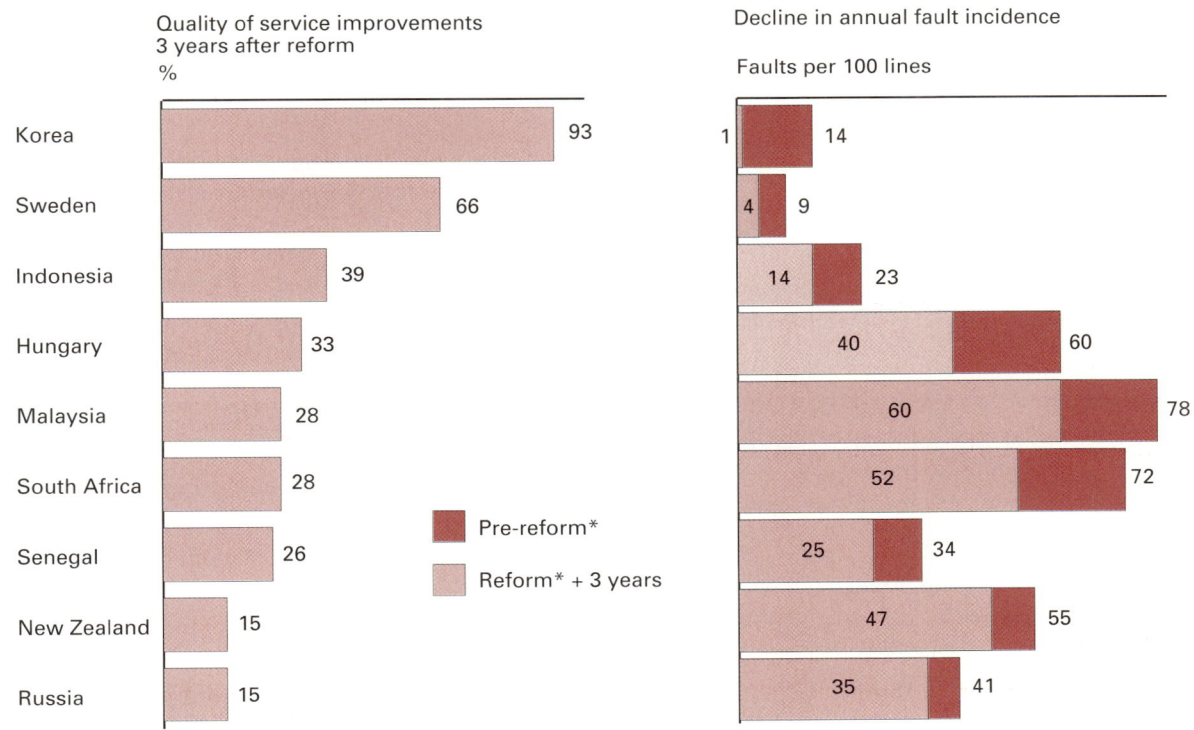

* Reform = telecommunications privatization and/or liberalization; pre-reform is year prior

Sources: ITU (2003a); OECD (2003); McKinsey analysis

Figure 2. Fixed Telephony Call Prices have Dropped Close to 50 percent in OECD Countries
Index 1990 = 100

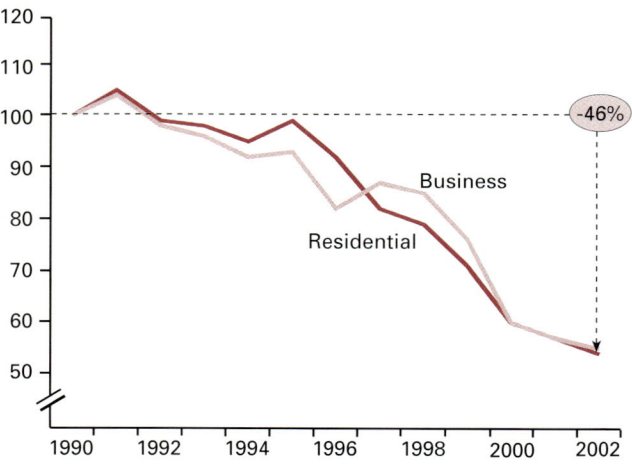

Fixed line call charges* in OECD countries

Business

Residential

-46%

* Call charge baskets are for domestic telephony, excluding monthly line rental, and international and fixed-to-mobile calls

Source: OECD (2003)

achieved at a high cost to the longer-term health of the industry by unintentionally discouraging innovation and investment.

From the earliest days of liberalization, the regulatory framework adopted in telecommunications was a result of compromises between the 'classical economic model,' focused on market efficiency and mitigating incumbent market power, and political and social objectives. Regulators often promoted these latter objectives at the expense of the economic imperatives of the market-based approach.

Several features of the traditional regulatory framework highlight how the model of market competition has been compromised:

- cross-subsidies continue, despite a relatively mature liberalization process, as regulators impose costly universal service obligations, restrictions on differentiating prices by geography (de-averaging), and strict limitations on access charges;

- regulators are generally reluctant to permit retail pricing flexibility for fixed incumbents, particularly in fixed-access services, regardless of whether other modes of competition (such as mobile) offer some degree of price discipline in selected market segments;

- regulators tend to intervene in the minutiae of wholesale contracts and negotiations to mitigate perceived incumbent advantages; and

- regulatory inertia continues to promote 'hit-and-run' competition, as low returns on network investments create disincentives for network-based competition,

virtually guaranteeing continued intervention in most markets.

The largest distortion lies in the unwillingness of regulators to allow local prices to rise. At the time of market liberalization, profitable services and customer segments were open to competitors. But there was another side to the bargain: as the ability to cross-subsidize was eliminated by market competition, the previously subsidized prices of less attractive services and segments were supposed to rise. Although formal price controls remained in place throughout the liberalization process, in many cases operators gained more flexibility to restructure their prices. In theory, this flexibility should have allowed prices to move towards levels permitted by market forces, with the least price-sensitive services rising above costs and the most price-sensitive (or competitive) services trending to costs.

Since the key service that was subsidized was access (through low monthly fees, and in some countries free local calls), its price should have risen as others fell, a process called re-balancing. As network access, or line rental, is a fairly inelastic service, consumers would continue to buy it even if the price rose substantially. In theory, this inelastic demand would mean that prices should have continued to rise well above costs. However, since line rental prices affect the mass of consumers—and often national inflation indices—its price was politically sensitive. For this reason, in many countries re-balancing has been allowed only to a limited extent (at most, to rises that reflect costs) or not at all.

Figure 3. Retail Line Rental Remains Significantly Below Recognised Costs In Most European Countries

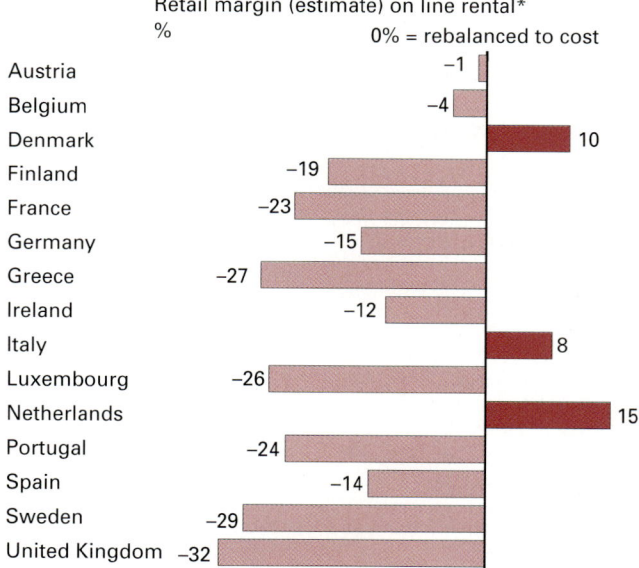

Retail margin (estimate) on line rental*
%   0% = rebalanced to cost

* Assuming 35% retail mark-up on fully unbundled loop charges (Aug 2004)

Sources: National regulatory authorities; company websites; McKinsey analysis

In the European Union, while most countries have officially rebalanced tariffs (Commission of the European Communities 2003), a closer examination of line rental prices shows a different reality. A comparison of retail line rental with underlying costs reveals that only a few countries brought prices in line (Figure 3). And no country has actually permitted prices for access to rise substantially above costs. Regulators have justified this by arguing again that operators can price access "affordably" and make money on other services. The effect, however, has been additional intervention in specific services, in direct contradiction of the pricing approaches suggested by economic theory. As a result, the traditional telecommunications regulatory framework continues to sustain persistent access deficits by incumbent operators, and competition in access by infrastructure players (particularly by cable operators) may have been unnecessarily held back.

Figure 4. **Net Universal Service Costs Vary Considerably by Country**

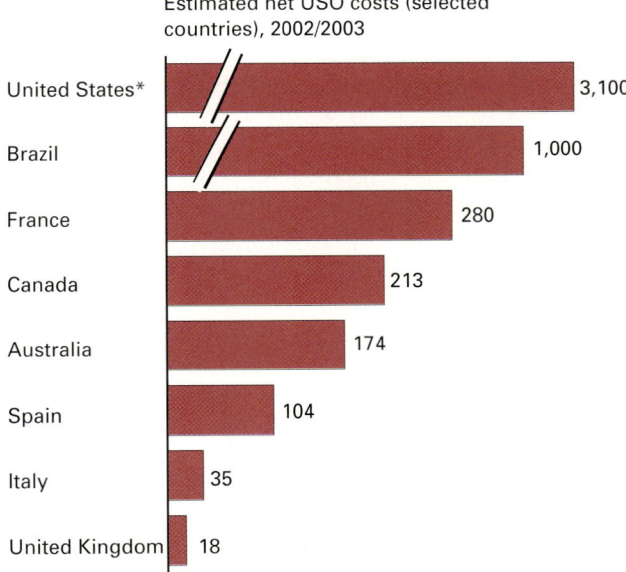

Estimated net USO costs (selected countries), 2002/2003

United States* — 3,100
Brazil — 1,000
France — 280
Canada — 213
Australia — 174
Spain — 104
Italy — 35
United Kingdom — 18

*Total for all regional bell operating companies

Source: National regulatory authorities

With policy makers and regulators still considering fixed telecommunications access and services as essential public services, incumbent operators have remained subject to universal service obligations and restrictions on geographic de-averaging of retail prices, despite extensive competitive entry. Figure 4 demonstrates how universal service costs can vary by country. Although funding mechanisms for universal service obligations are available as an option in most countries, in practice many incumbent operators bear the cost of ensuring that "unprofitable" customers have access to telephone services and of serving all citizens at an affordable price. In a recent survey of national regulators,

over one-third of the 130-odd regulators that responded acknowledged that the incumbent operator(s) in their country were expected to rely upon cross-subsidies between services to cover all or part of the cost of universal service (ITU 2003b).

Even where applied with an eye to creating economic incentives, price regulation can have unintended consequences. In the UK the telecommunications regulator, the Office of Communications (Ofcom), stated recently that the use of retail price caps on BT's services over a 17-year period of market liberalization had actually hindered the development of competition (Carter 2003). There are other examples of unintended consequences: regulation intended to promote both service-based and network-based competition can sometimes conflict with a desire for infrastructure. Regulation that creates attractive options for service-based competition (for example, carrier selection, wholesale leased lines and bitstream access) may delay or deter the development of network-based competition. Figure 5 illustrates the real impact, in both developed and emerging markets, that this difference in policy emphasis can have. And without the competing provision of network, there is no way of contesting the cost base from which the incumbent player provides its network.

The traditional regulatory framework has also compromised the classical economic model by promoting the entry of service-based competitors and resellers, which can survive only on the discount wholesale charges and retail price margin "arbitrage" opportunities created by regulation. Although policy makers and regulators recognize the importance of network-based competition in the sector, they tend to see service-based competition as a sustainable business model—especially given the recent economic downturn in the industry and the lack of financial funding in the sector. But, by allowing contradictions to persist within the regulatory framework, they create excessive "regulatory arbitrage opportunities" for new entrants and may be damaging the economic evolution of the industry over the long run by discouraging investments in alternative networks and technologies.

At the same time, under the current regulatory framework incumbent operators appear to be indefinitely subject to a regulatory burden that does not allow them the means or the incentives to engage in their own network investment. The cost-based pricing applied to the unbundled network elements means they lose money on these services that they are required to provide. They also argue that having to open up their networks to their competitors provides disincentives to invest in new technology in the local loop, such as fiber-to-the-building, because they would immediately have to give access to their competitors at discounted prices.

Figure 5. Regulatory Policy Focus Can Have Significant Impact on Sector Investment

Telecoms investments (selected countries) as % of GDP

Developed countries, 2001

Emerging markets, 2001

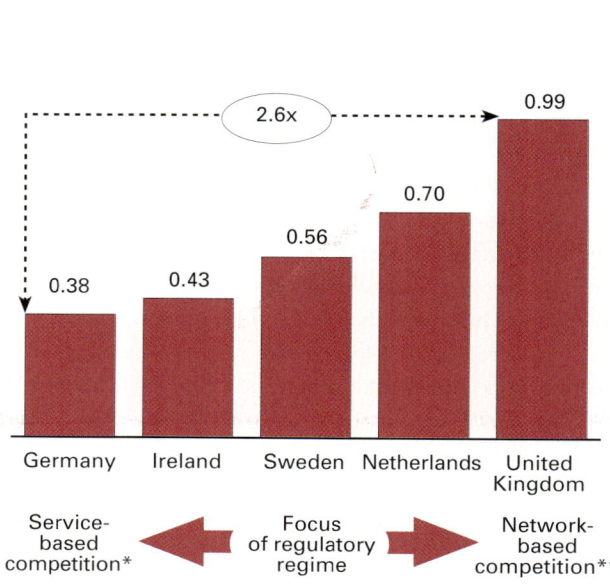

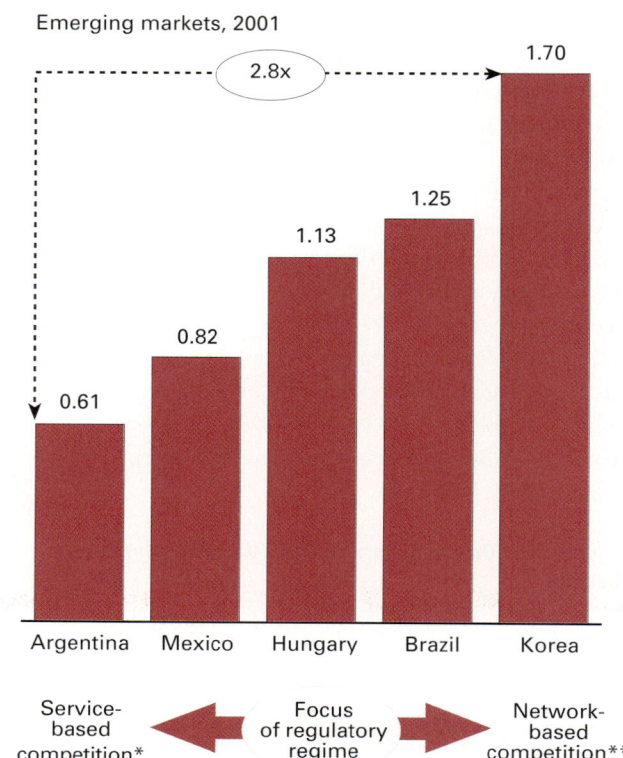

*Regulatory regime with licensing, interconnection and indirect access policies that favor entry
**Regulatory regime with licensing, interconnection and access policies that promote network investment or that create entry barriers that can be overcome only via network investment

Source: ITU (2003a); McKinsey analysis

Strict retail and wholesale price regulations are aimed primarily at transferring surplus to the customer while allowing the maintenance of existing networks. Increasingly, they are preventing incumbents from creating additional surplus to invest in the networks of the future—broadband and next-generation networks (NGNs).

The Nature of the Industry Poses Challenges to the Traditional Regulatory Framework

The telecommunications industry is capital-intensive and demonstrates strong economies of scale and scope. As shown in Figure 6, the telecommunications demands far more investments than many traditional "brick and mortar" industries. As a result, efficiency in the economic sense of the word—lower cost eventually translated into lower prices—is often achieved with a more limited number of network operators than with broad-based competition among smaller players. Figure 7 illustrates this feature in advanced mobile networks, where larger players enjoy more than a 50 percent network cost advantage over smaller players.

Figure 6. The Capital Intensity of Telecommunications Requires Policies That Focus on Investment Returns

Capital expenditure for selected US industries
% of revenues, average 1993–2003

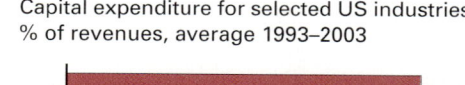

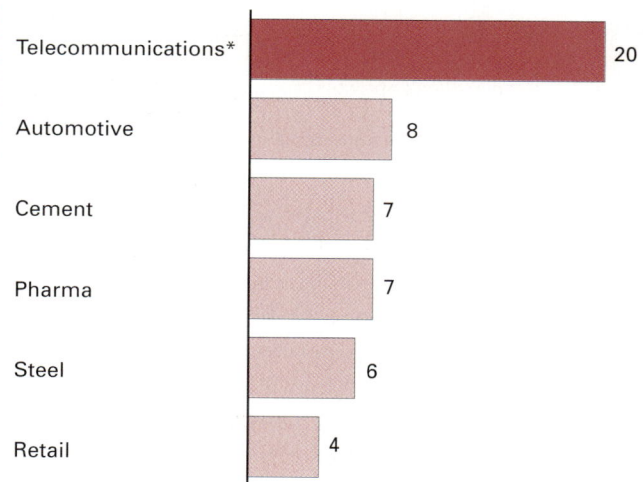

*All fixed and wireless companies; 22% for incumbent local exchange carriers

Sources: Standard & Poor's (2004); McKinsey analysis

Figure 7. **Operator Size is Critical to Explore Economies of Scale of 3G Networks**

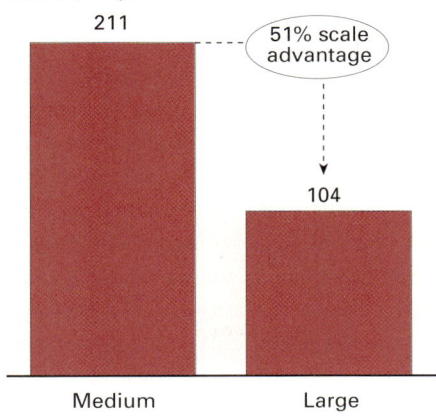

Cost of capacity for 3G networks
US$ thousands/Mbps

211

51% scale advantage

104

Medium   Large

Operator size

Sources: Bear, Stearns International Ltd. (2003); McKinsey analysis

Telecommunications infrastructure investments tend to have long funding and payback cycles. Such investment therefore is highly dependent on the likelihood of attractive returns: network operators will invest only if the investment uncertainty is manageable. The usual deciding factors are:

- the possible service offer to meet customers' existing and latent demand;
- the achievable price structure for existing and planned services; and
- the expected development of the subscriber base.

Since most innovation—whether process, product or technology—often requires the incentive of earning a monopoly "rent", for a limited period of time the room for pricing in a regulated market and the expected development of the subscriber base are the most critical factors for innovation in telecommunication services. If either is expected to change unfavorably, innovation may disappear completely. The rapid proliferation of new services seen in the mobile industry, as illustrated in Figure 8, demonstrates how commercial freedom and limited regulatory obligations can spark innovation and investments.

A difficulty with the traditional regulatory framework, which aims to mimic perfectly competitive markets by introducing service-based competition and establishing network access rules, is that the returns available to reward innovation are likely to be low—and subject to discretionary regulation and/or arbitrary political agendas.

Another feature of the industry that poses challenges for the current regulatory framework is that the positive

Figure 8. **The Mobile Industry Has Produced a Large Number of Innovations**

New technological innovations launched in mobile services or handsets*
NON-EXHAUSTIVE

1992–93	1994–95	1996–97	1998–99	2000–01	2002–03
					UMTS
				Bluetooth	Skins
				Video streams	TV tuner
			Blackberry	Java games	Edge
			Audio streaming	Downloads	IM
			i-mode	Symbian	Quad-band
		Full rate voice codec	Color display	Polyphonic	CDMA PTT
		Mobile payment	LBS	W-CDMA	Poly ring tone composition
		PTT	Tri-band	EMS	Digital ring tone service
		T9	WAP	CDMA 20001X	Mobile content
Voice mail	Prepaid	Dual-band	CDMA data service	HS-CSD	SD
CDMA	Contest SMS	Caller ID	Ring tone	Photo camera	MiDi FM
GSM	SMS	Clamshell phone	Changeable cover	GPRS	MMS

*First launch date indicated here, that is, launch could have taken place anywhere in the world

Sources: Press articles; McKinsey analysis

"network effect" for customers does not hold for the network operator. The physical and logical linkage of multiple telecommunications networks is extremely complex. It requires developing systems and processes that allow for the interoperability and interconnection of networks and the switching of providers, for example, number portability and wholesale line rental. Developing and implementing these systems and processes can impose significant costs on operators, many of which they cannot recover within the returns allowed under the regulation of retail and wholesale prices.

Access network competition and new services are creating more pressure for the traditional regulatory framework

The telecommunications industry is now in the midst of a period of significant technological change. The "access bottleneck" that has long plagued the industry is being removed by new access technologies, such as cable and wireless, and alternative platforms such as VoIP, in which voice is just another kind of data transmitted via the Internet. The combination of market liberalization and technology appears to have finally sparked competition in the last mile.

There are currently three critical trends:

- fixed-mobile substitution: people are increasingly more inclined to use mobile services than legacy fixed services;

- a portfolio of traditional voice and other services being offered via broadband access: the bundling of voice telephony, data, video and other value-added services onto multiple infrastructure platforms is blurring conventional definition of markets and thus how they should be regulated; and

- new competition by geographically focused access providers using a variety of access technologies from cable to wireless, from satellite to fiber.

Fixed-mobile substitution

In an increasing number of countries, mobile telephony markets are maturing, and mobile access is beginning to exceed fixed-line penetration. With the ubiquity of mobile networks and the erosion of the price premium of mobile services relative to fixed, increasingly large numbers of customers are using mobile phones for all or most of their communication. For example, global industry forecasts predict that by the end of 2006 mobile telephony will comprise over 50 percent of total voice revenues (Ovum 2004b); in many markets mobile telephony already represents over 50 percent of voice traffic (Beardsley et al 2004). This is particularly true in markets where fixed-line infrastructure was never broadly

developed. In other words, mobile has become a de facto substitute for fixed voice, although regulatory regimes do not acknowledge this basic fact.

ICT services-over-broadband

Broadband and the Internet protocol (IP) present a unifying platform for three converging industrial sectors: communications, software and broadcasting. Growth of VoIP services and broadband access are mutually reinforcing. VoIP services are more attractive over broadband access and will increase the attractiveness of broadband to some users, thereby promoting deeper broadband penetration. Consequently, increasing broadband penetration could swell the potential pool of VoIP users.

To date, VoIP take-up has been mostly in the business segment, driven by the cost savings offered on internal company calls. Industry surveys indicate that, in addition to further business-user growth, a significant take-up in residential usage seems very likely, as notable cost savings could make VoIP very attractive. Global industry forecasts predict strong growth for VoIP users, illustrated in Figure 9.

Figure 9. **VoIP-Telephony Will Grow Exponentially in the Near Future and will Capture a Significant Share of Fixed Lines**

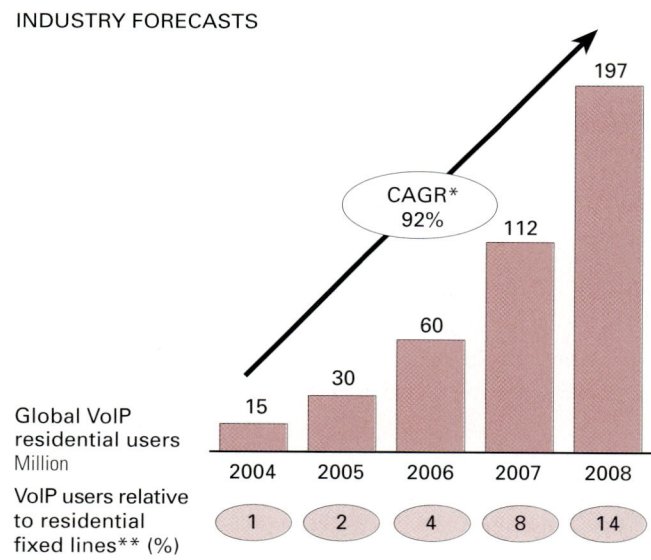

INDUSTRY FORECASTS

*CAGR = compound annual growth rate
**All residential lines capable of carrying VoIP services, including PSTN, upgraded cable, and fiber to the building

Sources: Ovum (2004a, 2004c); McKinsey analysis

Creative service bundling, combining fixed and mobile voice, video and pay television, and unified messaging and other productivity tools with broadband access, is emerging where allowed. The delivery of these services onto a variety of infrastructures and technologies from cable to wireless to satellite to fiber and traditional PSTN technologies has blurred the boundaries between industries and created a variety of substitute products—such as VoIP. And as more and more technologies—3G, WI-FI, WI-MAX, FTTX, to name but a few—are deployed, the competitive pressure and loss of market share on the operators of legacy networks, both fixed and mobile, only will increase.

New competition by geographically-focused access providers

The development of access technology has driven a rise in the focused entry of smaller scale access providers that have targeted urban areas. While in the past such business models have only been fully viable when targeted at the business segment, with its high long-distance revenues, advances in access technology and networks now mean that under certain conditions the residential market is becoming attractive as well. This is especially true for providing broadband access that serves as a platform for a wide range of services from voice to data and video.

The traditional regulatory framework at its limits

But many regulators still operate as if the last-mile bottleneck had yet to be broken—an approach that could have unforeseen consequences. First, the weakening of the incumbent operators' fixed-line business might not lead to more competition, just to a different type of market dominance, for example, by a mobile operator in emerging markets. Then there are also serious concerns about how to maintain and develop infrastructure if incumbents are undermined.

These trends are pushing the "classical" regulatory model to it limits through two sets of interrelated effects. On the one hand, for the first time ever the traditional incumbents face the real potential of significant losses of market share not just on the retail but the wholesale part of their business, with traffic migrating outright to alternative infrastructure providers. On the other hand, the rigidities and contradictions of the current regulatory model create significant additional uncertainty over the viability of investments into new infrastructure.

The implications of this can be quite profound. For example, IP has accelerated technological progress from outside the industry and is undermining the system of usage-based/per-minute charging. The unit costs per minute of legacy networks may increase, given a predominantly fixed cost base and a potentially significant migration of customers and traffic to alternative platforms. Having chosen to pursue retail and wholesale price regulation based on a broad principle of cost recovery,

regulators should recognize such increases in unit costs in setting the prices for future use of the legacy network and the dramatic discontinuity in wholesale and retail prices it could create.

Without such an adjustment, that is, a potentially significant increase in prices, the loss of market share by large established players has reduced the forward-looking potential rates of return on investments at current prices. The implication of this is that continuous investment in traditional fixed-line voice networks is becoming increasingly unattractive and, as a result, a shift to investment in new services and technologies may never happen.

This situation will be aggravated in developing countries, as they often are starting from a position of relative underdevelopment for the legacy fixed network and as, given lower penetration and usage levels, there is a smaller economic surplus to be distributed and to create investment incentives. The need to adapt the regulatory framework to changing market dynamics may be very critical in these countries.

Some signs of change

Most regulators do not seem to appreciate how technology is changing market dynamics. A few, though, have begun to review the situation and to consider necessary policy reforms:

- Under the EU's new regulatory framework for electronic communications, the European Commission has made a distinction between "relevant markets" and "emerging markets" where:

    - the narrow list of relevant markets are susceptible to *ex-ante* regulation by national regulators; and

    - emerging markets are to be left free from regulatory intervention so that their development is not stifled and competition not distorted;

- Ofcom (2004) launched a multi-phase strategic policy review and industry consultation in April 2004 to understand how changes in technology and patterns of consumer demand will shape the competitive landscape and to determine whether a new approach to economic regulation is needed; and

- the Federal Communication Commission's "Triennial Review," (FCC 2003) which is still the subject of much policy-making controversy, seeks to relieve costly-to-deploy fiber networks of unbundling regulations.

While these regulatory signals are encouraging—and, one hopes, will encourage other regulators to follow suit, the increasing lag between regulatory reform of technology and competitive developments leaves operators with continued uncertainty about the fiber they are putting in the ground and NGN equipment they are deploying today.

As outlined above, telecommunications markets are changing at a pace never seen before. Unless regulators can "get ahead of the curve," the trends outlined here could be detrimental to the economics of key players in the industry, thereby potentially distorting the market. Therefore, regulation needs to be adapted to allow operators to make decisions based on business economics.

Implications

We're living in a public policy paradox. We are asked to focus on the speedy development of new services but at the same time are facing some profound and unexpected regulatory barriers that hamper the implementation of these new services and networks.

(K. Karvala, Chairperson GSM Europe, 27 March 2003)

Broadband voice services are a new and emerging market. Our first task as regulator is to keep out of the way . . . As the market develops, we will ensure that consumers are appropriately informed and protected.

(Stephen Carter, Chief Executive Ofcom, September 2004)

It is imperative for regulators to look ahead and envisage how markets are likely to evolve given current industry trends and to compare this with how they want them to evolve. The traditional regulatory framework could be detrimental to the economics of key players in the industry and could therefore distort the market. Regulation needs to be adjusted to allow operators to make decisions based on business economics. This will contribute to ensuring a sustainable market and the continued "networked readiness" of society. This context suggests several implications for regulation of the sector.

Fix existing legacy regulation

The first imperative is to adjust the regulation of legacy assets to the new industry environment. This process has started in a few countries but so far it has been neither explicit nor consistent. The necessary adjustments include several elements. First, regulators need to explore how to reduce the scope of retail price controls and accommodate rising prices if a high proportion of traffic migrates off the legacy PSTN network because of fixed-to-mobile substitution and increased use of VoIP services. Second, to the extent possible, regulators should fund universal service and other obligations objectively and through demand-side subsidies instead of supply-side intervention. Third, they should ensure that greater symmetry is gradually achieved on network and customer access measures to recognize the presence of competing infrastructures such as mobile and cable networks—with a view to potentially scaling them down if a market is truly contestable, or installing proportionate requirements for

quality of service, emergency services, and so forth. Fourth, regulators should evolve their scope to be able to constructively regulate video, voice, and data services in an integrated manner.

When in doubt, do not regulate

Markets are not perfect; plenty of markets with static market failures do not merit intervention. Indeed, there are compelling reasons not to intervene in some situations. But the bias in mature markets, particularly in fixed telecoms, has been in favor of intervention: when in doubt, regulate. This has created disparities not only among fixed telecommunications operators (for instance, pricing freedom asymmetries) but also between modes of access (such as between fixed and mobile network build-out) or usage (the treatment of "plain old" voice versus VoIP). It has also opened up opportunities for strategic "regulatory gaming" by players. As a result, market dynamics are often not determined by markets but rather by regulatory outcomes.

The solution may involve a shift away from industry-specific regulation to general competition law. In markets where competition has been successfully introduced, there is no need for further regulation. In addition, newly emerging product or service markets should not be subjected to inappropriate upfront regulatory obligations, as the premature imposition of regulation may unduly influence or stifle the competitive conditions taking shape within them. Instead, regulators should rely on their ability to reassess emerging markets and intervene at a later stage.

Abandon the explicit or implicit focus on price as the key indicator of competitive intensity

While low prices are a good indicator of competition in markets with slow technological change and limited capital intensity, they are less appropriate in telecoms markets. Using regulation to enforce price reductions and transfer value to consumers works for existing services on installed networks and technology but it does not reflect the need for investments in new technologies and services. As a result, the current regulatory focus on continuously reducing prices—in fixed and, more recently, in mobile—threatens the long-run evolution of the sector. Networks are capital-intensive with long amortization periods. Operators, whether incumbents or attackers, will invest only if they are able to achieve an acceptable return on capital. In the long run, sufficient investment incentives for infrastructure-based competition are critical, which implies that prices may have to rise in some areas.

Consider the long-term evolution of the sector in evaluating specific regulation

Much regulation focuses on resolving immediate, sometimes relatively minor, problems between operators. It is developed in the course of periodic negotiation between the operators and the regulator. This approach, by its nature, fails to consider the impact on the long-term evolution of competition. For example, the long-run incremental cost approach that is so widely used for regulating interconnection charges by its very nature creates fundamental disincentives for new network investment by giving a new entrant easy "access" to the cost level of an efficient network and by fundamentally shifting most business risk associated with building and running a network to the incumbent operator.

Another de facto focus of policy makers and regulators in many countries has been to extract or transfer value away from the sector. Beyond a strong focus on price reductions, other decisions such as the sale of 3G mobile licenses via auction for over €100 billion clearly have not helped the financial health and capacity of the industry. Although the industry remains profitable, the high fixed-cost nature of the industry means that it is easier than one might think "to kill the goose that laid the golden egg."

Policy makers and regulators also need to carefully consider what objective they are trying to achieve. There is no one right answer. Developing countries need to consider their unique starting positions, and not make the mistake of copying regulatory regimes from countries with fundamentally different infrastructure and demand starting points. Other countries need to consider what will constitute success—for some this may mean ubiquitous high-speed broadband infrastructure, for others low prices or advanced services—and the regulatory regime ought to carefully consider how to make the appropriate trade-offs.

Conclusion

The most important task for telecoms regulators is to ensure a healthy industry that delivers long-term benefits for customers and the economy. While price competition and the accompanying need for efficiency are critical elements, innovation, infrastructure and service levels will be equally critical in the long run. Add to this multiplicity of often conflicting objectives the political pressures with which regulators must deal, and it is clear that they have to manage substantial and complex trade-offs. Given this context and the economic characteristics of network industries, regulators need to be aware of the shortcomings of existing regulatory tools, which are primarily geared to pushing price competition and leveraging existing infrastructures.

Competition at the retail level is now flourishing in many, though not all, customer segments and products. This follows the broad penetration of mobile services, the growth of alternative access providers in some territories, the increasing share gains of alternative long-distance operators, and the emergence of new technologies, such as VoIP, that have allowed new players to enter the sector. This competition is weakening the rationale for strict retail regulation in some fixed-line segments, such as voice. Left alone, these formerly regulated markets should start behaving like "normal" competitive markets in other industries, meaning prices may continue to fall or they may rise depending on the evolution and dynamics of competition. However, many regulators continue to equate lower prices with "effective" competition, driven in part by political considerations. When this bias leads to policies that target reductions of fixed-line retail prices in the long term, it may unwittingly damage long-term competition by discouraging innovation, infrastructure investments and service-level improvements. Where necessary, less market-distorting regulatory interventions, such as user subsidies to low-income families and rural small enterprises, would be favorable options.

Regulators must take a long-term view of industry evolution. This requires a substantial shift in focus away from the *ex-post* redistribution of rents from existing assets and networks towards the creation of the necessary preconditions to justify investments that will generate new rents. Only such a major shift will provide appropriate incentives to liberate funding for the next cycle of technology investments.

Some specific, practical implications of this are that regulators should adjust the way legacy assets are currently controlled, should cease to regard price as the key indicator of competitive intensity, should explicitly consider the likely long-term impact of specific regulation, and should avoid regulation altogether in cases where it is not necessary. They should also evolve their capabilities to be able to simultaneously regulate increasingly converged voice, video and data services.

References

Bear, Stearns International Limited. 2003. *Wireless Network Economics*. London: Bear, Stearns International Limited, 2 June.

Beardsley, S., I. Beyer von Morgenstern, L. Enriquez and C. Kipping. 2002. "Telecommunications Sector Reform— A Prerequisite for Networked Readiness." In G. Kirkman, P. Cornelius, J. Sachs and K. Schwab, eds., *Global Information Technology Report 2001–2002: Readiness for the Networked World*. New York: Oxford University Press. Online. http://www.cid.harvard.edu/cr/pdf/gitrr 2002_ch11.pdf

65

Chapter 4 Is a New Regulatory Framework for Telecommunications Needed for the Tewnty-first Century?

Beardsley, S., I. Beyer von Morgenstern, L. Enriquez and W. Verbeke. 2004. "Towards a New Regulatory Compact." In S. Dutta, B. Lanvin and F. Paua, eds., *Global Information Technology Report 2003–2004, Towards an Equitable Information Society*. New York: Oxford University Press.

Carter, S. 2003. "The Communications Act: Myths and Realities," speech of 9 October by Stephen Carter, Chief Executive, Office of Communications. Online. http://www.ofcom.org.uk/media_office/speeches_presentations/carter_20031009

Commission of the European Communities. 2003. *European Electronic Communications Regulation and Markets 2003: Report on the Implementation of the EU Electronic Communications Regulatory Package*. Brussels, 19.11.2003. COM (2003) 715 final. Online. http://europa.eu.int/eur-lex/en/com/cnc/2003/com2003_0715en01.pdf

FCC (Federal Communications Commission). 2003. *Report and Order and Order on Remand and Further Notice of Proposed Rulemaking* (FC 03–366). Washington, DC: FCC. Adopted 20 February 2003; released 21 August 2003. Online. http://www.fcc.gov/Daily_Releases/Daily_Business/2003/db0821/FCC-03-36A1.pdf

ITU (International Telecommunications Union). 2003a. *World Telecommunication Indicators Database (7th Edition)*. Geneva.

——. 2003b. *Trends in Telecommunication Reform: Promoting Universal Access to ICTs – Practical Tools for Regulators. 5th Edition, 2003*. Geneva: ITU.

OECD (Organisation for Economic Cooperation and Development). 2003. *OECD Communications Outlook 2003*. Paris: OECD. Online. http://www1.oecd.org/publications/e-book/9303021E.PDF

Ofcom. 2004. *Strategic Review of Telecommunications Phase 1 Consultation*. London: Ofcom, 28 April. Online. http://www.ofcom.org.uk/consultations/past/telecoms_review1/telecoms_review/?a=87101

Ovum Ltd. 2004a. *Broadband@Ovum advisory service forecasts*. London: Ovum Ltd., May.

——. 2004b. *Mobile@Ovum advisory service forecasts*. London: Ovum Ltd., July.

——. 2004c. *Consumer VoIP: the market opportunity*. London: Ovum Ltd, 11 August.

Standard & Poor's. 2004. *Compustat (North America) Database*. New York.

Chapter 5

Building a Sound Foundation for the Information Society

Elliot E. Maxwell, Center for the Study of American Government, Johns Hopkins University

with

Michael R. Nelson, Internet Society

Introduction

Government policy can and does have a major impact on the development of the Information Society and the information and communication technologies (ICT) infrastructure needed to support it. Many independent experts agree that inconsistent telecommunications policies and regulations (and related court decisions) in the United States have slowed investment in and deployment of advanced network infrastructure and services. Other countries, most notably the Nordic countries, have adopted clear policy frameworks that have fostered competition and encouraged the rapid build-out of advanced networks.[1] Since 2000, the Economist Intelligence Unit has undertaken a survey of the "e-readiness" of leading countries around the world. In its first survey the United States ranked well ahead of the rest of the world. By 2004, the United States had slipped to sixth place (see Table 1).

Table 1. **E-Readiness Country Rankings, 2001 and 2004**

	2001	2004
1	US	Denmark
2	Australia	UK
3	UK	Sweden
4	Canada	Norway
5	Norway	Finland
6	Sweden	US
7	Singapore	Singapore
8	Finland	Netherlands
9	Denmark	Hong Kong
10	Netherlands	Switzerland

Source: Economist Intelligence Unit (2001; 2004)

In almost every country, political leaders realize the importance of ICT for fostering economic development, providing better government services, and improving the lives of their citizens. The United Nation's two-part World Summit on the Information Society (WSIS) is just one reflection of the desire of national governments to find ways to foster the use of ICT and the Internet. In the past, policies related to ICT were almost entirely under the purview of telecommunications ministers; but today, economics ministers, finance ministers, foreign ministers and even prime ministers and presidents are involved in determining how the digital economy will develop.
In order to foster the growth of the Information Society, governments and the private sector need to address a wide range of challenges, including:

- establishing the rule of law so that competitive markets can flourish;

- fostering a competitive telecommunications sector;

- encouraging further deployment of the Internet and increased connection;

- reducing the cost of computer equipment (by eliminating tariffs and fostering competition);

- increasing the availability of a wide variety of affordable applications and content; and

- investing in education and training both of computer professionals and the general public.

These tasks are all critical and they are interrelated (see Table 2). For instance, it is very hard for competition to develop in the telecommunications sector if government regulation hinders the entry of new competitors or unnecessarily restricts the activities of established companies. And without a competitive telecommunications sector, the cost of connections to the Internet backbone can be 5, 10, or 20 times more expensive than it would otherwise be (see Figure 1).

Today, most countries are at a critical juncture in the development of the telecommunications sector. New technologies are dramatically changing the business model of traditional telephone companies, and the Internet is offering a host of new services to users. Unfortunately, in too many countries telecommunications policies are based on old technologies and old assumptions. As a result, they often—inadvertently or otherwise—hinder new competitors, discourage initiatives by existing companies, and slow the development and deployment of technologies including broadband, licensed and unlicensed wireless and digital cable TV networks. When they fail to

Table 2. **Layers of the Information Society**

Education and training
Software, e-business, and content
Computer hardware
Internet
Telecommunications networks
Rule of law (contracts, anti-corruption, etc.)

encourage competition they eliminate the most effective stimulus for investment and for the provision of new services, applications, devices, and content.

This chapter describes the need for government policy makers, the private sector, and civil society to reinvent regulatory policies in order to create a business environment that fosters investment and innovation while empowering and protecting consumers. It examines in detail how outdated telecommunications policies can stifle a competitive market in new digital services. Using the 1996 Telecommunications Act in the United States as a case study, the chapter highlights the problems caused when old models are imposed on new technologies.[2]

The Telecommunications Act of 1996

Over the last ten years, many governments have taken steps to liberalize their telecommunications sectors. The Basic Telecommunications Agreement at the World Trade Organization (URL) has helped to motivate these efforts. The Telecommunications Act of 1996 in the

Figure 1. **Worldwide Broadband Costs**

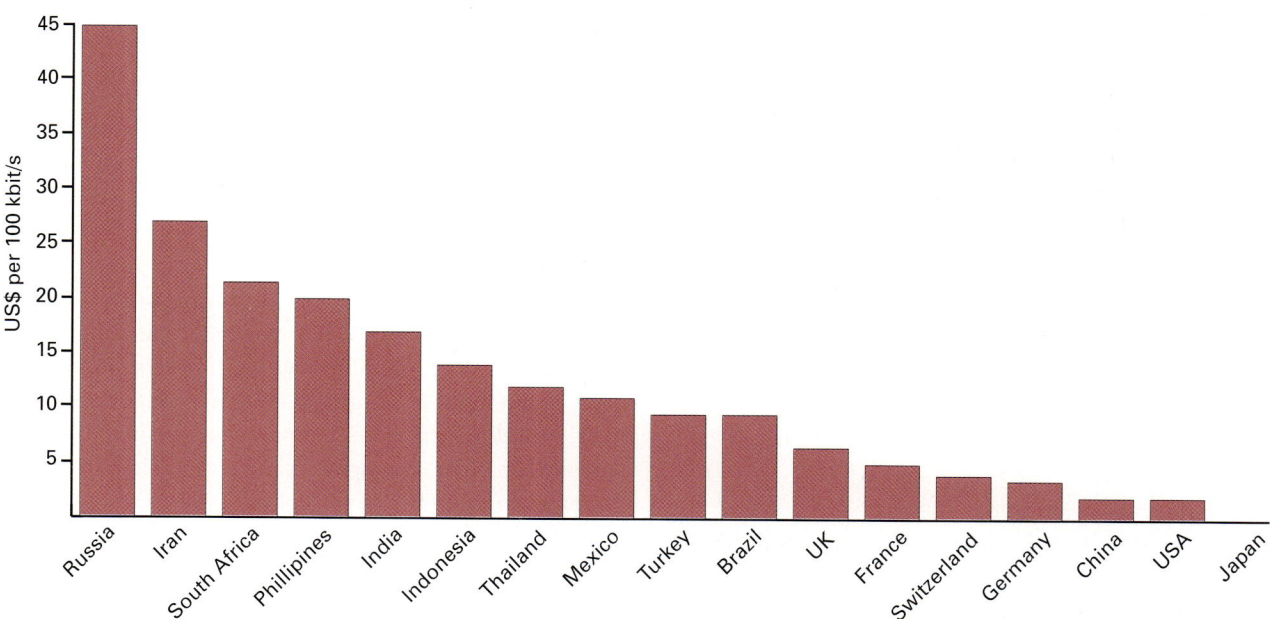

Source: ITU (2004b)

United States (FCC URL)was designed to open up the telecommunications sector to more competition and it has had some limited success in doing that. Yet, fewer than ten years after it passed, the Act is already outmoded.

The telecommunications landscape has changed dramatically since 1996. The bursting of the "telecommunications bubble" in 2000 devastated the industry, saw the collapse of dozens of companies, and destroyed trillions of dollars in shareholder value. In the United States, AT&T, the company synonymous with telecommunications for anyone over 35 years old, announced that it would no longer invest in traditional consumer telecommunications services (Richtel 2004), and a combination of wireless, e-mail, and instant messaging services cut the number of wireline long-distance minutes by over 50 per cent between 1998 and 2003 (Insight Research Corporation 2004). According to ITU statistics (ITU 2004a), in most countries the number of wireless subscribers now exceeds the number of wireline access lines. New unlicensed wireless services with names like WiFi and WiMax are emerging from industry standards-setting bodies and are being installed in laptops and deployed in hot spots or are being spread over cities and towns.

Today's regulatory debates in the US look both backward and forward. One of the most contentious is based on the provisions of the 1996 Act that set out the rights of, and costs for, new competitors to use parts of the existing networks of incumbent local-exchange companies (ILECs). Similar provisions to enable new competitors to efficiently enter the market have been included in regulations in Europe and elsewhere. But in the US these provisions have been the subject of litigation ever since the first regulations interpreting the Act were issued, and are still not settled.

At the same time regulators are trying to look into the future, asking how to regulate, if at all, voice services being offered over broadband Internet connections (VoIP) and other broadband-based applications utilizing the Internet protocol (IP). This debate has been triggered by a dramatic expansion in broadband connections since the turn of the century, with more than 27 million broadband connections in the US and, according to the ITU, over 100 million worldwide at the start of 2004 (ITU 2004b).

Given the seismic shifts in telecommunications since 1996, it is time to revisit the way we regulate the telecommunications market upon which so much economic, social, political, and cultural activity depends. To continue to make progress toward a more competitive and capable telecommunications sector, it is critical to create a new regulatory structure informed by the lessons of the past but not perpetuating earlier mistakes.

Today's telecommunications regulations reflect their origins in a world of monopoly telecommunications providers, not today's world of multiple competing platforms, expanded functionality, and increasing customer choice. Tomorrow's regulatory paradigm should be more consistent with the convergence of various service platforms and reflect more the competitive state of the marketplace. It should be more flexible and dynamic in order to accommodate change, more effective in preventing abuses of market power, more oriented to incentives, innovation and investment, and more efficient and equitable in achieving widely accepted social goals.

How We Got Here

Traditional telecommunications regulation was grounded in what for many years was taken as a given—that the provision of telecommunications services was a natural monopoly. The argument was that a single provider could provide services to customers at lower prices and more efficiently extend the network. The years immediately following the invention of the telephone saw the creation of many competing telephone companies, but governments eventually authorized a monopoly telecommunications provider. These monopoly entities became, over time, subject to a wide range of rules governing, among other things, the introduction and withdrawal of services, their prices, capital investment and returns, depreciation, accounting practices, organizational structure—even the staffing of customer service offices.

The grant of a monopoly franchise was based on what was thought to be the efficiencies of scale and scope offered by a single provider. The technology then employed by telecommunications providers inextricably tied the services that were to be offered to the telecommunications facilities that existed in a specific geographical area. In other words, a customer couldn't obtain a telecommunications service from anyone other than the single company that was authorized to build and operate the single telecommunications network facility in that locale. Regulators exercised jurisdiction over the company which operated the facilities in the geographical area over which the regulators had jurisdiction. The overall picture was relatively simple—customers, monopoly service provider, and regulator.

Over the last 40 years the notion that telecommunications is a natural monopoly has crumbled. Long-distance service in the US, for example, was an AT&T monopoly, permitted because of the perceived high capital cost of building a national long-distance network. The prices AT&T charged were averaged and did not reflect the costs of providing services in any particular area—and were vulnerable when new competitors sought to enter a particular market. When MCI began to serve large customers with high volumes of traffic between point A and point B utilizing point-to-point microwave services, it could offer a lower price than the averaged nationwide price offered by AT&T. A pricing structure filled with subsidies, and unrelated to the cost of providing a service, was vulnerable to competitors. Soon, what had been a monopoly market had become competitive. In a pattern that has been repeated time and again, new technology, combined with changes in law and regulation, and economically inefficient pricing systems, enabled competition in a narrow market. Competition would then spread to a broader market. New competitors, and those who benefited from their existence, would begin to advocate against monopoly and for even more competition.

To make a long story short, the AT&T divestiture followed. The potentially competitive long-distance market was separated from the then presumed "natural monopoly" local telecommunications market. In order to promote competition in the long-distance market, an "equal access" requirement was imposed on the local providers, ensuring that all long-distance competitors could interconnect in a non-discriminatory way with the local monopoly provider. At the same time the local monopoly provider was "quarantined" from the long-distance market to eliminate any incentive that it might have to discriminate.

An indirect challenge to the notion that local telecommunications service was a natural monopoly was emerging in another area. Wireless (radio-based) telephony was initially restricted in the US to a tiny number of local radio channels. The introduction of new technologies—"cellular" and digital modulation—into the mobile telephony market created the opportunity to serve many more customers. Over the last 20 years, wireless telephony in the US has changed from a local monopoly to a duopoly to the present system of multiple nationwide providers with large blocks of the radio-frequency spectrum. Wireless telephony has developed from an expensive, business-oriented, low-quality service which complemented local wireline voice service into a local and national voice and data service that business and residential users increasingly view as an acceptable substitute for local and long-distance wireline telephony.

Regulators have encouraged competition among wireless providers by expanding the amount of spectrum available, forgoing many of the regulations applied to the local wireline providers, and requiring number portability among wireless operators. The competition has led to lower prices, new service offerings, and a continuing increase in wireless subscribers.

Technological progress has also been changing other telecommunications platforms, leading to a convergence of platform capabilities and further discrediting the idea of

local telecommunications as a natural monopoly. Cable companies had originally installed coaxial cable to deliver video signals to homes where broadcast quality was poor. Over time, cable networks were extended and enhanced. Cable owners upgraded their coaxial cable facilities with optical fiber and new head-ends in order to deliver more, and higher quality, digital video channels and interactive services. But the network upgrades also gave the cable companies the opportunity to offer broadband Internet connections and VoIP in competition with local telephone companies

The Telecommunications Act of 1996 stimulated even more market convergence. It recognized the end of the natural monopoly era and abolished any remaining legal local telecommunications monopolies. Unfortunately this has not been the case in many other countries, particularly in the developing world, where governments continue to protect incumbent monopolies.

The 1996 Act had a positive effect on competition. But it barely changed legal and regulatory categories created over the preceding 60 years that are increasingly inconsistent with today's marketplace. These inconsistencies now stand as a barrier to even more competition and even greater innovation.

In the US and much of Europe most customers have access to multiple telecommunications platforms that do not rely on the facilities or services of the incumbent wireline operator. The vast majority of US households are passed by cable company facilities that are already, or soon to be, capable of providing broadband services and VoIP (Martin 2004). Wireless providers reach almost all US households—the Federal Communication Commission (FCC 2004) has found that 97 percent of the US population lives in counties with access to three or more wireless providers—and many of these licensed wireless providers are installing wireless broadband infrastructures, allowing them to offer a far greater range of services. All residential customers are connected to power lines which, in the future, may provide a fourth facilities-based broadband services provider (or a fifth or sixth or seventh depending on the number of licensed wireless broadband providers). New forms of wireless services are developing in the radio frequency spectrum's unlicensed bands, independent of existing telephone, cable, or terrestrial wireless companies' infrastructures. These unlicensed services began by offering very short-range Internet access connections in "hot spots", but are now evolving into longer-range, higher-speed substitutes for local telephone or cable access lines. These unlicensed wireless WiFi or WiMax services may provide additional broadband platforms well-suited to compete for broadband customers.

A Snapshot of Today's Competitive Landscape

Changes in technology coupled with regulatory requirements designed to promote competition have led to competitive entry in all sectors of the local telecommunications marketplace. Business customers, the most profitable, received the benefits of competition first. Residential customers increasingly have choices, with competitors focused on the most profitable and the easiest to serve. There has been much less competitive interest in those customers who are more expensive to serve, or who generate little revenue and less, if any, profit.

Overall, since the Telecommunications Act of 1996 was passed, some 50 million households have obtained voice service from providers other than the incumbent local-exchange carrier. While a large majority of those have been served by companies reselling service offered by the ILECs or leasing ILEC facilities, an increasing number have chosen alternatives completely separate from the incumbent provider. For the first time in years the ILECs are losing access lines customers (McKinsey and Company 2003: 7).

Cable companies pass 97 percent of US households and now provide voice service to over 2 million customers (Frost and Sullivan 2004). They plan to offer broadband Internet access and VoIP services ubiquitously by the end of 2006, partnering with companies such as AT&T and Sprint (Hu 2004; Gonsalves 2004). Goldman Sachs (2003) estimates that cable companies will serve 6 percent of residential households by 2007 and 20 percent of households by 2012.

Wireless services have totally replaced wireline services in approximately 5 percent of homes (USA Today 2003) and are viewed by many (particularly younger customers) as the primary choice for communications. Licensed wireless providers are likely to provide even greater competitive challenges to existing providers as new technology improves the functionality of wireless services and new devices allow customers to use both licensed and unlicensed wireless services, or to use previously incompatible wireless systems. Regulations that provide for wireline-to-wireless number portability have increased the competitive threat of wireless by eliminating a major barrier to switching; the absence of regulatory restrictions on wireless offerings has allowed wireless companies to roll out new offerings in a far more flexible manner than the ILECs.

Even if one assumed that competitors relying on ILEC services or facilities would disappear as new regulations were promulgated by the FCC, "off-net" providers such as cable or wireless companies are likely to continue to compete in the voice and broadband Internet access markets. These services are a source of incremental revenues from technologies that the off-net competitors have deployed to better compete in their core markets of video delivery and mobile voice services.

If the focus changes from voice services to broadband services the competitive picture is even more at odds with that of a continuing telecom monopoly. Cable providers provide approximately 60 percent of the broadband residential market. Competition in broadband is likely to heat up as cable continues its broadband deployment (broadband capable cable plant now passes 88 percent of US households) (NCTA 2004) and as wireless providers extend and improve their broadband offerings, which are now less capable than those offered by ILECs or cable companies. Provision of broadband Internet access via unlicensed wireless services may become more important as more spectrum is made available for unlicensed services and as private companies and public entities spread unlicensed broadband Internet access across the US in cities such as Philadelphia and in rural areas with low population densities where the costs of new wiring are prohibitive.

For today, the broadband market is dominated in the US by the cable companies and the ILECs rather than being robustly competitive. While forecasting the future in a rapidly changing market is difficult, it seems probable that the long-term competitive threats to the ILECs will be less intra-platform—from firms that resell ILEC services or lease ILEC facilities or that replicate the ILECs' existing platforms in competition with them—and more inter-platform from cable, licensed and unlicensed wireless, and powerline companies that do not rely on ILEC facilities. While the 1996 Act has many failings, it has brought us to the point where multiple platforms capable of providing voice and data services can compete for the more than 90 per cent of residential customers who today can be reached by an ILEC as well as a cable, powerline and multiple wireless companies.

The Rise of Broadband and IP

The rise of broadband services, whichever companies provide them, can be seen as an inflection point in the development of telecommunications regulation. In the past the ILECs either had a monopoly or were the leading, often dominant, providers of narrowband services, particularly voice. High-speed services were aimed at large customers, not the residential market. In today's market for residential broadband services, the ILECs neither have a monopoly nor are they dominant.

But it is not just the change in market share that is important. Broadband connections provide the opportunity for a platform provider, *or any other application provider*, to offer IP-based services, including VoIP. The introduction of broadband and IP services has led to the separation of facilities provision from the offering of applications. In VoIP, as with other IP applications, the owner of the underlying facilities need not be the provider of the application. In addition, the end-user customers pay for high-speed transport by subscribing to the broadband service so that the applications or content providers need not. In theory, an unlimited number of applications providers can compete, utilizing the facilities of whatever broadband providers serve an area.

The broadband platforms connected to the Internet theoretically allow anyone with a good idea to create an application that rides upon it, and make it quickly and widely available—a far cry from the traditional world of telecommunications with its highly centralized architecture and tight control over applications. Even if there are only one or two broadband facilities providers, there can be many applications providers offering voice and data services in competition with each other and with those of the facilities' owner. The more broadband platform providers there are, the better it is for applications providers, as the platform providers compete for the applications business to increase traffic on their facilities and obtain a higher return on their platform investment.

The separation of facilities and applications is a major change from the world in which today's regulation emerged. Moreover, voice applications are now being offered by companies very different from the ILECs or even such new entrants to the voice market as cable companies. Windows XP has, since 2002, included capabilities for voice communication. Yahoo, AOL, and MSN include voice chat capabilities. In the gaming world, voice applications already supplement the core gaming experience. Among the most successful new voice providers is European based Skype, a peer-to-peer service from the creators of KaZaa, providing free service among those who download the Skype software (over a million in the US and over 12 million worldwide) and low-cost voice services to reach Skype-less users. When voice applications are part of computer operating systems, messaging and peer-to-peer systems, and games, it is hard to imagine that regulators or platform providers will be able to identify and control these offerings. What has been commonly understood as telecommunications is becoming another shrink-wrapped application. Those whom regulators would have labeled telecommunications service providers call themselves manufacturers, software houses or game publishers.

The rise of broadband and IP applications such as VoIP is having another important effect on the telecommunications marketplace. As applications providers do not have the be facilities providers, they can roll out new offerings far more quickly, even on a national basis. Companies can enter new local telephone markets in far less time and with far less expense.

The lower cost and increased functionality of IP-based applications are increasing the competitive pressures on existing providers. AT&T and others have indicated that they will move to VoIP offerings due to their superior economics and strong feature sets; the cost and functionality advantages of IP offerings will likely increase over time if they follow well-known information technology performance curves. In addition, the presence of voice capabilities in software, messaging, and games brings a whole new world of creative application developers into what has been the limited and arcane world of voice engineering. While there are only a relatively small number of VoIP users today—perhaps a million—VoIP appears to have all the characteristics of a fundamentally disruptive technology for the telecommunications marketplace.

From a regulatory standpoint, the rise of broadband and IP applications has had another important impact. In the IP world, it is not necessary that a service provider be located in the same area as the facilities provider. While the local exchange carrier can be regulated because its facilities are in the same jurisdiction as the regulator, an IP-based applications provider may operate in another state or even in another country, potentially out of reach of the local regulator. Nor does the customer have to be located in the same regulatory jurisdiction. VoIP providers, unconstrained by geography, are already offering customers an area code of their choice, allowing the customer to reduce telecommunications costs by transforming what would have been long-distance calls into local calls. This is a far cry from the traditional world of telephony, where facilities, services and geography were closely tied.

One last point should be noted. In another change from the traditional marketplace, neither the facilities provider nor the regulator can assume that revenues from IP-based applications will support the investment in facilities. In the world of natural monopolies, 100 per cent of the revenues from applications would flow to the facilities provider and support investment in facilities. The facilities provider would have 100 per cent of the market. Now it is much more difficult to forecast the revenues that will be available to support facilities investment, making it much harder to build a business case for new investment. While economic theory suggests that more applications providers will increase the overall demand for transport services above the level of demand created by a single applications provider, it was far easier to plan and finance a facility when all the revenues (from both transport and applications) were available to support it.

The Shortcomings of Today's Telecommunications Legislation and Regulation

The very intensity of today's debates in the US about the regulatory definition and treatment of VoIP or cable modem service is illustrative of a critical problem. Telecommunications laws and regulation in the US developed as new distribution technologies emerged and new services were offered. Specific laws and regulations were created to address the policy issues raised by the new distribution technology or the new application, be it broadcast or cable TV or wireless telephony. As there was little or no overlap with existing distribution technologies or applications, the regulations weren't necessarily consistent. Each regulatory category balanced competing policy goals in a different manner. As new services emerged, regulators had to determine which regulatory category was appropriate, primarily based on the distribution technology. The choice of the "conduit" determined the regulatory "content". In the US a separate body of regulation contained in Title II of the Communications Act was applied to local and long-distance wireline common carriers; Title III incorporated rules for terrestrial wireless providers; Title VI contained the specific rules for companies offering cable television services. For many years the FCC was even organized around the differing technologies, with Broadcast, Cable, Common Carrier, and Wireless Bureaus.

These differences in regulatory treatment might have been acceptable in a monopoly era or when different technologies were used only to distribute different services. For example, it might have been acceptable to treat a coaxial cable-based cable TV company differently from the wireline telephone company because of differences in what was being transmitted, to whom, and by what means. This is no longer the case. The digitization of information means that voice, video and data can all be represented by bits. These bits can be distributed by twisted wire pairs, coaxial cable, optical fiber, or the radio frequency spectrum (in both bands requiring licenses, like cellular, or in bands that do not, such as via WiFi.) The attributes of services are no longer determined by the distribution technology, and the distribution technologies are relatively indifferent to the services carried. But the different categories with their different regulatory treatments remain.

Even given this convergence, when a new service is offered it must be placed in a specific regulatory category to be regulated accordingly. For example, now that cable television companies can offer broadband Internet access via cable modems, what regulatory treatment should be applied? The answer will determine not only the nature of the regulation but who the regulators are and, more importantly, the competitive position of the service versus similar services which are regulated differently.

Cable modem services are offered by cable television companies but don't resemble their traditional video delivery services. Should they be treated as cable services because they are provided by cable companies? They are functionally similar to digital subscriber line services (DSL) offered by wireline telecom companies, but if they are telecommunications services they would be governed by common carrier regulations including the requirement that they be offered on a non-discriminatory basis to non-affiliated companies with other competitive safeguards and regulatory requirements. Unlike common carrier telecom companies, cable companies can and do discriminate. The differences in regulatory treatment go on and on, each difference distorting the competition that policy makers have sought to encourage.

Both the courts and local and federal regulators have struggled with the categorization of cable modem services. The US Supreme Court recently announced that it would review the Ninth Circuit Court of Appeal's categorization of cable modem services as presumptively telecommunications services. The FCC has concluded that cable modem services are presumptively information services with a telecommunications component that should be regulated under Title I rather than Title II, which governs common carriers. The FCC more recently tentatively concluded that DSL services are also information services subject to Title I regulation, and has opened a proceeding to determine what regulatory treatment should eventually be applied to both cable modem and telco DSL services

The effect of an FCC decision to regulate both services under Title I would be to achieve regulatory parity by substantially deregulating them—something that has considerable attraction given the competition between the two services and the outmoded nature of much regulation. But the deregulation would come at the cost of closing a previously open system in which non-affiliated providers could obtain broadband transport and bundle it with any number of IP applications. This could reduce competition in the applications marketplace. And it would place both services into a regulatory category where, if anti-competitive problems arose, the FCC's powers to regulate under US law are subject to the greatest challenge.

There are many other examples of the difficulties caused by the incorporation of outdated regulatory categories in the 1996 Act. A telephone company wishing to offer video services in competition with a cable TV company has very different regulatory requirements; if it were to compete by providing streaming video services it is not even clear what the requirements would be. Wireless services that are increasingly being substituted for wireline local and long-distance services are regulated differently under Title III from those they replace, which are regulated under Title II.

Services placed in different categories may be subject to different regulators and be treated differently, even if they are functionally equivalent. In yet another example, prices for functionally identical access services vary widely from state to state and between the states and the federal jurisdiction.

In an era when communications are increasingly digitized, this "stove-pipe" or vertical regulation with different rules for each transmission medium is illogical and increasingly unstable (see Figure 2). As voice, data, video and image are presented in digital form, there appears to be little basis for treating them differently based on the identity of the company making the offering or the transmission medium employed. The technology platforms of cable companies, telephone companies, and wireless companies have "converged" in capabilities; the services that they offer have likewise "converged" and increasingly overlap. The Telecommunications Act of 1996 paid little attention to this convergence, or the rise of the Internet, leaving in place regulatory categories inherited from a very different past. Convergence was left to swell into a tsunami breaking over the stove-pipes.

Figure 2. **Old Vertical "Stove-Pipe" Model**

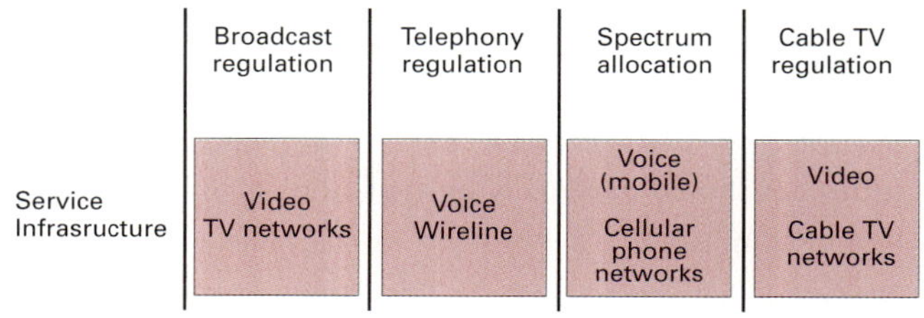

Nor did the 1996 Act adequately anticipate the inter-platform competition that has emerged. It justifiably sought to encourage the entry of long-distance companies and others into the local voice telecommunications market and to stimulate video competition between cable TV and telephone companies. But its focus on narrowband voice services and traditional video delivery seems almost quaint in an increasingly broadband and IP- driven marketplace.

Regulation which has been overtaken by technological or marketplace changes, or which is unnecessary or counterproductive in creating conditions in which competition can thrive, or which fails to protect consumers, can have profoundly negative effects. To continue to regulate as if we were still living in an era of natural monopolies, or to fail to respond in a timely way to increasing competition, has high costs. It creates great uncertainty for those who wish to enter the market. It encourages new competitors and incumbents to structure their offerings based on regulatory categories as opposed to

customer needs or market economics. It places heavy burdens on those firms subject to more extensive regulation, particularly as they seek to respond to competition. Given the importance of regulatory as opposed to market decisions, it invites calls for favorable treatment by those who serve as carriers of last resort as well as by those who claim to be bringing competition to a market. By treating similar services in dissimilar ways it undermines the very marketplace competition it seeks. It discourages investment by regulated firms or their offering of new services due to fears that regulators will use revenues generated by that investment to subsidize other services. While firms in unregulated markets engage in service trials and market first to early adopters, regulated firms can be discouraged from market experiments by regulations that require that new services or technologies be deployed ubiquitously. Regulations may impose additional costs by dictating specific organizational structures or requiring non-standard regulatory financial accounts in addition to those based on generally accepted accounting principles.

This is not to argue that all regulation is harmful or outmoded. Nor is it to argue that there is effective competition in the various telecommunications markets. There is a need for regulation to create and maintain the conditions for competition and to prevent abuses of market power where it genuinely exists. There is also a need for regulators to implement the social judgements embodied in legislation, such as how to ensure some form of universal connectivity for the benefit of all those who use the network and for the society that utilizes communication in its most important economic and social processes. But the time has come to review and update regulatory structures that are now impeding the further development of telecommunications in the US.

Telecom Regulation that Fits Today's Telecommunications Marketplace—And Tomorrow's

The FCC (like many other national regulators) is not indifferent to the inequity of treating functionally equivalent services differently and has taken some positive steps to remedy the problem. Increasingly, the FCC is applying the same requirements to firms offering functionally equivalent services over wireline, cable, or radio waves—if not prevented from doing so by legislative requirements or other policy needs. For example, the FCC is considering proposals to reform both the system of payments that carriers make to one another (inter-carrier compensation) and the universal service support mechanism to treat similarly situated carriers in a similar fashion. The regulatory treatment of broadband services offered by cable and wireline companies is also under review, with one concern being regulatory parity. But, given the existing state of the law, there are important limits on the FCC's freedom to reform.

The Commission has also responded to changes in competitive markets by removing some regulations in specific markets when competition has been convincingly demonstrated. But these steps at the federal level are few and far between, and the process is neither swift nor sure. Nor are state responses to changes in increasingly competitive markets. This is not surprising, given the political sensitivity of issues that must be addressed to create competitive markets such as rebalancing (read "raising") local residential telephone rates where they have been subsidized and do not cover costs. States have lagged behind the FCC in eliminating implicit subsidies in their rate structures, delaying even further the establishment of conditions necessary for effective competition.

The history of legislative efforts to change telecommunications regulation does not inspire confidence. But it is hard to argue against the need for new legislation after watching the FCC struggle to put square services in round regulatory holes or to apply yesterday's tools to today's problems.

The goals of such legislation are likely to be similar to those in the 1996 Act. New legislation should promote intra-platform competition among providers who employ the same technology platforms and inter-platform competition among firms using different platforms. It should be technologically neutral so as to accommodate inevitable technological changes. It should facilitate the identification of market power and focus on its control, both by establishing conditions under which competition can flourish and by preventing its abuse. It should rest on a preference for market forces rather than regulation but provide sufficient authority to regulators to deal with market failures as well as to forbear from regulating. It should encourage the availability of advanced technologies, starting with today's comparatively low-speed broadband offerings, for all Americans. It should provide mechanisms for ensuring that those most likely to be underserved have affordable access in a way that does the least damage to the achievement of a competitive telecommunications marketplace. It should seek to create a positive environment for innovation and investment. And it should reflect the value of "openness" which has served as the foundation for the extraordinary growth of the Internet.

While the goals of new legislation are similar, the nature of the regulatory framework must be different. Many academic experts and telecommunications policy analysts believe that a layered model of regulation, reflecting the layered model of the Internet, is more appropriate than the regulation inherited from the railroad commissions of the

nineteenth century and the regulatory categories that grew like barnacles upon the 1934 Communications Act as new distribution technologies emerged. Compared with today's stove-pipe regulations, a layered or "horizontal" model better reflects recent changes in technology, particularly the increasing digitization of communications and the convergence of the capabilities of various technology platforms. Such a layered model more easily accommodates recent changes in competition, particularly the importance of inter-platform competition and the increasingly central role of the Internet.

Proponents of a layered model generally include: a layer for the physical media over which bits are transported, such as copper wires, optical fiber, coaxial cable, or wireless; a logical layer for the operation and control of the physical media; a layer for applications such as voice services or video delivery that ride upon the physical layer; and a layer for the content included in the applications (see Table 3). The boundaries of the layers are not firm and unchanging, but a shift from vertical to horizontal regulation allows the regulators to focus on the detection and control of market power.

Table 3. **New Horizontal Layered Model**

Content (broadcasting programming, Web content, etc.)
Applications (Web, video, voice, instant messaging, etc.)
Transport (Internet, circuit-switched network, other)
Infrastructure (copper wire, fiber, wireless, satellite)

Each layer can be analysed separately to address whether a participant has market power in that layer. If so, appropriate regulation can be employed to foster competition or prevent abuse in that layer. Regulation would be crafted based on the specific conditions or policy issues within that layer. A layered system would allow the regulators to concentrate on those layers which exhibit bottlenecks or other sources of market power. Open interfaces between the layers would serve as a barrier to the extension of market power from one layer to another.

Layers which are relatively open and in which there is robust competition would not be subject to economic regulation, although non-economic regulations to meet social goals might still apply. In the physical layer—which appears to be the least competitive due to the high capital costs of building networks—economic regulation would play a more central role. But the regulator would focus on controlling market power and preventing its extension to other layers where it could impede competition. Such a check would be particularly important if a transport provider were vertically integrated, participating in the higher layers.

Effective control of market power in the physical layer would justify permitting transport providers to participate in all or any of the layers. Allowing vertical integration seems more likely to increase competition in the other layers than the quarantine approaches that were used in the US to block local telephone companies from entering the closely related long-distance market. Firms participating in the physical layer may well have competencies related to the higher layers and, absent a compelling reason such as a convincing demonstration of their possession of market power in the physical layer and their ability to impede competition in other layers, they should be presumed to be able to integrate vertically.

Adoption of a layered form of regulation in the US would mark a dramatic change. It would make it easier to harmonize US and European telecommunications regulation, which is likely to become more important given that service providers in Europe are likely to be offering services from Europe to the US and vice versa.

Support for layered regulation is widespread even among fierce competitors in the telecom marketplace. Where these competitors differ most profoundly is in their analyses of the extent of market power in the physical layer, and what should be done to prevent anti-competitive effects in other layers.

A detailed analysis of market power in the physical layer in the US is beyond the scope of this chapter. What can be said in brief is that there has been a dramatic increase in competition in the physical layer as described above. There are multiple voice providers and two dominant broadband providers today. More are likely to enter. As to what should be done to prevent broadband transport providers—whatever the medium used— from negatively affecting competition in the applications and content layers, the Internet provides some useful lessons.

Four Openness Principles that Should be Central to Telecommunications Regulation

The openness of the Internet has contributed profoundly to its success. It is built upon open standards. Many of its functions are powered by open source software. Anyone can connect to it with any device capable of using the TCP/IP set of protocols, and can reach others connected to the Internet anywhere. There is no controlling party at the Internet's center to authorize applications, so the hundreds of millions of human beings with Internet access can create and post applications and content which are immediately available around the world.

That same openness has not marked traditional wireline telecommunications or other platforms such as cable TV or wireless networks. In the US litigation and regulatory action were required to guarantee consumers the right to attach devices of their own choosing to the telecommunications network, subject to rules designed to prevent network harm. Similarly the introduction of "enhanced" (data) services by telephone companies took decades in the US. One source of delay was inherent in the centralized control over the network exercised by telephone companies, which required exhaustive coordination before a new service could be offered. On the other hand, regulatory concerns over the entry of monopoly telephone companies into the unregulated data processing markets slowed the process to a snail's pace.

With this legal and regulatory experience as a backdrop, and with the success of the Internet as a guide, there are four principles that should be embodied in regulation to help prevent the exercise of market power to thwart competition and innovation.

The first "openness" requirement is for the availability of interconnection for all networks—one of the few regulatory requirements upon which all competitors seem to agree. While all agree in principle, the issues of how and what to pay for interconnection and transport will require continuing regulatory attention.

A second and much more contentious requirement is that unaffiliated entities be guaranteed non-discriminatory access to broadband transport by any broadband service provider regardless of the technology that such provider employs. Broadband service providers would be allowed to bundle transport and applications into broadband services, but non-affiliated entities would have non-discriminatory access to unbundled broadband transport and could then create their own bundled services.

This hardly resembles the extensive unbundling of network elements required under the 1996 Act. It is limited to broadband transport because that is required to create bundled services that could be used by non-facilities providers to compete against today's limited number of broadband facilities providers.

One of the major hurdles for even considering this requirement has been the issue of price. Setting the price for broadband transport is a difficult problem for regulators, particularly when there is no well-functioning market in existence. The government could set the price, but past attempts by regulators to price unbundled network elements have resulted in continuing litigation and prices that varied widely across jurisdictions.

An alternative course would be to allow the market to operate on its own. Given the size and sophistication of the participants, it might be possible to rely on normal contract negotiations. But the resistance of some broadband facilities providers to making such transport available suggests the need for a way to bring negotiations to a successful conclusion.

In order to prevent a broadband platform provider from using its market position anti-competitively and unduly delaying the process, commercial arbitration with strict time limits for a decision would be imposed if price negotiations failed. The decision would need to cover contracts for a sufficient period of time—perhaps three years—to allow for both parties to implement their plans. Most participants would probably prefer to avoid the risks of arbitration, which would tend to moderate their negotiating positions. While initially there would be relatively little market information to guide the commercial arbitrator, there will be some pricing data from those carriers that have agreed to lease broadband transport. More would be available as arbitrations were concluded.

A third rule would ensure that customers could utilize any device that did not harm the network. This would encourage innovation in the equipment sector and make the underlying networks more valuable to users. There may be a good rationale for more extensive restrictions on attaching devices, such as a rule against connecting devices used for theft of service, but any such proposed rules should be reviewed carefully by regulators. In the past, cable companies have restricted connection of home networks and WiFi equipment to cable modems on the grounds that they enabled theft of service. While this practice by cable companies appears to have ceased, the right to attach non-harmful devices to networks was hard won and worth applying to all broadband services.

A final "openness" rule would ban unreasonable discrimination in the use of applications and in access to content. Users should be able to create and employ applications, and provide and access content as they choose; such rules against unreasonable discrimination are similar to the rules applied today to common carriers in the US (but not cable TV providers.) In the past, some providers have imposed "acceptable use" contracts forbidding customers from hosting websites or providing other services; others have filtered packets to prevent file sharing.

Such a rule would not ban all discrimination but *unreasonable* discrimination. There are sure to be different business arrangements between different parties; such differences would not be defined in advance as unreasonable.

Many arguments are likely to be raised in order to justify various forms of non-identical treatment. An argument might be made, for example, that the management of bandwidth could provide a justification for discriminating among applications. It might be reasonable for a cable TV company to charge different prices for using different amounts of bandwidth. But it would be highly suspect to use bandwidth management to justify an outright ban on streaming video offered by a non-affiliated company that used cable broadband transport to compete directly with the cable company's video delivery business. Similarly it would likely be highly suspect for a cable TV company to degrade cable-based connections to non-affiliated content providers who compete with the cable company's content offerings. To be even-handed by way of examples, it would be highly suspect for a telephone company to degrade the network connections of alarm providers who use telco transport services to compete with alarm services offered by the telephone company.

The successful implementation of such openness policies would render unnecessary other structural remedies or limits on vertical integration or bundling. It would foster competition today when cable companies and telephone companies dominate broadband services. Providers of broadband platforms subject to these rules, and those entities that obtain transport from them, would be free to participate in all layers and to be as creative as they can be in fashioning new bundled services. One small but positive effect of this would be improved IP applications, such as VoIP, that benefit from close coordination with the platform provider. From a broader standpoint, the participation of the largest number of capable players in all layers should stimulate competition and innovation. In a virtuous cycle, increased competition should stimulate new applications and encourage the production of additional content, which should increase the value of the underlying platforms and support further infrastructure investment.

While there is considerable support for the principles of non-discrimination regarding access to devices, applications and content, there certainly will be opposition to rules implementing them. Cable companies have, in the past, opposed requirements for non-discriminatory access to broadband transport; some ILECs that previously have argued for requiring cable companies to provide open access to the cable platform have recently argued that the ILECs should be relieved of their obligation to provide non-discriminatory access to transport in the interest of regulatory parity.

A number of arguments are likely to be made in opposition to regulations based on the openness principles. One is that, given preferences for deregulation and the use of market forces, no new regulatory requirements should be imposed without evidence of anti-competitive activities by broadband platform providers. But fears of anti-competitive behavior are not speculative and evidence not hard to find. Cable companies have denied transport to unaffiliated Internet service providers (ISPs), restricted streaming media applications, and banned home network and WiFi connections to cable modems. New cable technology standards include provisions that allow the blocking of access or degradation of connections to unaffiliated providers.

A related argument is that such a requirement is unnecessary because, as economic theory suggests, platform providers will provide transport to all comers as it is in their long-term interest to maximize revenues from use of their transport facilities. But, at least in the short run, cable providers have resisted just such a course. More recent economic analysis makes a similar argument about the long-term advantages of providing transport to all, but acknowledges that firms might resist if regulators set a price that that doesn't cover costs, if transport providers mistakenly believe they could obtain super-competitive profits from their application affiliates, or if providing transport interfered with legitimate forms of price discrimination. With a negotiated price subject to commercial arbitration, the first possibility would be eliminated. Super-competitive profits obtained by favoring an affiliate would indicate anti-competitive behavior and should not be allowed to justify a denial of transport. Cable companies can and do engage in many forms of legal price discrimination, most commonly in offering tiered packages of pay-programming, but maintaining such price discrimination by, for example, not offering transport to non-affiliated ISPs who might offer streaming video should not be acceptable.

Another argument is that non-discrimination rules would reduce innovation by barring different contractual arrangements between broadband platform providers and their customers. In effect, it might be argued, every arrangement would have to be like a tariffed offering, available to all. But differing arrangements for different customers have existed in the common-carrier world for generations and would be as readily available today. The rule would not bar discrimination, but unreasonable discrimination.

Another argument against the openness requirements is that they would discourage investment in broadband facilities. That surely would be the case if the prices were set by regulators and didn't cover the costs plus a reasonable return on investment—but this would not be the case if there were commercial negotiations. Perhaps the argument might be taken a step further. If the investment would provide new capabilities and was highly risky, the investing firm might seek to capture all the returns on the investment in order to justify it. But having other non-affiliated customers for broadband transport at a price set by commercial negotiation should provide the company with additional revenues. The increased innovation resulting from multiple applications providers should also increase the demand for the new facilities.

What economic theory would predict, and what is being proven in the broadband services market today, is that the presence of inter-platform competition—whether between cable companies and ILECs in the broadband Internet or voice or video markets, or between cable companies and direct broadcast satellites in the video delivery market— provides the most effective incentive for infrastructure investment. Without below-cost regulatory pricing or regulatory appropriation of revenues to subsidize other services, any effect of the openness requirements on investment incentives should be swamped by the competitive incentives to invest.

Another argument is that having one "open" platform— that of the ILEC—is sufficient. All things being equal, the open model should be more attractive. But asymmetric regulation inevitably distorts competition. More importantly, what if the existing common carrier requirements were eliminated and the ILEC chose— perhaps wrongly—to emulate the closed cable model? The "closing" of today's open wireline platform could dramatically reduce competition in the applications arena, just when applications are being separated from facilities and becoming a potential competitive check on the limited number of broadband service providers.

A variant of this argument is that there is intensifying inter-platform competition today, making such a requirement unnecessary. This argument has more power. But what exists today in the broadband transport market is not full-scale competition. Cable TV and telephone companies control the vast majority of the market. Broadband wireless services—licensed or unlicensed—are not yet comparable in capabilities or as widely available as either cable modem or DSL services. Satellite providers have been retreating from competing in the broadband Internet access market, and broadband powerline is just emerging. While it appears highly likely that there will be viable third and fourth and fifth broadband transport providers, this is not yet the case.

While full-blown inter-platform competition is still emerging, it would be useful to impose a sunset date on any open-access requirement as well as a performance standard for its elimination. Thus, the requirement could sunset in, for example, ten years, or would be removed when there are three or four or more broadband providers with no more than one counted from each distribution technology. An alternative performance standard that would demonstrate an effective market for broadband transport would be if, for example, 20 percent of the broadband transport were purchased by unaffiliated application providers.

In the final analysis a balancing decision must be made. Are the costs of requiring non-discriminatory access to broadband transport and enshrining the prohibitions against unreasonable discrimination in regulations greater than the benefits? The costs of the requirements are low. ILECs have demonstrated that such rules can be followed under much more intrusive regulatory schemes. Replacing regulatory pricing with commercial negotiations removes the most obvious source of problems. Cable companies in the US subject to specific open-access merger requirements have demonstrated that it can be done successfully in the cable environment even if cable's different architecture limits the number of unaffiliated parties that can purchase broadband transport. The recent Canadian decision to mandate cable access and to adopt a set of regulations governing it also shows that it is possible.[3]

Weighing in on the benefits side most heavily are the positive benefits for competition, particularly competition from those who lease broadband transport, bundle it with applications, and compete against vertically integrated broadband transport providers in all layers. Demonstrated threats of unreasonable discrimination can be prevented. Regulation, including pricing regulation and regulation of vertical integration and bundling, can be reduced overall. Innovation theory suggests that prospects for innovation are increased as the platform providers can no longer exercise as much control over innovations in applications and content. And all those who use telecommunications service as inputs in their own productive activities should have more and better choices.

Spectrum Policy

Until recently, most policy makers viewed spectrum policy as arcane and relevant only to the limited number of services that depended on access to the airwaves such as over-the-air broadcasting, mobile telephony, and so on. Given digitization of information and wireless transmission, and the resulting remarkable increase in the capacity of the airwaves to efficiently transmit all forms of information, spectrum policy has moved to the center of competition policy.

Policy makers over the past several years have benefited from a vigorous debate about spectrum and a fundamental rethinking of the proposition that spectrum is a scarce resource. Continually improving information technology has led to "smart" radios, "smart" antennas and improved signal processing. These improvements, in turn, allow much more intensive use of already allocated spectrum without harmful interference. As if by magic, these same trends have enabled the creation of what might be called "synthetic" spectrum through the use of "underlays" and "overlays" in existing allocations. "Open spectrum" is replacing the more traditional and static view of scarce spectrum broken up into blocks usable only for a specific service by an often limited number of specified users.

In order to gain all the benefits that wireless services can bring, national regulators should continue to expand civilian access to the spectrum, enabling even greater intra-platform competition among licensed wireless services, as well as heightened inter-platform competition. Additional "unlicensed" spectrum should be made available, as it is in these unlicensed bands that some of the most remarkable innovation has taken place. The various forms of WiFi and WiMax employed in these bands are providing another potential inter-platform competitor. Regulators should encourage the development of ultra-wideband communications—low-power communications that can be spread over vast amounts of spectrum—and new contractual arrangements for spectrum, that will allow even more intensive spectrum use with appropriate deference to the need to protect existing users from harmful interference.

The prospect of wireless services providing inter-platform competition is particularly important in rural areas. The high cost of building fixed wireline or cable facilities in low-density areas has resulted in fewer competitive choices for rural consumers.

Creative wireless pioneers have used the unlicensed bands for Internet access and other services in rural areas where the cost of new landline investments is prohibitive; satellite companies offer less capable connections but can offer them anywhere.

Such wireless service offerings in rural areas should be encouraged. Spectrum use in rural areas is, fortunately, much less intense, opening the possibility of creative solutions for making more spectrum available, including new arrangements for access to spectrum for competitive purposes on either a primary or a secondary basis. (The entry of powerline providers into rural telecommunications markets should also be encouraged as they do not have to build entirely new facilities.)

Rather than insulating rural telecommunications providers from competition, regulatory policies should encourage it. In the long term this is the most efficient way of providing choices for consumers in high-cost areas and for bringing down the total cost of ensuring access to advanced telecommunications throughout the country.

Economic Regulation Revisited

In the world of monopoly providers of telecommunications, economic regulation sought to prevent the abuse of monopoly power and encourage efficiency through oversight of virtually every aspect of the monopolist's business. Regulation was to provide a substitute for the effects of competition. Network plans were reviewed, the addition of lines or the cessation of service approved. Costs were allocated among various services, and prices authorized. Tariff filings and specialized accounting reports allowed regulators to review offerings and operations; rate of return proceedings provided the basis for the monopolist's capital structure. The organizational structure of the firm was subject to regulation and could be changed to avoid anti-competitive actions.

As intra- and inter-platform competition has increased, the need for pervasive economic regulation has diminished. In many circumstances competition can now replace regulation. Effective competition should eventually lead to the end of regulatory controls on prices and cost allocations, service offerings and deployment, organizational structures and specialized accounting and depreciation practices, and so on.

Determining whether there is effective competition is a complex and difficult process. But it is critical to the transition that must be made from a world of detailed economic regulation to a world in which competition, rather than regulators, controls the behavior of firms. What is needed for this change is a road map and a schedule, so that all parties to the process can know what questions must be addressed, can present probative evidence before an impartial fact-finder, and can obtain a decision in a timely manner.

National regulators should lay out such a path away from pervasive regulation. Reasonable timetables should be set which allow for appropriate deliberation but not interminable proceedings. Regulatory processes can be prolonged by the participants if they believe it is in their economic interest to do so; given the high stakes for the firms involved, fixed timetables are probably necessary to ensure that decisions, at least by regulators, are made in a timely way.

This chapter has focused on the importance of broadband and how it should be treated so that the coming broadband world will be competitive and innovative. Yet, even with the dramatic rise in broadband connectivity, in the near term most customers will continue to rely on an existing narrowband telecommunications network by choice or by necessity. Here too regulatory reform should begin immediately.

Reform should begin with what might be the single most sensitive telecommunications issue—that of the price of local voice service. In a world where consumers can choose among multiple voice providers there is little economic justification, for example, for regulating one firm's retail price for voice telephone service absent convincing evidence of market power. If there are adequate and sufficient substitutes, consumers will vote with their checkbooks. The number of customers who literally have no other choice for their narrowband communications is likely to be very small in many countries given the near-ubiquitous coverage of wireless companies.

Where there are alternatives, ILECs should able to set retail prices for their local voice service. Prices should be allowed to rise, with annual caps on upward movement of rates for several years in order to prevent rate shock. If rates do rise, other providers will increasingly take customers from the incumbents.

In evaluating the potential impact of such changes it should be remembered that, even though local rates have increased slightly over the past decade in the US as rates have been rebalanced, telephone subscription rates have also increased. Universal service systems should provide a safety net for those most likely to be negatively affected.

Other regulations can and should be eliminated. The goal should be to foster competition, particularly for broadband, and, over time, to encourage customers of narrowband communications to migrate to more capable and cost-effective broadband networks. At present, ILECs have limited incentives to invest in their legacy wireline networks. They should be authorized to employ whatever technologies they believe can most efficiently serve their narrowband customers and to price them efficiently. They should be allowed to organize the provision of these services in the way they determine to be most cost-effective.

In the long run, encouraging competition in both urban and rural areas is the most effective check on increased prices or the abuse of market power. Affirmative actions to facilitate competition, such as requiring intra- and inter-platform number portability, would help. The universal service system should provide the mechanism for protecting low-income customers and those in high-cost areas—not hidden subsidies or other actions that distort or discourage competition.

Universal Service and Taxation

Since the early days of telecommunications there have been discussions about the best means of encouraging the extension of telecommunications networks and serving customers, particularly in high-cost areas. The issue is still central to debates about regulation, although now it is couched in terms of providing some form of affordable universal connectivity or universal service, given the importance of access to telecommunications for participation in the economic and social life of the nation.

In the days of monopoly providers, the costs of providing service to everyone—those expensive to serve and those less so—were incorporated into the overall costs of the monopolist. The firm would then be allowed to charge rates that would cover these costs and provide a return on investment. The rates charged to different customers did not correspond to the different costs of serving them. Business customers subsidized residential users (and still do with business rates roughly double those of residential customers for functionally equivalent services). High-volume users subsidized low-volume users, residential customers who bought value-added services like call-waiting subsidized those who didn't, long-distance users subsidized local users. As markets have become more competitive these differences between the costs incurred and the prices charged served to attract new competitors.

In the US the 1996 Telecommunications Act sought to eliminate subsidies buried in rates so that the prices for various services would better reflect costs, sending the correct economic signals and encouraging efficient competitive entry. Subsidies to support universal service were to be made explicit. While much has been done in this area at the federal level, the process is by no means complete with even more to be done at the state level. To the extent that retail rates remain regulated, implicit subsidies must be removed and rates rebalanced if effective competition is to be achieved.

Even today, when there are proliferating choices for local voice service, the ILECs remain the carriers of last resort, required to provide affordable service to all who request it. But increased competition has led to the loss of their more profitable customers and has caused profit margins to shrink, undercutting the ILECs' ability to serve less profitable customers without external support. At the same time universal service funds in the US—the source of external support—are under pressure, raising serious questions about the capacity of today's universal service system to continue to fund universal service as presently defined.

How should universal service, particularly for low-income consumers and those in high-cost areas, be defined and supported? Some principles regarding universal service are almost self-evident. Support should be targeted to those who most need it. What is supported should be clearly defined and should reflect what is broadly purchased in the marketplace. The means of providing support should not distort the competition which regulation seeks to foster.

The best solution is that long preferred by economists. Rather than relying on implicit subsidies within telecommunications prices or industry-wide contribution programs that may not encompass all the relevant players, it would be better to simply recognize the social, economic, and political importance of universal connectivity and support it via direct governmental funding. Such funding is as appropriate as the funding of other social programs. It could be distributed in a variety of ways, the most attractive being the provision of funds directly to customers via vouchers which they could use to help them purchase particular services from providers they choose.

If the funding of universal connectivity is still to be internalized into the telecommunications industry, it should reflect the same principles that underlie the layered model of regulation. The universal service contributions from the various companies need to be rationalized with those similarly positioned treated in a similar fashion. Recent universal service proposals that base contributions on the number of telecommunications lines or telephone numbers are a step in that direction. In the same way, if funds are not provided directly to end user customers, there should be no preferences as to which companies can draw from the universal service fund. Incumbents and newcomers should be treated in a similar, technologically neutral, fashion.

One point should be made about telecommunications and taxation. Given the key role that telecommunications now plays, it is striking that the telecommunications industry is among the most highly taxed—perhaps *the* most highly taxed. Because it provides an essential input for other productive activities, the telecommunications industry should not be used to collect excise taxes to provide general revenues. A rationalization and lowering of taxes on telecommunications would serve to stimulate the substitution of information processing and telecommunications for other processes that consume more resources or provide less efficient means to accomplish important economic and social goals. Reducing taxes would also reduce the price of broadband, thereby stimulating greater broadband subscription.

Other Regulatory Requirements

Numerous other forms of regulation are applied to the telecommunications industry. An increasing challenge is to determine whether regulation is still required to accomplish particular goals and, if so, how to advance such goals in the most efficient manner with the least impact on the competitive marketplace. The goals, in general, should be set by political rather than regulatory bodies. Broad-based industry groups, with public interest representatives, should be utilized, to the extent possible, to propose efficient and effective means for accomplishing these goals.

For example, traditional wireline carriers have provided their customers with the capability to connect to emergency 911 services. Over the last several years the licensed wireless industry has been mandated, at considerable cost, to provide emergency 911 services as well. Will new competitive entrants—from cable companies to unlicensed wireless providers to VoIP companies—face the same requirements? Will they have to replicate the systems in place now or will they be allowed to implement innovative systems that meet the social goals but in different ways? Broadly based advisory groups can play a key role in addressing such issues.

Similar questions arise with respect to access to telecommunications services for individuals with disabilities. The 1996 Act reflected a societal decision that guaranteed such access. Some new competitive service offerings are less hospitable to the needs of the disabled community. Will they face the same requirements or can their providers be challenged to find acceptable innovative solutions in this area as well?

Less widely discussed are the issues of robustness, resiliency and restoration of the telecommunications network. In the past the very limited number of local and long-distance firms made it possible for the government and industry participants to cooperate in planning for network restorations, particularly in the event of an emergency. In the post 9/11 environment in the US it is necessary to contemplate the possibility of truly large-scale outages, such as those that might be caused by terrorists using weapons of mass destruction. It is therefore increasingly important but increasingly difficult to bring together today's wider range of market participants to determine how to improve the robustness and resilience of the telecommunications network as a whole, and to anticipate and plan for such contingencies.

Conclusion

A great deal of attention is being paid by national governments around the world to the regulatory treatment of IP services, including VoIP. But we need to look at the more fundamental issues of how, and to what extent, we regulate the entire telecommunications industry. What form of regulation is appropriate in a world of converging platforms and IP services? Can today's regulation be justified in the presence of multiple competing platform providers or where there is no longer an unbreakable tie between facilities ownership and service provisioning, or a match between regulatory jurisdiction and the location of service providers? Given the growth of competition, where, if anywhere, does market power exist? How will regulation accommodate further technological change as functionality and service move to the desktop from central offices and head-ends?

The intellectual foundation of today's regulation is, at best, under siege. The next challenge is to provide a process for reform and a design for regulation that reflect a world of convergence, not stove-pipes. We must break out of the confines of today's regulatory categories that emerged to deal with disparate distribution technologies with different capabilities, because these technologies now provide functionally equivalent services while being regulated differently. We need to find more effective and efficient ways to promote effective competition, particularly inter-platform competition, and be prepared to reduce or remove regulation as competition develops. We need to prevent anti-competitive acts and draw upon the lessons of the Internet promoting open systems such as ensuring that all of the disparate platforms can be interconnected. We need to find ways to implement principles of openness—

openness regarding platforms, network attachments, applications, and content—by preventing unreasonable discrimination. We need to move away from structural remedies or quarantines and encourage the greatest possible innovation by all market participants. As we move increasingly to a broadband world we need to ensure that all broadband platforms provide transport in a non-discriminatory manner to non-affiliated providers so that anyone can provide bundled services. We need to think more about incentives than control, investment rather than costs, the power of innovation rather than the threats from change.

We also need to re-examine how to accomplish societal aims such as the universal availability of advanced technologies while ensuring that the most vulnerable are not disconnected. Other social goals also merit re-examination, involving not only how they might be most efficiently and effectively accomplished but also whether the present arrangements should apply to the plethora of new and varied market participants.

The challenge to create a new regulatory paradigm and pass legislation that meets the needs of policy makers and is broadly acceptable to consumers and the industry alike is enormous. Among experts there is broad agreement on the need to rethink how, and whether, we regulate, and to update our regulatory institutions. We need to begin that process soon if we are to build a strong foundation for the Information Society.

Notes

1   The World Economic Forum's Networked Readiness Index includes a Political and Regulatory Environment subindex designed to assess whether a country's policies are fostering the growth of the networked economy (see Chapter 1). In the 2004–2005 rankings, the United States is ranked 10th, after Denmark, the United Kingdom, Singapore and Finland among others.

2   Much of the chapter is adapted from a longer paper by Elliot Maxwell supported by the Economic Policy Institute.

3   Canadian Radio-Television and Telecommunications Commission, Telecom Decision CRTC 99-8, 6 July 1999 and Telecom Decision 2003-47, 14 July 2003.

References

Economist Intelligence Unit. 2001. *The Economist Intelligence Unit/Pyramid Research e-readiness Rankings*. London: Economist Intelligence Unit. Online. http://www.ebusinessforum.com/index.asp?layout=rich_story&doc_id=367&categoryid=&channelid=&search=e%2Dreadiness

———. 2004. *The 2004 e-readiness Rankings*. London: Economist Intelligence Unit and IBM Institute for Business Value. Online. http://graphics.eiu.com/files/ad_pdfs/ERR2004.pdf

FCC (Federal Communications Commission). URL. Online. http://www.fcc.gov/telecom.html

——. 2004. FCC 04-216. Online. http://www.neca.org/wawatch/wwpdf/100104_5.pdf

Frost and Sullivan. 2004. *North American IP Cable Telephony Market*. Frost and Sullivan.

Goldman Sachs. 2003. *Telecom Services: United States*. 7 July. Online. http://www.vonage.com/media/pdf/res_07_07_03.pdf

Gonsalves, A. 2004. "Cable Operator Mediacom Partners With Sprint." *Information Week*. 25 August. Online. http://informationweek.networkingpipeline.com/news/33200001

Hu, J. 2004. "AT&T strikes VoIP deals with cable." CNET *News.com*. 19 August. Online. http://ecoustics-cnet.com.com/AT38T+strikes+VoIP+deals+with+cable/2100-7352_3-5316842.html

Insight Research Corporation. 2004. "The Wireless Impact on Long Distance Markets as Seen by the Sale of AT&T Wireless." Press release. Boonton, NJ. 26 February. Online. http://www.insight-corp.com/pr/02_26_04.asp

ITU (International Telecommunication Union). 2004a. "Cellular subscribers." Online. http://www.itu.int/ITU-D/ict/statistics/at_glance/cellular03.pdf

——. 2004b. *The Portable Internet*. ITU Internet Reports. Geneva: ITU.

McKinsey and Company. 2003. *The State of the Industries: Past, Present, and Future*. New York: McKinsey and Company. Online. http://www.mckinsey.com/practices/mediaentertainment/home/content/pdfs/McK%20Media%202003%20State_of_Industries.pdf

Martin, K. 2004. Presentation at the 22nd Annual Institute on Telecommunications Policy and Regulation. FCC. 3 December. Online. http://ftp.fcc.gov/commissioners/martin/documents/presentation120304.pdf

NCTA (National Cable and Telecommunication Association). 2004. "Broadband Services". Online. http://www.ncta.com/Docs/PageContent.cfm?pageID=37

Richtel, M. 2004. "The Diminishing Bell: The Industry; Bells Win a Battle, Not Necessarily the War." *New York Times*, 23 July, p. 1

USA Today. 2003. "For many, their cell phone has become their only phone." 24 March. Online. http://www.usatoday.com/tech/news/2003-03-24-cell-phones_x.htm

WTO (World Trade Organization). URL. Online. http://www.wto.org/english/tratop_e/serv_e/telecom_e/telecom_e.htm

Chapter 6
Next Generation IT Outsourcing:
Profits or Perils?

Mark Melford, Booz Allen Hamilton

Myles Wright, Booz Allen Hamilton

Suvojoy Sengupta, Booz Allen Hamilton

Introduction

Outsourcing is now an established fixture in the organizational models of the twenty-first century. At first confined to peripheral business activities such as cleaning, transport or legal services, outsourcing now encompasses business functions that are closer to the "core", such as manufacturing, customer management . . . or IT.

Outsourcing offers strategic and economic benefits that are too compelling to ignore. When it works, outsourcing decreases costs, increases flexibility and discipline, enhances expertise, and provides the freedom to focus on core business capabilities. However, the worlds of business and government are replete with examples of outsourcing decisions gone wrong. And as suppliers become more tightly integrated into the fabric of a company's basic business operations, the risks attending the failure of these relationships escalate.

IT outsourcing is at the sharp end of this movement. Many companies' IT activities are so close to "core"—that is, so closely intertwined with basic business processes—that an IT outsourcing decision becomes a strategic choice. IT outsourcing is now a developed market where global suppliers are offering differentiated and more flexible service propositions, resulting in increased complexity of procurement processes and relationship models.

Outsourcing in general and IT outsourcing in particular are increasingly not just make-or-buy decisions; they are make-or-break decisions. This chapter is aimed at executives facing an imminent or potential IT outsourcing decision. It addresses the questions:

- What is the track record of outsourcing so far, and what lessons can be learned from it?

- What are the common pitfalls and how can they be avoided?

- How to do it right? What are the key success factors in making your IT outsourcing decision?

Working with clients across a range of industries, Booz Allen Hamilton has identified five common myths about good outsourcing practice, perceptions which can lead to failure of the outsourcing deal. We have also observed the common characteristics of successful outsourcing decisions, and crystalized six steps which successful companies tend to follow as they identify what should and should not be outsourced and manage new supply relationships.

The Recent History of IT Outsourcing

With all aspects of running a company becoming more competitive and complex, it is little wonder that companies are refocusing their efforts on what they do best and outsourcing the rest. But what exactly have they been outsourcing? And how successful have they been?

Services typically outsourced have been those deemed "non-core", at least until recently. In a manufacturing context, those services have included cleaning, transport, catering, maintenance, and training, among others. In other industries, functions such as legal, financial, human resources and website development are regularly farmed out. In general, the less strategic the activity, the more likely it is to be outsourced. But as outsourcing takes hold, its reach is spreading into areas closer and closer to the core.

IT outsourcing in the context of the overall outsourcing movement

This chapter addresses the issues around IT outsourcing specifically. But to do so, it is important to recognize that many of the dynamics, myths and truisms that apply to IT outsourcing apply to the outsourcing movement overall. How exactly does IT outsourcing fit in?

IT outsourcing typifies the trend to outsource systems and processes close to the core. In many industries, such as financial services or other information-related sectors, business activities rely heavily on IT. This means that IT is increasingly considered a core activity. Even in more "traditional" industries, such as the manufacturing sector, many IT systems are considered business-critical due to the impact that a breakdown might have on business

operations. Therefore an IT outsourcing decision process must be supported by a deep analysis of business issues and must be approached as a potential *business process* outsourcing decision.

The distinction between *system* and *process* is an important one, as it represents a greater degree of autonomy and responsibility on the part of the supplier, as well as greater value added. An airline may outsource its ticketing systems—a significant step—but to outsource the very process of ticket sales is a considerable step further. In this context, the outsourcing of IT systems and processes can be seen alongside the outsourcing of other support services (which may in themselves be "core" or strategic), such as finance, IT or human resources. A recent Booz Allen Hamilton survey found that, of these three areas, outsourcing of IT was the most advanced, having the highest percentage of companies outsourcing selected—or even "most"—processes (see Figure 1). It's also worth noting that almost 60 percent of companies surveyed were outsourcing functions to some degree.

But what exactly *is* IT outsourcing, and what does it mean to outsource "selected" processes? With the growth of specialized services and more mature service offerings from vendors, the question facing companies now is rarely "should we outsource IT?" The challenge is usually to identify the right "bundles" of IT activity to outsource and to be clear upfront about the business reasons for outsourcing. Functions such as user support or Internet services are among the commonest "bundles" outsourced, whereas functions such as IT strategy planning, policies, governance, and so forth are rarely outsourced (Figure 2).

Figure 1. **IT Outsourcing vs Outsourcing of Other Functions**

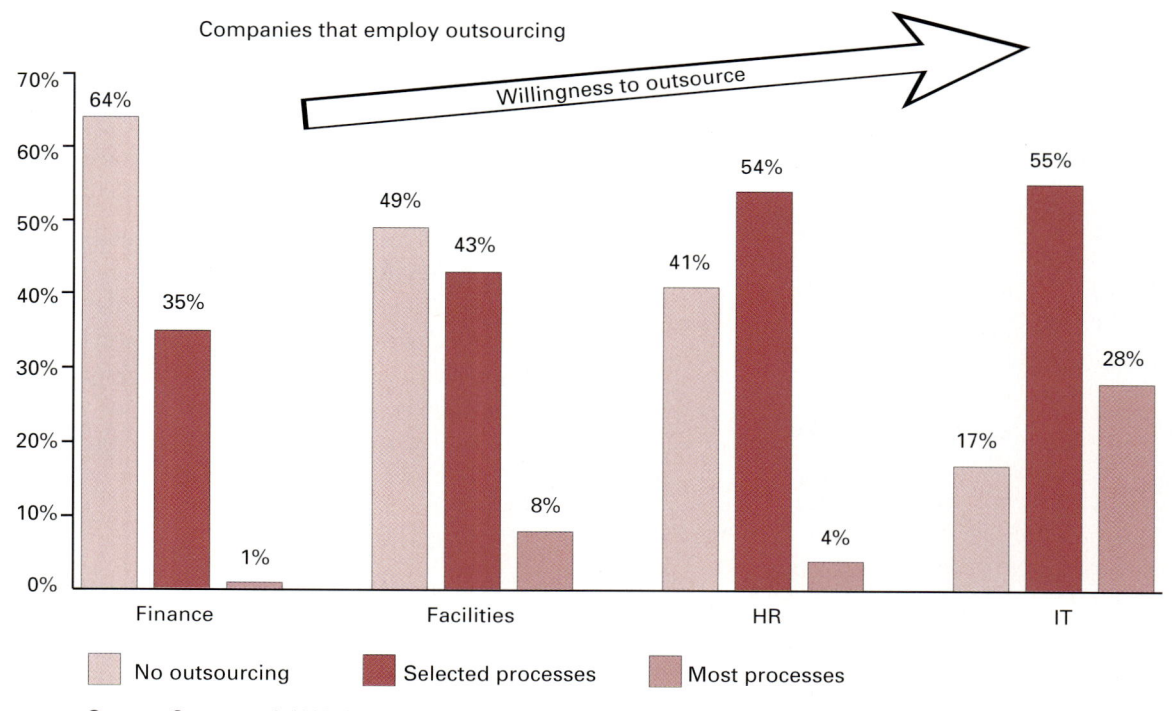

Source: Couto et al. (2004)

Figure 2. **IT Activities Outsourced**

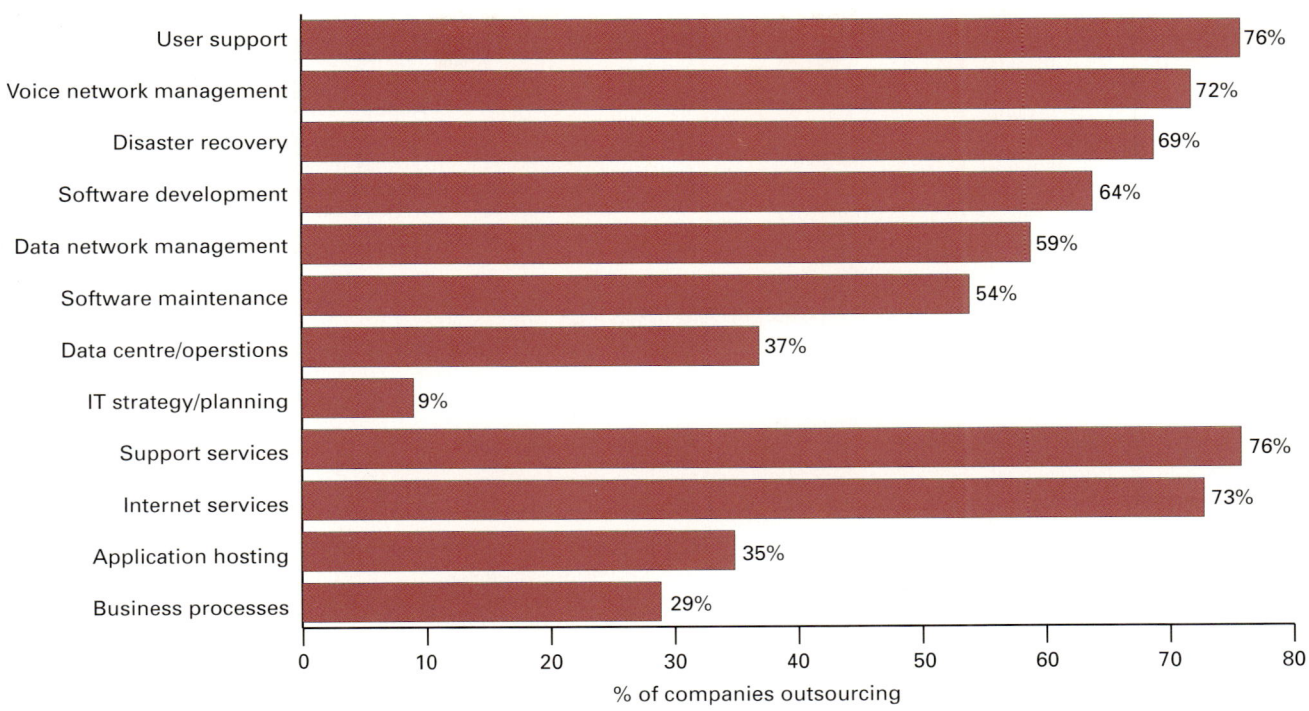

Source: The Conference Board

What are the potential benefits of IT outsourcing?

What is driving this phenomenon? The initial and superficial perception was that outsourcing was simply a means to lower cost and add flexibility in cases of rapidly fluctuating demand. The first wave of IT outsourcing savings were based on capturing factor cost differences, that is, getting work done by cheaper resources at variable cost. This also drove the first push to move activities to cheaper locations such as India.

Since the turn of the present century, a more sophisticated set of rationales has emerged. In an accelerated, global competitive market that rewards focus, companies have sought to contract those segments of their supply chain that are not adding optimal value. "Vertical" thinking (own or control every link in the supply chain) has given way to "virtual" thinking (create a flexible web of supply relationships and focus exclusively on what one does best), and companies are rushing to implement the new, networked business models that result. When it works, outsourcing offers companies compelling strategic and financial advantages, including lower costs, greater flexibility and discipline, enhanced expertise, and the freedom to focus on core business capabilities.

In 2004, cost remains the greatest among a wide set of motivators for outsourcing (Figure 3). However, there are telling differences between the rationale for smaller and larger firms: smaller firms are motivated by cost, access to

technology and flexibility; large firms are likewise motivated by cost-reduction objectives, but also by other factors such as improving business focus, internal reorganization, freeing up internal resources and accessing expertise not available in-house.

How successful have companies been?

In short, the report card is mixed. In fact, recent research suggests that fully a third of the companies that have tried outsourcing failed to realize cost savings, with some even suffering significant cost increases! (Figure 4).

In terms of cost, a 2001 international review of more than 200 outsourcing studies highlighted the mixed nature of the results achieved. On average, firms estimated a 15 percent reduction in costs, but almost a third of the companies surveyed believed their costs had remained constant or had actually *increased*. More recent research (people3 2003) shows that in 2003 this pattern of success was unchanged: 18 percent of companies did not achieve any cost reductions, and almost 10 percent experienced an increase in costs from their IT outsourcing contracts! Only 24 percent of companies reported a cost saving greater than 20 percent.

Furthermore, there is strong reason to believe that most studies conducted to date have overestimated the benefits generated from outsourcing. Many simply ask respondents to *estimate* the savings they have achieved rather than requiring a more rigorous economic analysis. In coming

Figure 3. **Benefits of Outsourcing**

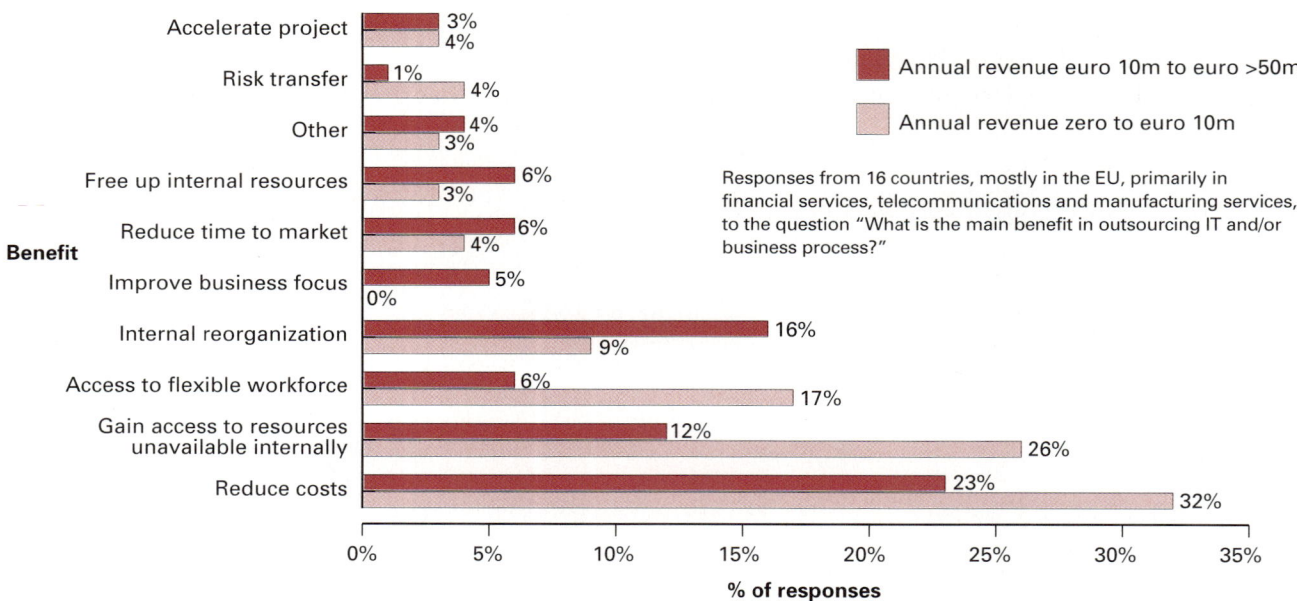

Source: Trestle Group (2004)

up with these estimates, many companies fail to consider the full economic impact of their outsourcing decisions. They frequently cite only the absolute reduction in costs since the relationship began rather than looking at those savings from a net present value (NPV) perspective, including associated costs and effects on production levels. Companies also frequently ignore the fixed costs that do not go away when a task is outsourced. As well, they ignore the administrative costs incurred in managing a new supplier relationship. Based on our experience, fixed and transition expenses can account for up to 30 percent of the total costs related to an outsourced function.

More broadly, our work with clients worldwide reveals a catalogue of errors in the management of the outsource contracts themselves. Many IT outsourcing deals have encountered problems that not only led to renegotiation of the contracts but in many cases a rethinking of IT sourcing strategies and organization. Often, the root cause can be traced to the beginning, when many companies failed to adequately redesign IT costs, processes, organization and demand management before making the outsourcing decision.

- Many outsourcing deals have evolved in such a way that the *price* for the IT services greatly exceeds industry

Figure 4. **Percentage of Studies Reporting Cost Savings from Outsourcing**

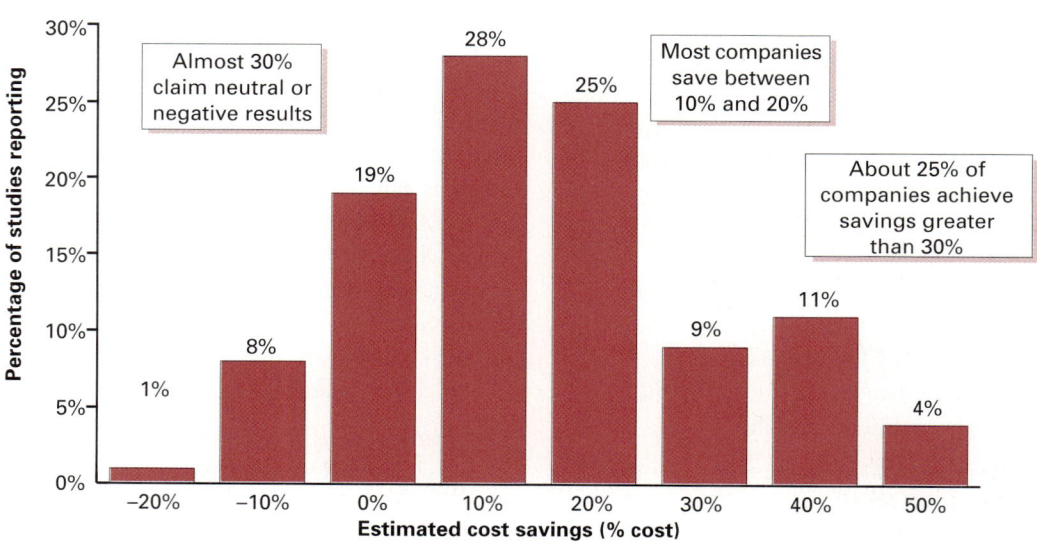

Sources: Industry Commission: Booz Allen Hamilton analysis

benchmarks, especially if the agreements were not structured to both manage internal demand and capture the naturally declining costs of suppliers.

- Most companies suffer from limited *leverage* as they have minimal insight into the true costs and drivers, and have few options to change suppliers. Further, their organization does not have the *skills* to manage the more sophisticated outsourcer.

- Finally, many companies have gone too far in their "partnerships" with suppliers, leading to inadequate IT *governance* over the relationship, uncontrolled demand, higher prices paid for services, and ultimately an acrimonious relationship between customer and supplier.

As companies look to outsource activities closer to their core, the consequences of such failures escalate; indeed, they threaten the very viability of the organization.

Unsurprisingly, after the initial surge and disappointments, the market has gone through a period of cooling off. Between 2002 and 2003, the value of global IT outsourcing contracts actually fell by 3 percent to US$55 billion (NelsonHall 2004).

In 2004, IT outsourcing is emerging a more mature sector as companies become more sophisticated in their approach.

- The total value of large deals (more than €40 million) has been increasing again, particularly in Europe, where the total value of outsourcing contracts rose from €13 billion 2002 to €24 billion in 2003 (The TPI Index Europe 2004)—levels so far for 2004 are up on 2003.

- Vendors are improving their performance. Most have recognized that factor cost substitution is not a sustainable competitive basis. In order to remain competitive the key players are striving to reduce their structural costs through a commitment to improving IT operations and development process discipline (most outsourcing majors now boast of quality standard certifications with ISO9002, CMMI[1] and Six Sigma[2]) and investment in automation technologies. The majority of long-term outsourcing contracts today include clauses stipulating year-on-year productivity gains that the vendor must realize and subsequently share with the outsourcing company.

- Satisfaction with outsourcing also seems to be recovering. In fact, it is *dissatisfaction* levels which are generally more telling: in 2004 these had dropped to 10 percent, against 23 percent in 2002 (DiamondCluster 2004).

Next-generation outsourcing

The IT outsourcing services market is moving towards more mature and innovative service offerings, not just relying on salary cost differentials between developed and developing economies. Here are three emerging but clear trends that will shape IT outsourcing in the next two to four years:

- *Multisourcing*: the days of the monolithic IT outsourcing arrangements are over. Large corporations are outsourcing a range of IT services from different providers. This increases the flexibility of the overall arrangement and reduces dependence on any single provider. The lessons from early outsourcing experiences in vendor management are being leveraged now to manage more complex multi-vendor relationship models

- *Utility computing*: the new paradigm in enterprise computing, utility computing holds the promise of computing capacity becoming ubiquitous and usage-based. Thus, the corporation of the future need not invest in expensive assets such as data centers, servers and applications, but can get computing power "on tap" and pay for what it uses—much like gas and water. The enabling technologies are coming to market now; for example, IBM is investing heavily in its On Demand services. Utility computing can have a dramatic effect on the IT outsourcing market, as it can change the existing asset ownership-based model of IT into a more flexible usage-based model

- *Global delivery models*: while the economic advantages of moving IT activities offshore are well established, a purely offshore delivery model has proved to be less than optimal. This is largely due to differences in culture or language between countries and also due to the need for support closer to end-users. Large global outsourcing providers are developing innovative delivery models which incorporate the optimal combination of activities performed at on-site, onshore, offshore and near-shore locations. This is aimed at enhancing the level of on-site service while leveraging the cost advantages of offshoring. Outsourcing providers have developed specific offerings such as EDS' Bestshore and Cap Gemini Ernst & Young's Rightshore. Both offerings seek to optimize the delivery model globally.

In short, the game is changing and the stakes have never been higher. An IT outsourcing decision is increasingly make-or-break for the company. Next-generation outsourcing is different not only in terms of the sophistication of suppliers and service offerings but above all in terms of the sophistication of the way in which "customer" companies are making and executing against their outsourcing decisions. What then are the characteristics of successful approaches? How do the successful companies do it?

Myths and Truths about Successful Outsourcing

Many companies have already reaped significant benefits from IT outsourcing, and others are poised to capture that value. The imperative to enhance performance is still there; if anything, the bar has been raised. And there is no turning back. While outsourcing is here to stay and the benefits are many, companies should embark on these initiatives with their eyes open. More specifically, Booz Allen client work indicates that organizations can benefit from outsourcing and offshoring and create value if they avoid falling prey to common misconceptions—that is, outsourcing myths—and if they adhere to certain critical best practices when setting up outsourcing arrangements.

Our goal in exploring both outsourcing best practices and myths is to *ensure that companies can capture value by successfully implementing outsourcing and offshoring strategies*.

Common outsourcing myths

Organizations can improve their chances of success by rejecting outsourcing truisms or myths that have gained widespread currency and often determine how companies implement outsourcing deals. The following myths, often advocated by suppliers or sometimes even by internal company staff, lead to the failure of outsourcing core and non-core agreements:

Myth 1. Partnerships lead to successful outsourcing arrangements.

Myth 2. A highly competitive bidding process will ensure the best deal.

Myth 3. Outsourcing reduces management complexity.

Myth 4. A broken process should not be outsourced until fixed.

Myth 5. Everyone is outsourcing whole functions.

A closer examination of each myth makes it clear that popularity is no substitute for veracity and value.

Myth 1: Partnerships lead to successful outsourcing agreements. The idea that partnership is the key to outsourcing success is a common fallacy. While strong and open relationships are critical, relationships must be based on well-defined contracts that capture all aspects of the relationship. Organizations must insist upon transparent agreements that explicitly define exit clauses and include clear risk-reward structures—all within detailed contractual frameworks. Outsourcing suppliers often seek to leave elements of the contract vague in order to close deals and to structure them in their favour.

Furthermore, strong relationships between senior executives, while important, are no substitute for detailed contracts, as these relationships rarely serve to resolve critical issues, particularly those with economic implications. Recently a major European banking group outsourced a critical customer facing process based on a "handshake" between CEOs. When issues surfaced and customer service levels began to drop, the partners struggled to resolve them. Due to lack of clear guidelines and a detailed contract, it took nearly a year and a half for customer service to return to pre-outsource levels, and transaction costs increased by almost 50 percent.

Myth 2: A highly competitive bidding process will ensure the best deal. While loosely defined partnerships can scuttle outsourcing efforts, so can overly competitive bidding processes that push suppliers to the lowest cost. Clearly, competition is important to ensure good pricing, but suppliers must make a sustainable margin. When assumptions are unrealistic, suppliers often try to make up the shortfall by designing inflexible agreements and increasing charges when requirements or volumes change.

Indeed, overly aggressive bidding can thwart economic gain. A large company sought to outsource a critical customer-facing activity and negotiated aggressively with all bidders to achieve a lower-than-industry cost structure. As customer demands changed and processes required adjustment, the outsourcer invoked several costly change clauses. Furthermore, the agreement did not sufficiently account for required upgrades in the underlying technology. As a result, several key changes designed to improve market share were delayed.

As organizations seek to outsource increasingly complex core activities, instead of focusing on overly competitive bidding processes they should focus on developing transparent cost-plus structures, ensuring a sustainable margin for suppliers (see Figure 5). Firms should also supplement these transparent agreements with risk-reward objectives.

Myth 3: Outsourcing reduces management complexity. Many companies mistakenly assume that outsourcing will reduce management complexity and simplify running an organization. In reality, management complexity increases when activities are outsourced, for one primary reason: while activities can be outsourced, management understanding of these activities and their processes cannot. To manage and achieve positive change within an organization, management must understand in detail how processes are executed and why.

A large retail bank recently sought to implement a new customer strategy focused on basic customer experiences, such as account application through receipt of debit cards, cheques and the first transaction. Parts of the account opening process had been outsourced for several years, including data entry, call-centre support, and sending customer chequebooks. But most bank managers had lost touch with how the processes were operating. As a result, these managers had to make a significant effort to relearn

Figure 5. Understanding How the Outsource Supplier Makes Money (client example)

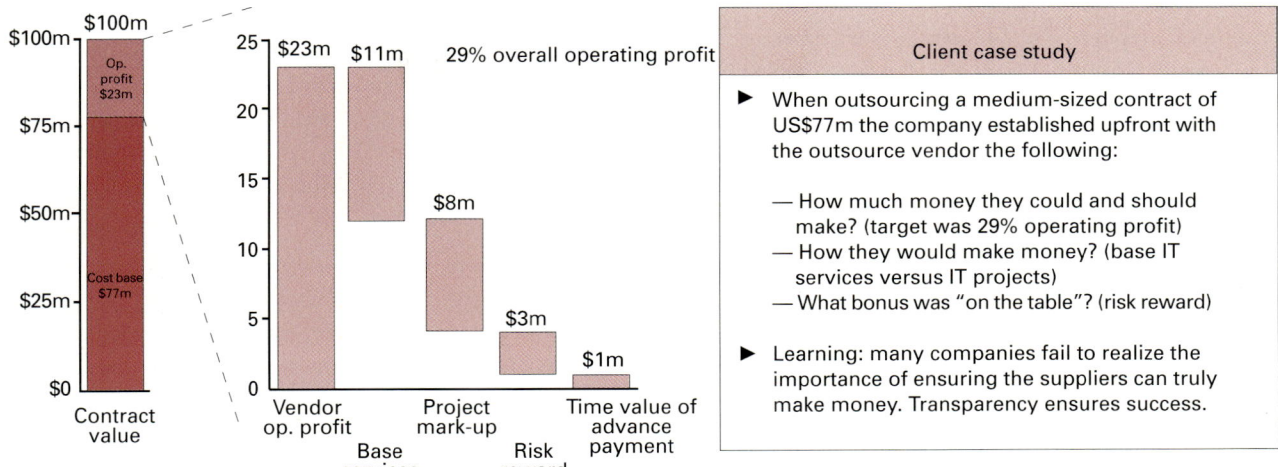

the bank's processes so that they could plan and implement the new customer strategy.

One of the advertised benefits of outsourcing is that it frees up management time to focus on critical value-creation activities. In reality, unless managers understand the workings of the outsourced processes, they will need to spend a significant amount of time managing outsourcer interfaces instead of focusing on high-value activities.

Myth 4: A broken process should not be outsourced until fixed. It is not necessary to fix broken processes before outsourcing: it is important to understand the broken processes and the cost of fixing them. In fact, companies can enjoy a competitive advantage during negotiations if they understand the processes and the potential benefits of re-engineering them. Who fixes the processes is less relevant.

Cost savings of 10–30 percent are possible through process improvements. Companies should invest in the requisite upfront analysis and then work with the outsourcer to identify savings and agree upon a clear action plan to capture and share them. Business process re-engineering assumptions and savings should be part of the negotiation, and each supplier should be asked to submit a realistic plan to implement the savings.

Myth 5: Everyone is outsourcing whole functions. Not all activities can be outsourced. In fact, most companies are struggling to outsource basic technology successfully.

In a recent Booz Allen study of CFOs of global companies, we identified only a handful of pioneers who were outsourcing whole functions. While most companies are outsourcing processes, they engage in this practice selectively. Even within functions, such as IT, which is among the most readily outsourced, only about 28 percent of the respondents are outsourcing it fully. Other non-core activities like finance and human resources are even further behind.

One of the reasons companies are lagging in outsourcing core functions is that regulators can place stringent requirements on them when they outsource these activities. To meet these regulatory demands, organizations must prove they can manage the operational risk of outsourced activities, and only those with deep outsourcing skill and experience will be able to pass this test.

Six pragmatic outsourcing best practices

Outsourcing challenges are significant, and the most successful organizations will be those that can rise to them. Here we outline *six pragmatic outsourcing best practices* to help organizations overcome the key challenges to next-generation outsourcing.

1. Manage complex relationships with multiple internal and external suppliers.

2. Establish clear service parameters, such as definitions, service level agreements, and cost baselines.

3. Overcome lack of in-house negotiation experience in structuring complex outsourcing arrangements.

4. Avoid overly optimistic expectations about the time required to outsource.

5. Structure the agreement flexibly to handle changes in volumes and requirements.

6. Practise financial engineering carefully so that fundamental economies are not traded away to achieve short-term objectives.

Organizations that bring a disciplined and rigorous approach to implementing these best practices will be best-equipped to meet the complex and complicated challenges of outsourcing—and reap the financial rewards. Here's a more comprehensive assessment of each practice:

Manage complex relationships with multiple internal and external suppliers. Most next-generation outsourcing deals involve complex relationships with multiple internal and outsourced suppliers. The IT domain is no different. To be successful, companies must plan for this complexity by developing new organizational structures and processes and by ensuring that staff have the right skills to manage the new constellation of relationships. Failing to prepare sufficiently ensures savings will be lost and the overall health of the outsourcing arrangement will decline.

Recently, a large European bank sought to outsource nearly half of its €4 billion business processes and technology activity. The initiative failed because the bank did not sufficiently prepare both the outsourcer and the other internal and external suppliers.

A key element of the failed initiative was outsourcing mortgage processing. The bank failed to achieve cost and service targets because it could not manage and coordinate the ten-plus suppliers involved in the overall mortgage process. Five of the suppliers were internal in separate business units, and provided mainframe, data input, and bank customer number maintenance. Another internal group provided IT infrastructure, such as networks and desktops, to all participants. Similar complexity existed with external suppliers to the mortgage outsourcer. Customer service quality costs increased threefold and the time required to make changes to the process increased significantly. The contract was nearly cancelled, and a senior executive was diverted from other critical activities to implement new organizational structures and processes and to hire experienced staff.

Many companies outsourcing business processes today have already outsourced major components of technology management. Therefore, companies must decide whether to maintain current technology outsourcing agreements or bundle them into new agreements with the outsourced business processes. This requires that management critically analyse how current activities and services are being executed, and then decide whether to cancel or change these arrangements to reduce relationship complexity.

Establish clear service definitions, service-level agreements, and cost baselines. IT outsourcing deal success hinges on clear service definitions of what will be outsourced. Furthermore, service-level agreements must be clearly defined with robust cost and volume projections that cover the length of the proposed agreement. Unfortunately, most organizations do not explicitly define service levels or the costs of internally delivered activities.

Sufficient time and resources must also be provided to allow teams to create—usually from scratch—current cost and service levels and also to project demand. For example, a major European insurance provider spent twelve months mapping processes and defining current

and projected costs and volumes before moving activity to an Indian outsource provider.

Overcome lack of in-house experience negotiating and structuring complex outsourcing arrangements. Outsourcing suppliers negotiate many deals each year and have professional deal teams that focus on generating value. Companies must field similarly skilled teams with process re-engineering, financial and legal skills. Most important is experience in negotiating contracts.

A common error is to have current process owners negotiate the deal. This practice often results in missed opportunities to reduce cost, to improve service quality, or to fundamentally change the way an activity is executed. Furthermore, many companies rely on the internal "sourcing" department to negotiate deals. Although these staff are experts at negotiating contracts for discrete commodities such as photocopiers, technology and paper, negotiating complex outsourcing agreements is not their forte.

A major European credit card company outsourced and offshored IT development and certain business processes. The deal was negotiated by the IT and operations director, supported by the sourcing department. Only nine months into the outsourcing contract, several penalty clauses were breached, resulting in a 50 percent cost increase. At issue was the actual type of work required by the business users, which was significantly different from that agreed to in the negotiation (see Figure 6).

The sourcing team, inexperienced in outsourcing major contracts, failed to negotiate key clauses related to volumes and type of work. Instead, it focused on negotiating a 5 percent reduction in the hourly cost of the outsourcer's staff.

Figure 6. **Differences between Forecast and Actual Workload for a Credit Card Company (client example)**

Strategic	*Turnaround*
— IT projects — New system replacements — "Right to match and like for like market test"	— Consultancy services as required on a preferred-supplier basis
Forecast: 22–40% of total *Actual: ~10% of total*	*Forecast: no commitment* *Actual: ~5% of total*
Factory	*Support*
— Enhancements to in-scope systems — In-scope package amendments	— Support for all in-scope systems — Support for all new in-scope systems
Forecast: 27–55% of total *Actual: ~10% of total*	*Forecast: 22–25% of total* *Actual: ~35% of total*

Figure 7. Forecasting Multiyear Volumes and Costs (client example)

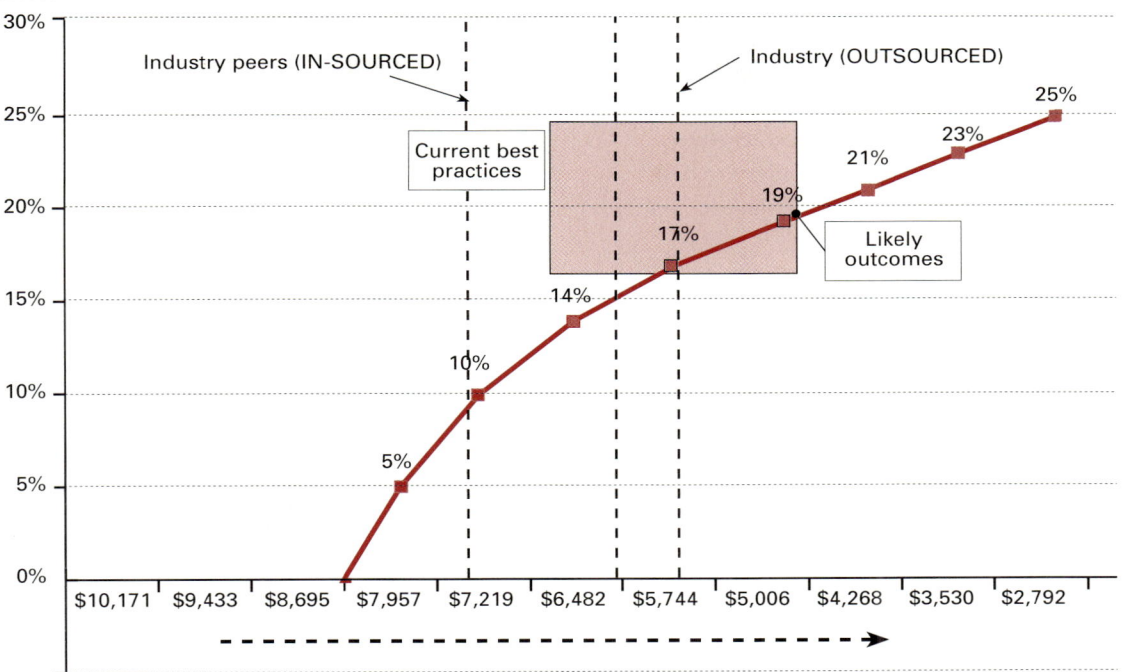

Annual ROI

Note: Discount rate 4%, four-year contract

Avoid overly optimistic expectations about the time required to outsource. Success demands that companies make realistic estimates of the time required to project costs and service levels, negotiate, transfer legal ownership, and implement an outsourcing deal. Most companies grossly underestimate timeframes to outsource, and as a result often devote insufficient time to key elements of implementing outsourcing agreements.

In addition, outsourcing suppliers often push to shorten the analysis and estimation phase in an effort to bury critical issues such as cost projections and service levels—key elements of the agreement. Furthermore, internal pressures to deliver value can quickly lead to unrealistic timeframes.

Structure the agreement flexibly to handle changes in volumes and requirements. Most outsourcing agreements fail to acknowledge that the new arrangement will have consequences for both the requirements and workload of the business; it is unlikely that business requirements and volumes two years hence will be the same as current levels.

Best practice organizations spend significant time forecasting future scenarios (see Figure 7) and design flexible contracts for a changing organization. Figure 7 is the output of a model that a client used to capture costs, volumes and and potential return on investment across a number of scenarios. Building a flexible forward–looking model is important for assessing and tracking progress against the outsourcing objectives.

Practise financial engineering carefully so that fundamental economics are not traded away to achieve short-term objectives. Most outsourcing agreements require significant change costs and upfront investment, and can take up to two or three years to break even. However, most firms cannot wait this long. Financial engineering techniques, such as funding upfront costs through supplier financing, venture capital and even bond issues, can help companies address these issues and achieve their short-term objectives such as reducing cost:income ratios and headcount, and ensuring that costs are changed from fixed to variable. While financial engineering can be a deal enabler, companies must ensure that the cost benefits of the arrangement are transparent over the length of the contract.

Outsourcing companies must be especially wary of the financial engineering offered by suppliers and must first ensure that deal economics are sound and that transparency is evident. Caution is critical because suppliers often use financial engineering to mask deal economics by turning the focus from overall cost reduction and service improvement to short-term financial savings. Booz Allen has analysed several large outsourcing contracts and found that at first glance the overall deals were positive, with savings early in the contracts. However, once the entire life of each contract was analysed, the economics of the deals were not as compelling.

One European financial services company recently worked with a venture capital group to finance the outsourcing of a core business process. The venture capital group funded necessary investments and even guaranteed savings several years into the future. This allowed the company to record savings immediately and achieve cost:income and headcount reductions promised to investors without sacrificing long-term economic benefits. Although important to achieving short-term objectives, financial engineering was performed only after the deal economics were agreed upon, ensuring transparency for all parties.

Despite the significant challenges of implementing next-generation outsourcing and offshoring, organizations can enjoy success by adhering to the above best practices.

Conclusion

IT outsourcing is here to stay and can create significant value, despite significant challenges. Booz Allen's experience suggests that organizations that succumb to the allure of myth or fail to embrace the best practices set out above will fail to create value. However, organizations that enter into these arrangements with their eyes open will achieve success with next-generation outsourcing and offshoring and will ensure their own competitive advantage.

Notes

1 Software development companies can be benchmarked against best practices embodied in the Capability Maturity Model Integration (CMMI) defined by the Software Engineering Institute at Carnegie Mellon University.

2 Six Sigma is a rigorous information/analytics driven methodology to measure and improve a company's operational processes and systems by identifying and preventing "defects" in manufacturing and service-related processes.

References

Couto, V., J. Schaedler, P. Pigorini and C. McNeese. 2004. *The New CFO Agenda: Global G&A Survey Insights and Implications*. Booz Allen Hamilton. Online. http://www.roberthalfmr.com/html/downloads/BoozAllen_CFOAgenda.pdf

DiamondCluster. 2004. *2004 Global IT Outsourcing Study*. Diamond Cluster International, Inc. Online. http://www.diamondcluster.com/press/News/Others/2004%20Global%20IT%20Outsourcing%20Study_single.pdf

NelsonHall. 2004. *Global BPO Contract Analysis & Future Opportunities*. Slough, UK: NelsonHall.

people3. 2003. *Embarking on a Successful IT Outsourcing Journey: Refocusing your Human Capital Management Strategy*. Schomberg, ON: Gartner Inc. Company

The TPI Index Europe. 2004. *An Informed View of the State of the Global Outsourcing Market*. Technology Partners International, Inc. 21 July. Online. http://www.tpi.net/pdf/2Q04%20TPI%20Index%20Presentation%20Europe%20Final.pdf

Trestle Group. 2004. *Summer 2004: Outsourcing Survey Results*. Trestle Group. Online. http://www.trestlegroup.com/downloads/common/TGR_Summer2004_Survey_Results.pdf

In addition, internal Booz Allen Hamilton client examples were used in this chapter.

Chapter 7

Case Study: Rearing Taiwan's ICT Industry

F. C. Lin, Institute for Information Industry

Introduction

From motherboards to monitors, from PCs to PDAs, Taiwanese companies produce a very substantial share of the devices that now pervade workplaces and homes. This chapter examines the factors that have led to Taiwan's emergence as a leading producer of information and communication technology (ICT) products, tracing the governmental and socio-economic forces that have propelled the small island to new heights of success and will continue to do so.

In addition to the factors that have pushed Taiwan on to the global stage, the Taiwanese government has initiated numerous projects to promote the role of technology in the public sphere, the business world, and the daily lives of its citizens. Yet the potential application of these projects extend beyond the domestic front and, combined with Taiwan's contribution toward the global penetration of ICT, Taiwan is set to play an even bigger role in helping to narrow the global digital divide.

Taiwan gained its status as one of the four "Asian tigers" from its enormous foreign reserves and its long-term average annual economic growth rate, one of the world's highest. Although the island was confronted with economic challenges in the mid- to late 1980s, a subsequent transition to technology-intensive industries sustained continued growth and helped Taiwan weather downturns such as the Asian financial crises of 1997. Technology-intensive industries, mostly in ICT, now make up over half of Taiwan's economy, compared with less than a quarter in the late 1980s.

Taiwanese manufacturers collectively produce well over half the global supply of the devices that make up the core of the worldwide ICT industry and infrastructure: notebook PCs, motherboards, thin-film transistor liquid crystal display (TFT LCD) panels, LCD monitors, network interface cards, hubs/switches, wireless local area network equipment, digital subscriber line (xDSL) modems, cable modems, and analog modems. This is the result of the Taiwanese government's clear policy direction for the ICT industry and of the private sector's vibrant entrepreneurial spirit.

Some of the government's efforts to boost the ICT industry assisted the establishment of organizations with a promotional role, as well as science parks. Incentives were devised to attract multinational corporations and, later, Taiwanese expatriates from Silicon Valley. Furthermore, the government took steps to enrich the local manpower pool, resulting in an explosion of science and technology graduates. A highly energetic private sector helped accelerate development, fuelled by easier access to venture capital.

Despite Taiwan's outstanding achievements, its ICT industry is facing formidable challenges due to its midstream position in the global value chain. Taiwanese players generally have had limited success in penetrating the higher-value ends of the value chain that have long been dominated by major brands or established technology leaders. At the same time China is taking over more mid-chain activities while leveraging the power of its home market to penetrate the higher value-added ends of the chain. Meanwhile, global competition is intensifying, and emerging market opportunities are harder to spot.

In light of the difficulties imposed by the wider environment, and with the aim of boosting Taiwan's presence at the higher ends of the industrial value chain, the government has enacted the Challenge 2008 National Development Plan, a six-year plan that is intended to bring about political, financial and fiscal reform. Challenge 2008 is comprised of sub-plans, such as digital infrastructure, international innovation and R&D base establishment, and industrial value strengthening, which are expected to play a role in shaping the future development of the Taiwanese ICT industry.

The Evolution of Taiwan's ICT Industry[1]

The first economic miracle: 1953–1986

Since the implementation of its first four-year plan in 1953, Taiwan has successfully transformed itself from an agricultural to an industrial economy. Over the past 50 years, Taiwan has recorded an average annual growth rate of roughly 8 percent, the highest in the world. At the same time, Taiwan has sharpened its export competitiveness and has accumulated the world's third largest holding of foreign exchange reserves, after Japan and China. This achievement may be called the first miracle in Taiwan's economic development, earning Taiwan's status as one of the four Asian tigers, alongside South Korea, Singapore and Hong Kong.

In 1986 and 1987, under pressure from the mounting trade surplus, the New Taiwan dollar appreciated sharply against the US dollar, from US$40 to $25. This posed a severe challenge to exporters and forced many traditional industries to relocate overseas. Meanwhile, the domestic economy was facing soaring wage and land costs, rising demand for environmental protection, and intensifying competition from emerging south-east Asian economies. Many economists began to question whether Taiwan would be able to remain competitive and sustain its fast pace of growth.

The second miracle: 1987–2000

Taiwan was eventually able to overcome the challenges and successfully complete its industrial restructuring. From 1987 to 2000 it continued to record moderate annual growth rates, even during the onset of the Asian financial crisis. Exports increased from US$53.7 billion in 1987 to $148.3 billion in 2000, while per capita income rose to $14,118. Deemed Taiwan's second miracle prior to its transition to a knowledge-based economy, this achievement can be attributed mainly to the successful development of technology-intensive industries, particularly in the IT sector.

Between 1986 and 2000, Taiwan's industrial structure underwent a rapid transformation, with technology-intensive industry doubling its share of the economy, from 24 percent to 48 percent, mainly due to the contribution from the IT industry.

Overview of Taiwan's ICT industry

Back in 1986, the ICT industry's production value amounted to only US$15.2 billion, which was 16 percent of Taiwan's total manufacturing output of $95.2 billion. By 2000 it had increased more than sixfold, to $101.4 billion, amounting to 37 percent of the total manufactured output of $270.2 billion.

Within the ICT industry, the fastest-growing segment was computer hardware, whose output rose more than ten times, from US$2.1 billion to $23 billion, during the 15 years 1986–2000. The integrated circuit (IC) industry also grew at a rapid rate to reach US$22.4 billion in 2000.

Owing to the remarkable development of its ICT industry, Taiwan has become a key supplier to major international firms. In terms of output, Taiwan is among the top four largest producers of IT hardware. Although China has recently inched ahead of Taiwan, it is notable that Taiwanese firms on the mainland produce more than 70 percent of China's IT output.

As they have become more globalized, Taiwanese firms have actively pursued an international division of labor. Taiwan's total IT hardware output reached US$57 billion in 2003, of which more than 70 percent was produced overseas. By leveraging cross-strait and international resources, Taiwan has become the largest producer of more than a dozen of the world's ICT products.

With tremendous flexibility and a unique original equipment manufacturing/original design manufacturing (OEM/ODM) business model, Taiwan has become one of the most important partners of the world's major ICT producers. Taiwan's share of the global supply of most ICT products is expected to continue to increase.

The Foundations of Growth

Strong government leadership

The success of Taiwan's IT industry can be attributed to the government's clear policy direction and the private sector's vibrant entrepreneurial spirit.

Figure 1. **Taiwanese ICT Industry Production Value, 2001–2003**

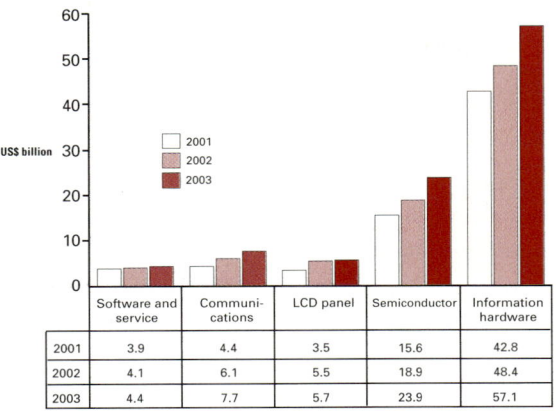

	Software and service	Communi- cations	LCD panel	Semiconductor	Information hardware
2001	3.9	4.4	3.5	15.6	42.8
2002	4.1	6.1	5.5	18.9	48.4
2003	4.4	7.7	5.7	23.9	57.1

Sources: Data collected in May 2004 from the Market Intelligence Center (URL) and the Industrial Economics and Knowledge Center (URL)

Since the 1970s, the development of the science and technology industry has been firmly established as a main formal objective of government policy. In 1973 the Industrial Technology Research Institute (ITRI) was set up, and in 1979 the government launched a science and technology development plan. Also in 1979, the government took the important initiative of establishing the Institute for Information Industry (III) to promote the IT industry. One year later the Hsinchu Science-based Industrial Park was established and the government formally designated IT as a strategic industry for priority development.

Socioeconomic context

The social and economic environment was also conducive to the development of the IT industry. Early on, by offering tax incentives and setting up export processing and industrial zones, Taiwan was able to persuade both foreign electronics manufacturers and local enterprises to invest in production facilities. This influx laid the foundations of the Taiwanese electronics industry.

When IBM introduced the personal computer in 1981, and the world entered the PC era, the government was able to take advantage of the well-established electronics industry to move quickly into this promising sector. The authorities promptly instructed research institutions to develop PC-compatible models and transfer the technology to private industries for mass production, thereby laying the groundwork for Taiwan's IT hardware industry. Subsequently, by offering incentives to lure back Taiwanese expatriates from Silicon Valley to the Hsinchu Science-based Industrial Park to work alongside former employees of ITRI and III, the government was able to foster, one after another, the emergence of such industries as semiconductors, networking equipment, flat-screen display, wireless communications and digital content.

Manpower development

Taiwan has long placed an emphasis on the nurturing of talent in science and technology to promote industrial restructuring. The island's pool of science and technology graduates at the college level and above surged from

Figure 2. **Taiwanese Global Share of ICT Products, 2003**

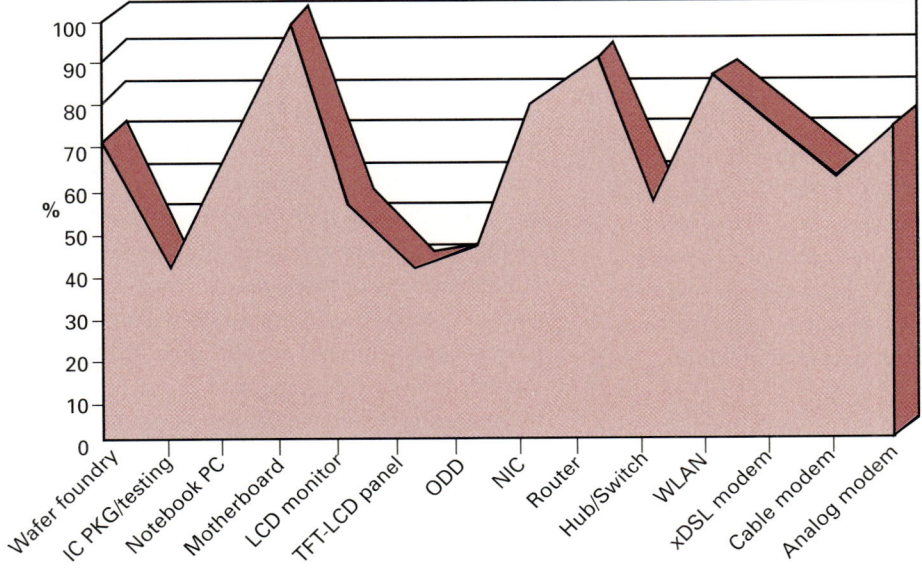

Note: All production shares other than wafer foundry and IC packaging and testing are based on volume.

Sources: Data collected in May 2004 from the Market Intelligence Center (URL) and the Industrial Economics and Knowledge Center (URL)

38,100 in 1981 to 81,400 in 1991 and 138,600 in 1999. Year after year, the input of these highly trained individuals has kept Taiwan's industrial development moving rapidly forward and upward.

The return of expatriate Taiwanese experts has also been important. From 1960 to 1994, the number of Taiwanese studying abroad rose steadily year by year. Since then it has fallen slightly, but still remains substantial. Most of the students study in the United States, with a large proportion in engineering courses. After graduation, most of them stay in the United States to work, particularly in the high-tech industries of Silicon Valley.

When the specialized knowledge of Taiwanese expatriates matched the needs of emerging industries back home, often they would return in waves to start new ventures. It was largely through these people that such close links of technological cooperation were forged between Silicon Valley and the Hsinchu Science Park.

Coalescing high-tech clusters

Another notable feature of Taiwan's high-tech industry is the formation of clusters along the lines of Silicon Valley, which has been highly advantageous to the development of the entire industry. In northern Taiwan, the cluster of firms in the 70–90 kilometers between Taipei and the Hsinchu Park covers the whole spectrum of IT-related production activities. At present, to balance development between Taiwan's regions the government is actively promoting another high-tech belt between two new science-based industrial parks in Tainan and Luchu. The new cluster in the south is expected to become a major global base for manufacturing 12-inch wafers and TFT LCD.

The development of venture capital

Venture capital (VC) and the development of high-tech industry go hand-in-hand. Taiwan's first VC company was set up with government encouragement in 1984, and by the end of 2003 there were 241 in operation. In terms of investment activities, Taiwan's VC business is second only to that in the United States. VC has been indispensable to the success of Taiwan's high-tech small and medium enterprises (SMEs).

A highly energetic private sector

The vigor of Taiwan's private sector, especially its SMEs, is particularly striking. This private-sector vitality has certainly made an important contribution to Taiwan's IT development. Its most influential aspects in this respect are as follows:

- techno-entrepreneurship creates SMEs with cutting-edge technologies that are potential stars of emerging industries;
- a great number of overseas Taiwanese professionals who possess profound industry experience return to Taiwan, not only contributing to and bringing in advanced expertise but also building a wide and solid network between Taiwan and the US;
- stock options provide incentives to encourage developers and engineers to devote themselves to both design and manufacturing process innovation;
- the highly flexible and agile business operations of Taiwanese SMEs provide an advantageous position to respond to the rapidly changing market;
- a high savings rate provides abundant capital to sustain SMEs;
- long-term cooperation between Taiwanese and major international ICT players has established a solid basis of mutual trust for sharing knowledge and working as a team, which are essential to the further growth of both parties in creating and capturing new business opportunities.

The Future of Taiwan's ICT Development[2]

Current challenges

Its longstanding role as the original design outsourcing partner for the world's major ICT brands has brought the Taiwanese ICT industry to its present tremendous scale. Yet Taiwanese players remain absent from the higher value-added links of the value chain. Segments dominated by major brands, such as initiating industry standards, technological innovation and branding, are all areas in which the Taiwanese ICT industry has generally had limited success. Furthermore, the value activities that have been the province of Taiwanese makers—design verification testing (DVT), production verification testing (PVT), process improvement, product design enhancement, global logistics, channel management, and after-sales service—are progressively being absorbed into China. Chinese players are also aggressively making inroads into brand management and standards formulation by leveraging the power of their domestic market.

Yet even the more valuable ends of the value chain offer increasingly scant prospects. The products and processes comprising Taiwan's current industry portfolio have reached a specific phase of product life cycle, compressing the value-added formerly available at the corners of the value chain. The Taiwanese ICT industry is therefore faced with marginal profits, minimal differentiation and heavy competition, while at the same time emerging

Figure 3. Taiwan's Position in the Global ICT Value Chain

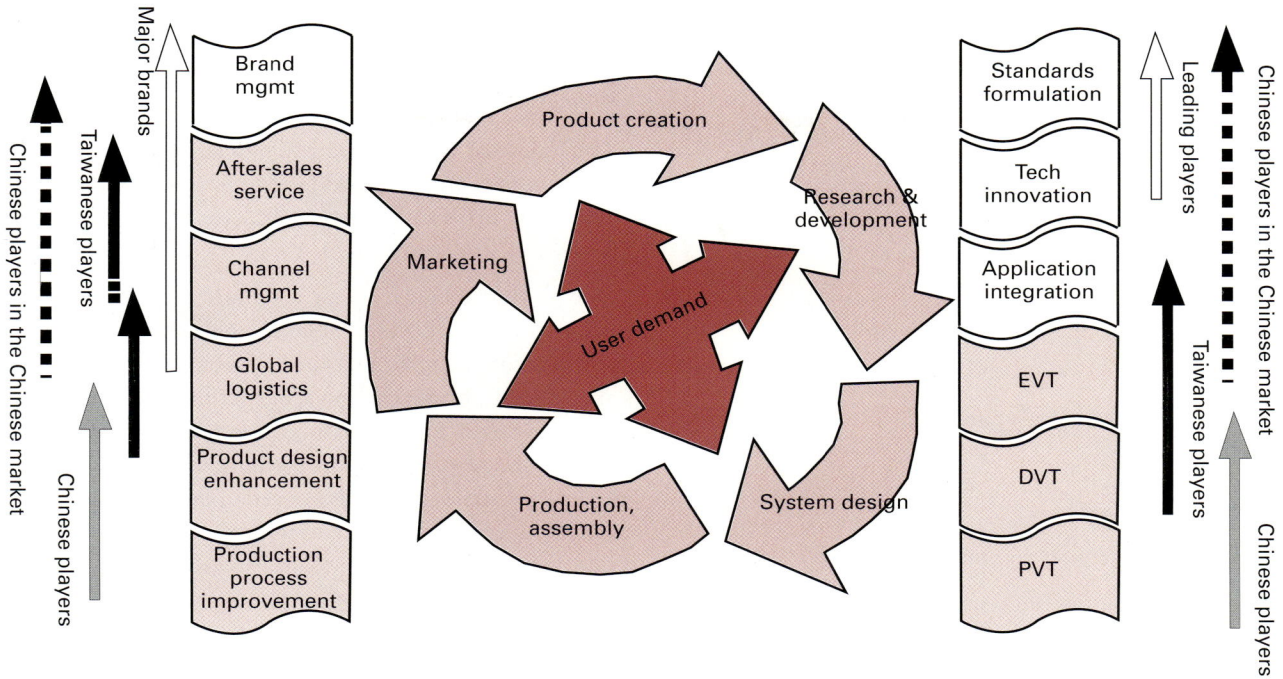

Source: Data collected in November 2003 from the Market Intelligence Center (URL)

market opportunities are still vague and uncertain for this ICT little giant.

Meanwhile, the island is beset by other external macroeconomic threats. Global competition is growing increasingly intense, industrial chains are being reorganized, and other countries are aggressively competing for talent and funds. Taiwan is also faced with the impact of its enormous neighbor, China, whose cheap labor, low-cost land, and huge potential market are luring Taiwan's industries to its shores. On the internal front, the quality and discipline of the labor force, which in the past enabled Taiwan to so successfully develop manufacturing industries, is no longer sufficient to meet present or future industrial needs.

Figure 4. Diminishing Value-Added of Taiwan's Current Industry Portfolio

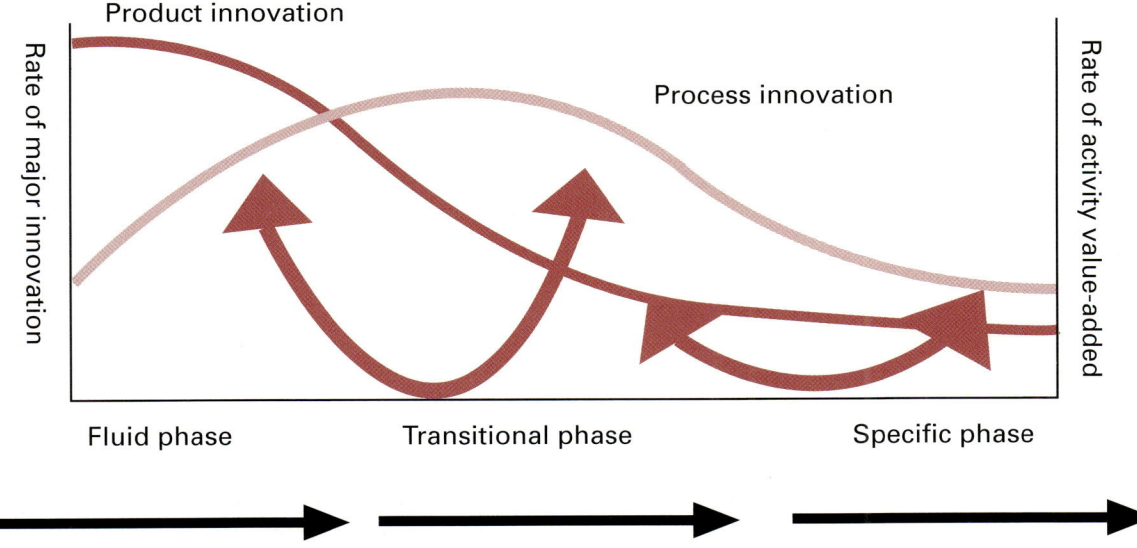

Source: Data collected in November 2003 from the Market Intelligence Center (URL)

Meeting the challenges ahead

In light of the difficulties imposed by the wider environment, and with the aim of boosting Taiwan's presence at the higher ends of the industrial value chain, the government has enacted the Challenge 2008 National Development Plan, a six-year plan that is intended to effect political, financial, and fiscal reform (Executive Yuan 2002). Challenge 2008 is comprised of ten individual sub-plans, to promote:

- Manpower cultivation
- Cultural and creative industry development
- International innovation and R&D base establishment
- Industrial value strengthening
- Tourism
- Digital infrastructure
- Operations headquarters establishment
- Island-wide transportation construction
- Environmental improvement
- Community development

The promotion of international innovation and R&D base establishment, industrial value strengthening, and digital infrastructure (e-Taiwan) will have the most immediate impact on the Taiwanese ICT industry.

International innovation and R&D base plan

The chief objective of the international innovation and R&D base plan is to raise R&D spending to 3 percent of GDP within six years, and to turn Taiwan into a major hub for innovation and R&D in Asia.

The plan will endeavor to attract international R&D personnel and introduce R&D resources from around the world. This will include recruiting overseas high-tech personnel and students, encouraging Taiwanese students to study abroad, internationalizing domestic universities, and building a global academic Internet. Key-industry technical colleges for semiconductor and digital content will be established, and cooperation among industry, schools, and research institutions will be encouraged in order to cultivate industry personnel.

R&D loans worth NT$50 billion (US$1.53 billion) will also be provided to boost innovation and R&D activity, and innovation and R&D centers will be established to create competitive advantage in special fields. These efforts would also include attracting multinational firms to set up regional centers in Taiwan, and the establishment of specialized centers for genetic research, software design, mobile telecommunications R&D, nanotech applications R&D, precision machinery R&D, and environmental protection technology incubation. Key industrial

technology research and the establishment of core industrial technologies such as telecommunications, chips, nanotech, and biotech will also be promoted

Industrial Value Strengthening Plan

The Industrial Value Strengthening Plan is intended to heighten industrial value-added and to mold Taiwan into a global production and supply center for high-value-added products.

One measure includes raising a venture capital fund of NT$100 billion (US$3.1 billion) to enlarge funding channels for emerging industries. Assistance will also be provided for the development of core industrial technologies including electronics and IT, optoelectronics, communications, machinery, textiles and biotechnology. Hybrid technologies are also covered, including micro-mechatronics, intelligent transportation systems (ITS) and vehicle systems, nanotechnology, navigational electronics and semiconductor processing equipment. Industrial parks will also be built to house research for domains as diverse as IC design, medicine, horticulture, and recycling technologies.

Key industries will be furthered, including higher-value-added traditional industries as well as the semiconductor, display, digital content, biotechnology, information application, logistics, healthcare, and environmental industries. The government will also provide incentives for investment in international distribution channels and brands. This includes providing guidance and assistance in setting up international marketing companies and developing international marketing channels, and attracting multinational corporations to expand industrial and sales cooperation with domestic firms.

The government will also work to upgrade labor by establishing a nationwide vocational training network, helping the disadvantaged and unemployed obtain vocational training, and strengthening on-the-job and second-skill training.

e-Taiwan

The goal of e-Taiwan is to develop the island into Asia's most digitized country, with six million households using broadband by 2008. The vision and prospects of e-Taiwan are to employ ICT to set up a high-efficiency government, strengthen industrial competitiveness, create an intelligent transportation environment, and build a high-quality information society in order to accelerate Taiwan's progress toward a knowledge-based economy, materializing the ideal of Taiwan as a high-tech service island and one of the most advanced e-nations in Asia, and realizing the vision of a superior-quality e-society.

Upon this foundation, the proliferation of electronic applications will also be promoted on a wide scale, with components for infrastructure, e-Society, e-Industry, e-Government, and e-Opportunity.

The infrastructure establishment is to sustain the continuous development of a national information infrastructure and build an Internet security environment in which to construct an integrated broadband network covering cable, wireless, mobile, and fixed telecom networks.

e-Society encompasses nationwide digital learning and archiving, digital entertainment promotion, and an intelligent transportation system to enrich cultural information, to upgrade the quality of learning and entertainment, and to enhance the quality of transportation.

e-Industry seeks to forge a joint-design system among key industries and knowledge management for agricultural industries and SMEs, as well as to strengthen international cooperation and exchange in e-commerce in order to develop a manufacturing and service center with high added value, enhancing the quality of the supply chain and logistics management mechanisms.

e-Government will work to create a single online portal for integrated government services. e-Government will also provide a G2B2C electronic documentation exchange, a virtual conference system for government agencies, and an integrated communication system for management of emergencies.

In order to provide digital information and information accessibility for everyone, e-Opportunity will focus upon providing equal opportunity, increasing overall industry competitiveness, and fostering exchange with the international community, thus enabling all citizens to utilize ICT and so improve their lives. In the process, the government would also balance the development of urban and rural areas and provide equal opportunities for everyone.

Conclusion

Based on years of experience in large-scale, cost-effective production, Taiwanese manufacturers have played a major role in generating price points that have accelerated the global penetration of information and communication devices. In addition to this global contribution in terms of

hardware by Taiwanese enterprises, the Taiwanese government has been working to narrow the digital divide at home and internationally. Taiwan's Challenge 2008 National Development Plan has been implemented to close the gap between certain demographics and residents in remote areas, and has reached out to assist domestic enterprises as well as entities abroad. So far, these measures have contributed significantly to advancing e-adoption; the Taiwanese government continues to look for partners abroad with which to share these successful models and help bridge the digital divide on an international scale.

Notes

1 This section draws extensively on Market Intelligence Center (2004a, 2004b) and Industrial Economics and Knowledge Center. Along with the next section it also draws extensively on Ministry of Economic Affairs (URL), Chung Hua Institution for Economic Research (URL), and Taiwan Research Institute (URL).

2 This section draws extensively on Chyn (2003).

References

Chung Hua Institution for Economic Research, Taiwan. URL. Online. http://www.cier.edu.tw

Chyn, C. 2003. "Repositioning Taiwan in Global ICT Industry Value Chain". PowerPoint presentation. Taipei.

Executive Yuan. 2002. *Challenge 2008—National Development Plan*. Taipei: Council for Economic Planning and Development. Online. http://www.cepd.gov.tw/2008

Industrial Economics and Knowledge Center. URL. Taipei: Industrial Technology Research Institute. Online. http://www.iek.itri.org.tw/eng

Market Intelligence Center. 2004a. The *Taiwanese Software Industry, 2003–2004 and Beyond*. Taipei: Institute for Information Industry.

———. 2004b. *The Taiwanese Information Hardware Industry in 2003 and Beyond*. Taipei: Institute for Information Industry.

———. URL. Taipei: Institute for Information Industry. Online. http://mic.iii.org.tw/english

Ministry of Economic Affairs Taiwan. URL. Online. http://www.moea.gov.tw

Taiwan Research Institute. URL. Taipei. Online. http://www.tri.org.tw

Part 2
Country Profiles

How to Read the Country Profiles

The Country Profiles section presents the rankings of the 104 countries considered in the *Global Information Technology Report 2004–2005*. It provides a quick picture of the level of ICT development of each country by grouping information in the following sections:

1. Key macroeconomic and ICT indicators such as population, gross domestic product (GDP) per capita, and Internet users per 100 inhabitants.[1]

2. Overall Networked Readiness Index (NRI) ranking for 2004–2005, which gives immediate insight into overall ICT competitiveness; one can compare this rank to that of the NRI 2003–2004 and NRI 2002–2003 if the country was ranked for those years.

3. Three component indexes, each consisting of a list of variables. Detailed rankings for the country presented can be found for each of the variables listed and taken into consideration for the current NRI study.

This information, which identifies key areas of relative over- and under-performance, provides a rapid understanding of a country's ICT competitiveness. For example, the rankings of the variables of venture capital availability and the state of cluster development in the Environment Component Index enable the reader to identify key parameters contributing to the country's performance.

The inferences that can be derived from the ranking of a given country can be augmented by closer inspection of the relative performance of other countries. Each variable is preceded by a variable number—for example, 1.01, or 2.03—which provides a link to the Data Tables where the performance of all 104

countries in terms of that variable is presented.

In this way, by becoming acquainted with the performance of two countries that are similar and models such as Finland, Singapore, and Korea, one can quickly assess the two countries' strengths and weaknesses, and key areas requiring development.

Note

1 Data sources for these indicators are as follows.

Population: UNFPA State of World Population 2003. Online. http://www.unfpa.org/swp/2003/english/ch1/index.htm; Economist Intelligence Unit.
GDP per capita: The World Bank, International Comparison Program (ICP). Online. http://web.worldbank.org/WBSITE/EXTERNAL/DATASTATISTICS/ICPEXT/0,,pagePK:62002243~theSitePK:270065,00.html
Internet users per 100 inhabitants: International Telecommunication Union, 2003.

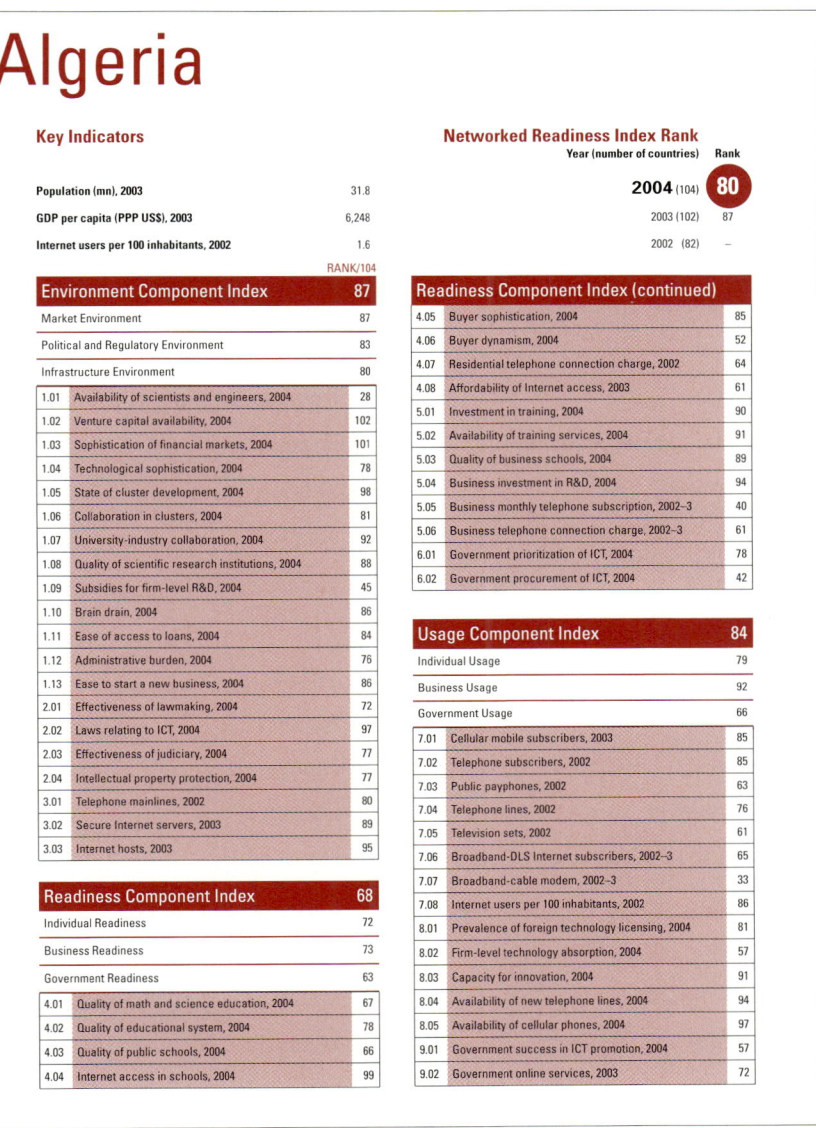

Algeria

Key Indicators

Population (mn), 2003	31.8
GDP per capita (PPP US$), 2003	6,248
Internet users per 100 inhabitants, 2002	1.6

Networked Readiness Index Rank

Year (number of countries)	Rank
2004 (104)	**80**
2003 (102)	87
2002 (82)	—

Environment Component Index	**87**
Market Environment	87
Political and Regulatory Environment	83
Infrastructure Environment	80
1.01 Availability of scientists and engineers, 2004	28
1.02 Venture capital availability, 2004	102
1.03 Sophistication of financial markets, 2004	101
1.04 Technological sophistication, 2004	78
1.05 State of cluster development, 2004	98
1.06 Collaboration in clusters, 2004	81
1.07 University-industry collaboration, 2004	92
1.08 Quality of scientific research institutions, 2004	88
1.09 Subsidies for firm-level R&D, 2004	45
1.10 Brain drain, 2004	86
1.11 Ease of access to loans, 2004	84
1.12 Administrative burden, 2004	76
1.13 Ease to start a new business, 2004	86
2.01 Effectiveness of lawmaking, 2004	72
2.02 Laws relating to ICT, 2004	97
2.03 Effectiveness of judiciary, 2004	77
2.04 Intellectual property protection, 2004	77
3.01 Telephone mainlines, 2002	80
3.02 Secure Internet servers, 2003	89
3.03 Internet hosts, 2003	95

Readiness Component Index	**68**
Individual Readiness	72
Business Readiness	73
Government Readiness	63
4.01 Quality of math and science education, 2004	67
4.02 Quality of educational system, 2004	78
4.03 Quality of public schools, 2004	66
4.04 Internet access in schools, 2004	99

Readiness Component Index (continued)	
4.05 Buyer sophistication, 2004	85
4.06 Buyer dynamism, 2004	52
4.07 Residential telephone connection charge, 2002	64
4.08 Affordability of Internet access, 2003	61
5.01 Investment in training, 2004	90
5.02 Availability of training services, 2004	91
5.03 Quality of business schools, 2004	89
5.04 Business investment in R&D, 2004	94
5.05 Business monthly telephone subscription, 2002–3	40
5.06 Business telephone connection charge, 2002–3	61
6.01 Government prioritization of ICT, 2004	78
6.02 Government procurement of ICT, 2004	42

Usage Component Index	**84**
Individual Usage	79
Business Usage	92
Government Usage	66
7.01 Cellular mobile subscribers, 2003	85
7.02 Telephone subscribers, 2002	85
7.03 Public payphones, 2002	63
7.04 Telephone lines, 2002	76
7.05 Television sets, 2002	61
7.06 Broadband-DLS Internet subscribers, 2002–3	65
7.07 Broadband-cable modem, 2002–3	33
7.08 Internet users per 100 inhabitants, 2002	86
8.01 Prevalence of foreign technology licensing, 2004	81
8.02 Firm-level technology absorption, 2004	57
8.03 Capacity for innovation, 2004	91
8.04 Availability of new telephone lines, 2004	94
8.05 Availability of cellular phones, 2004	97
9.01 Government success in ICT promotion, 2004	57
9.02 Government online services, 2003	72

List of Countries

Algeria

Key Indicators

Population (mn), 2003	31.8
GDP per capita (PPP US$), 2003	6,248
Internet users per 100 inhabitants, 2002	1.6

RANK/104

Environment Component Index	87
Market Environment	87
Political and Regulatory Environment	83
Infrastructure Environment	80

1.01	Availability of scientists and engineers, 2004	28
1.02	Venture capital availability, 2004	102
1.03	Sophistication of financial markets, 2004	101
1.04	Technological sophistication, 2004	78
1.05	State of cluster development, 2004	98
1.06	Collaboration in clusters, 2004	81
1.07	University-industry collaboration, 2004	92
1.08	Quality of scientific research institutions, 2004	88
1.09	Subsidies for firm-level R&D, 2004	45
1.10	Brain drain, 2004	86
1.11	Ease of access to loans, 2004	84
1.12	Administrative burden, 2004	76
1.13	Ease to start a new business, 2004	86
2.01	Effectiveness of lawmaking, 2004	72
2.02	Laws relating to ICT, 2004	97
2.03	Effectiveness of judiciary, 2004	77
2.04	Intellectual property protection, 2004	77
3.01	Telephone mainlines, 2002	80
3.02	Secure Internet servers, 2003	89
3.03	Internet hosts, 2003	95

Readiness Component Index	68
Individual Readiness	72
Business Readiness	73
Government Readiness	63

4.01	Quality of math and science education, 2004	67
4.02	Quality of educational system, 2004	78
4.03	Quality of public schools, 2004	66
4.04	Internet access in schools, 2004	99

Readiness Component Index (continued)	
4.05	Buyer sophistication, 2004	85
4.06	Buyer dynamism, 2004	52
4.07	Residential telephone connection charge, 2002	64
4.08	Affordability of Internet access, 2003	61
5.01	Investment in training, 2004	90
5.02	Availability of training services, 2004	91
5.03	Quality of business schools, 2004	89
5.04	Business investment in R&D, 2004	94
5.05	Business monthly telephone subscription, 2002–3	40
5.06	Business telephone connection charge, 2002–3	61
6.01	Government prioritization of ICT, 2004	78
6.02	Government procurement of ICT, 2004	42

Usage Component Index	84
Individual Usage	79
Business Usage	92
Government Usage	66

7.01	Cellular mobile subscribers, 2003	85
7.02	Telephone subscribers, 2002	85
7.03	Public payphones, 2002	63
7.04	Telephone lines, 2002	76
7.05	Television sets, 2002	61
7.06	Broadband-DLS Internet subscribers, 2002–3	65
7.07	Broadband-cable modem, 2002–3	33
7.08	Internet users per 100 inhabitants, 2002	86
8.01	Prevalence of foreign technology licensing, 2004	81
8.02	Firm-level technology absorption, 2004	57
8.03	Capacity for innovation, 2004	91
8.04	Availability of new telephone lines, 2004	94
8.05	Availability of cellular phones, 2004	97
9.01	Government success in ICT promotion, 2004	57
9.02	Government online services, 2003	72

Angola

Key Indicators

Population (mn), 2003	13.6
GDP per capita (PPP US$), 2003	2,319
Internet users per 100 inhabitants, 2002	0.3

Networked Readiness Index Rank

Year (number of countries)   Rank

2004 (104)   **101**

2003 (102)   99

2002 (82)   —

RANK/104

Environment Component Index		**104**
Market Environment		104
Political and Regulatory Environment		98
Infrastructure Environment		95
1.01	Availability of scientists and engineers, 2004	104
1.02	Venture capital availability, 2004	104
1.03	Sophistication of financial markets, 2004	104
1.04	Technological sophistication, 2004	101
1.05	State of cluster development, 2004	104
1.06	Collaboration in clusters, 2004	104
1.07	University-industry collaboration, 2004	104
1.08	Quality of scientific research institutions, 2004	100
1.09	Subsidies for firm-level R&D, 2004	94
1.10	Brain drain, 2004	77
1.11	Ease of access to loans, 2004	100
1.12	Administrative burden, 2004	75
1.13	Ease to start a new business, 2004	99
2.01	Effectiveness of lawmaking, 2004	76
2.02	Laws relating to ICT, 2004	98
2.03	Effectiveness of judiciary, 2004	90
2.04	Intellectual property protection, 2004	104
3.01	Telephone mainlines, 2002	95
3.02	Secure Internet servers, 2003	94
3.03	Internet hosts, 2003	102

Readiness Component Index		**99**
Individual Readiness		93
Business Readiness		98
Government Readiness		89
4.01	Quality of math and science education, 2004	104
4.02	Quality of educational system, 2004	104
4.03	Quality of public schools, 2004	104
4.04	Internet access in schools, 2004	103

Readiness Component Index (continued)		
4.05	Buyer sophistication, 2004	104
4.06	Buyer dynamism, 2004	99
4.07	Residential telephone connection charge, 2002	79
4.08	Affordability of Internet access, 2003	92
5.01	Investment in training, 2004	98
5.02	Availability of training services, 2004	104
5.03	Quality of business schools, 2004	104
5.04	Business investment in R&D, 2004	102
5.05	Business monthly telephone subscription, 2002–3	94
5.06	Business telephone connection charge, 2002–3	89
6.01	Government prioritization of ICT, 2004	89
6.02	Government procurement of ICT, 2004	83

Usage Component Index		**101**
Individual Usage		97
Business Usage		102
Government Usage		89
7.01	Cellular mobile subscribers, 2003	101
7.02	Telephone subscribers, 2002	98
7.03	Public payphones, 2002	81
7.04	Telephone lines, 2002	96
7.05	Television sets, 2002	98
7.06	Broadband-DSL Internet subscribers, 2002–3	65
7.07	Broadband-cable modem, 2002–3	49
7.08	Internet users per 100 inhabitants, 2002	97
8.01	Prevalence of foreign technology licensing, 2004	98
8.02	Firm-level technology absorption, 2004	103
8.03	Capacity for innovation, 2004	104
8.04	Availability of new telephone lines, 2004	101
8.05	Availability of cellular phones, 2004	102
9.01	Government success in ICT promotion, 2004	79
9.02	Government online services, 2003	90

Argentina

Key Indicators

Population (mn), 2003	38.4
GDP per capita (PPP US$), 2003	11,586
Internet users per 100 inhabitants, 2002	11.2

RANK/104

Environment Component Index	88
Market Environment	90
Political and Regulatory Environment	96
Infrastructure Environment	49

1.01	Availability of scientists and engineers, 2004	50
1.02	Venture capital availability, 2004	95
1.03	Sophistication of financial markets, 2004	76
1.04	Technological sophistication, 2004	50
1.05	State of cluster development, 2004	75
1.06	Collaboration in clusters, 2004	89
1.07	University-industry collaboration, 2004	88
1.08	Quality of scientific research institutions, 2004	83
1.09	Subsidies for firm-level R&D, 2004	92
1.10	Brain drain, 2004	70
1.11	Ease of access to loans, 2004	99
1.12	Administrative burden, 2004	92
1.13	Ease to start a new business, 2004	95
2.01	Effectiveness of lawmaking, 2004	102
2.02	Laws relating to ICT, 2004	85
2.03	Effectiveness of judiciary, 2004	97
2.04	Intellectual property protection, 2004	88
3.01	Telephone mainlines, 2002	54
3.02	Secure Internet servers, 2003	50
3.03	Internet hosts, 2003	32

Readiness Component Index	81
Individual Readiness	69
Business Readiness	46
Government Readiness	99

4.01	Quality of math and science education, 2004	80
4.02	Quality of educational system, 2004	84
4.03	Quality of public schools, 2004	72
4.04	Internet access in schools, 2004	64

Readiness Component Index (continued)	
4.05	Buyer sophistication, 2004	62
4.06	Buyer dynamism, 2004	80
4.07	Residential telephone connection charge, 2002	52
4.08	Affordability of Internet access, 2003	34
5.01	Investment in training, 2004	66
5.02	Availability of training services, 2004	64
5.03	Quality of business schools, 2004	29
5.04	Business investment in R&D, 2004	75
5.05	Business monthly telephone subscription, 2002–3	49
5.06	Business telephone connection charge, 2002–3	51
6.01	Government prioritization of ICT, 2004	103
6.02	Government procurement of ICT, 2004	89

Usage Component Index	59
Individual Usage	51
Business Usage	70
Government Usage	74

7.01	Cellular mobile subscribers, 2003	67
7.02	Telephone subscribers, 2002	57
7.03	Public payphones, 2002	11
7.04	Telephone lines, 2002	54
7.05	Television sets, 2002	66
7.06	Broadband-DSL Internet subscribers, 2002–3	44
7.07	Broadband-cable modem, 2002–3	44
7.08	Internet users per 100 inhabitants, 2002	47
8.01	Prevalence of foreign technology licensing, 2004	72
8.02	Firm-level technology absorption, 2004	78
8.03	Capacity for innovation, 2004	85
8.04	Availability of new telephone lines, 2004	52
8.05	Availability of cellular phones, 2004	60
9.01	Government success in ICT promotion, 2004	104
9.02	Government online services, 2003	17

Australia

Key Indicators

Population (mn), 2003	19.7
GDP per capita (PPP US$), 2003	29,143
Internet users per 100 inhabitants, 2002	48.2

Networked Readiness Index Rank

Year (number of countries)   **Rank**

2004 (104)	**11**
2003 (102)	9
2002 (82)	15

RANK/104

Environment Component Index	9
Market Environment	20
Political and Regulatory Environment	6
Infrastructure Environment	5

1.01	Availability of scientists and engineers, 2004	20
1.02	Venture capital availability, 2004	8
1.03	Sophistication of financial markets, 2004	7
1.04	Technological sophistication, 2004	13
1.05	State of cluster development, 2004	37
1.06	Collaboration in clusters, 2004	31
1.07	University-industry collaboration, 2004	18
1.08	Quality of scientific research institutions, 2004	9
1.09	Subsidies for firm-level R&D, 2004	15
1.10	Brain drain, 2004	30
1.11	Ease of access to loans, 2004	19
1.12	Administrative burden, 2004	33
1.13	Ease to start a new business, 2004	18
2.01	Effectiveness of lawmaking, 2004	5
2.02	Laws relating to ICT, 2004	8
2.03	Effectiveness of judiciary, 2004	10
2.04	Intellectual property protection, 2004	11
3.01	Telephone mainlines, 2002	16
3.02	Secure Internet servers, 2003	5
3.03	Internet hosts, 2003	5

Readiness Component Index	11
Individual Readiness	8
Business Readiness	13
Government Readiness	20

4.01	Quality of math and science education, 2004	18
4.02	Quality of educational system, 2004	7
4.03	Quality of public schools, 2004	12
4.04	Internet access in schools, 2004	12

Readiness Component Index (continued)	
4.05	Buyer sophistication, 2004	8
4.06	Buyer dynamism, 2004	8
4.07	Residential telephone connection charge, 2002	31
4.08	Affordability of Internet access, 2003	16
5.01	Investment in training, 2004	11
5.02	Availability of training services, 2004	12
5.03	Quality of business schools, 2004	7
5.04	Business investment in R&D, 2004	19
5.05	Business monthly telephone subscription, 2002–3	24
5.06	Business telephone connection charge, 2002–3	32
6.01	Government prioritization of ICT, 2004	36
6.02	Government procurement of ICT, 2004	13

Usage Component Index	16
Individual Usage	20
Business Usage	19
Government Usage	16

7.01	Cellular mobile subscribers, 2003	27
7.02	Telephone subscribers, 2002	27
7.03	Public payphones, 2002	23
7.04	Telephone lines, 2002	17
7.05	Television sets, 2002	22
7.06	Broadband-DSL Internet subscribers, 2002–3	32
7.07	Broadband-cable modem, 2002–3	22
7.08	Internet users per 100 inhabitants, 2002	12
8.01	Prevalence of foreign technology licensing, 2004	5
8.02	Firm-level technology absorption, 2004	15
8.03	Capacity for innovation, 2004	24
8.04	Availability of new telephone lines, 2004	34
8.05	Availability of cellular phones, 2004	37
9.01	Government success in ICT promotion, 2004	31
9.02	Government online services, 2003	11

Austria

Part 2 Country Profiles

Key Indicators

Population (mn), 2003	8.1
GDP per capita (PPP US$), 2003	29,972
Internet users per 100 inhabitants, 2002	41.5

Networked Readiness Index Rank

Year (number of countries)	Rank
2004 (104)	**19**
2003 (102)	21
2002 (82)	16

RANK/104

Environment Component Index — 19

Market Environment	22
Political and Regulatory Environment	18
Infrastructure Environment	18

1.01	Availability of scientists and engineers, 2004	22
1.02	Venture capital availability, 2004	31
1.03	Sophistication of financial markets, 2004	33
1.04	Technological sophistication, 2004	29
1.05	State of cluster development, 2004	27
1.06	Collaboration in clusters, 2004	17
1.07	University-industry collaboration, 2004	14
1.08	Quality of scientific research institutions, 2004	23
1.09	Subsidies for firm-level R&D, 2004	12
1.10	Brain drain, 2004	26
1.11	Ease of access to loans, 2004	36
1.12	Administrative burden, 2004	8
1.13	Ease to start a new business, 2004	35
2.01	Effectiveness of lawmaking, 2004	30
2.02	Laws relating to ICT, 2004	18
2.03	Effectiveness of judiciary, 2004	16
2.04	Intellectual property protection, 2004	15
3.01	Telephone mainlines, 2002	24
3.02	Secure Internet servers, 2003	16
3.03	Internet hosts, 2003	14

Readiness Component Index — 21

Individual Readiness	14
Business Readiness	16
Government Readiness	35

4.01	Quality of math and science education, 2004	11
4.02	Quality of educational system, 2004	10
4.03	Quality of public schools, 2004	10
4.04	Internet access in schools, 2004	13

Readiness Component Index (continued)

4.05	Buyer sophistication, 2004	23
4.06	Buyer dynamism, 2004	22
4.07	Residential telephone connection charge, 2002	18
4.08	Affordability of Internet access, 2003	24
5.01	Investment in training, 2004	17
5.02	Availability of training services, 2004	11
5.03	Quality of business schools, 2004	11
5.04	Business investment in R&D, 2004	22
5.05	Business monthly telephone subscription, 2002–3	23
5.06	Business telephone connection charge, 2002–3	17
6.01	Government prioritization of ICT, 2004	34
6.02	Government procurement of ICT, 2004	31

Usage Component Index — 15

Individual Usage	16
Business Usage	20
Government Usage	13

7.01	Cellular mobile subscribers, 2003	13
7.02	Telephone subscribers, 2002	15
7.03	Public payphones, 2002	34
7.04	Telephone lines, 2002	24
7.05	Television sets, 2002	16
7.06	Broadband-DSL Internet subscribers, 2002–3	22
7.07	Broadband-cable modem, 2002–3	8
7.08	Internet users per 100 inhabitants, 2002	17
8.01	Prevalence of foreign technology licensing, 2004	57
8.02	Firm-level technology absorption, 2004	19
8.03	Capacity for innovation, 2004	13
8.04	Availability of new telephone lines, 2004	18
8.05	Availability of cellular phones, 2004	6
9.01	Government success in ICT promotion, 2003	29
9.02	Government online services, 2003	7

Bahrain

Key Indicators

Population (mn), 2003	0.724
GDP per capita (PPP US$), 2003	17,789
Internet users per 100 inhabitants, 2002	24.6

Networked Readiness Index Rank

Year (number of countries) Rank

2004 (104) **33**

2003 (102) —

2002 (82) —

RANK/104

Environment Component Index		35
Market Environment		30
Political and Regulatory Environment		37
Infrastructure Environment		46
1.01	Availability of scientists and engineers, 2004	77
1.02	Venture capital availability, 2004	32
1.03	Sophistication of financial markets, 2004	23
1.04	Technological sophistication, 2004	31
1.05	State of cluster development, 2004	51
1.06	Collaboration in clusters, 2004	66
1.07	University-industry collaboration, 2004	69
1.08	Quality of scientific research institutions, 2004	86
1.09	Subsidies for firm-level R&D, 2004	78
1.10	Brain drain, 2004	12
1.11	Ease of access to loans, 2004	16
1.12	Administrative burden, 2004	24
1.13	Ease to start a new business, 2004	19
2.01	Effectiveness of lawmaking, 2004	55
2.02	Laws relating to ICT, 2004	34
2.03	Effectiveness of judiciary, 2004	46
2.04	Intellectual property protection, 2004	34
3.01	Telephone mainlines, 2002	46
3.02	Secure Internet servers, 2003	36
3.03	Internet hosts, 2003	61

Readiness Component Index		31
Individual Readiness		34
Business Readiness		57
Government Readiness		14
4.01	Quality of math and science education, 2004	62
4.02	Quality of educational system, 2004	48
4.03	Quality of public schools, 2004	40
4.04	Internet access in schools, 2004	30

Readiness Component Index (continued)		
4.05	Buyer sophistication, 2004	35
4.06	Buyer dynamism, 2004	25
4.07	Residential telephone connection charge, 2002	28
4.08	Affordability of Internet access, 2003	37
5.01	Investment in training, 2004	50
5.02	Availability of training services, 2004	68
5.03	Quality of business schools, 2004	93
5.04	Business investment in R&D, 2004	88
5.05	Business monthly telephone subscription, 2002–3	11
5.06	Business telephone connection charge, 2002–3	28
6.01	Government prioritization of ICT, 2004	25
6.02	Government procurement of ICT, 2004	16

Usage Component Index		32
Individual Usage		37
Business Usage		39
Government Usage		30
7.01	Cellular mobile subscribers, 2003	35
7.02	Telephone subscribers, 2002	37
7.03	Public payphones, 2002	39
7.04	Telephone lines, 2002	44
7.05	Television sets, 2002	25
7.06	Broadband-DSL Internet subscribers, 2002–3	30
7.07	Broadband-cable modem, 2002–3	60
7.08	Internet users per 100 inhabitants, 2002	33
8.01	Prevalence of foreign technology licensing, 2004	14
8.02	Firm-level technology absorption, 2004	43
8.03	Capacity for innovation, 2004	92
8.04	Availability of new telephone lines, 2004	26
8.05	Availability of cellular phones, 2004	25
9.01	Government success in ICT promotion, 2004	35
9.02	Government online services, 2003	35

Bangladesh

Key Indicators

Population (mn), 2003	147
GDP per capita (PPP US$), 2003	1,786
Internet users per 100 inhabitants, 2002	0.2

RANK/104

Environment Component Index — 97

Market Environment		93
Political and Regulatory Environment		94
Infrastructure Environment		101
1.01	Availability of scientists and engineers, 2004	60
1.02	Venture capital availability, 2004	98
1.03	Sophistication of financial markets, 2004	91
1.04	Technological sophistication, 2004	93
1.05	State of cluster development, 2004	42
1.06	Collaboration in clusters, 2004	91
1.07	University-industry collaboration, 2004	97
1.08	Quality of scientific research institutions, 2004	92
1.09	Subsidies for firm-level R&D, 2004	90
1.10	Brain drain, 2004	100
1.11	Ease of access to loans, 2004	86
1.12	Administrative burden, 2004	93
1.13	Ease to start a new business, 2004	80
2.01	Effectiveness of lawmaking, 2004	77
2.02	Laws relating to ICT, 2004	101
2.03	Effectiveness of judiciary, 2004	82
2.04	Intellectual property protection, 2004	100
3.01	Telephone mainlines, 2002	99
3.02	Secure Internet servers, 2003	99
3.03	Internet hosts, 2003	104

Readiness Component Index — 101

Individual Readiness		102
Business Readiness		101
Government Readiness		73
4.01	Quality of math and science education, 2004	92
4.02	Quality of educational system, 2004	95
4.03	Quality of public schools, 2004	97
4.04	Internet access in schools, 2004	101

Readiness Component Index (continued)

4.05	Buyer sophistication, 2004	81
4.06	Buyer dynamism, 2004	69
4.07	Residential telephone connection charge, 2002	104
4.08	Affordability of Internet access, 2003	88
5.01	Investment in training, 2004	101
5.02	Availability of training services, 2004	103
5.03	Quality of business schools, 2004	96
5.04	Business investment in R&D, 2004	98
5.05	Business monthly telephone subscription, 2002–3	86
5.06	Business telephone connection charge, 2002–3	103
6.01	Government prioritization of ICT, 2004	56
6.02	Government procurement of ICT, 2004	96

Usage Component Index — 95

Individual Usage		95
Business Usage		96
Government Usage		86
7.01	Cellular mobile subscribers, 2003	100
7.02	Telephone subscribers, 2002	101
7.03	Public payphones, 2002	104
7.04	Telephone lines, 2002	98
7.05	Television sets, 2002	88
7.06	Broadband-DSL Internet subscribers, 2002–3	65
7.07	Broadband-cable modem, 2002–3	60
7.08	Internet users per 100 inhabitants, 2002	103
8.01	Prevalence of foreign technology licensing, 2004	80
8.02	Firm-level technology absorption, 2004	83
8.03	Capacity for innovation, 2004	102
8.04	Availability of new telephone lines, 2004	103
8.05	Availability of cellular phones, 2004	78
9.01	Government success in ICT promotion, 2004	84
9.02	Government online services, 2003	79

Belgium

Key Indicators

Population (mn), 2003	10.3
GDP per capita (PPP US$), 2003	28,396
Internet users per 100 inhabitants, 2002	32.8

Networked Readiness Index Rank

Year (number of countries)	Rank
2004 (104)	**26**
2003 (102)	24
2002 (82)	22

RANK/104

Environment Component Index		24
Market Environment		25
Political and Regulatory Environment		24
Infrastructure Environment		25
1.01	Availability of scientists and engineers, 2004	24
1.02	Venture capital availability, 2004	24
1.03	Sophistication of financial markets, 2004	15
1.04	Technological sophistication, 2004	18
1.05	State of cluster development, 2004	31
1.06	Collaboration in clusters, 2004	8
1.07	University-industry collaboration, 2004	16
1.08	Quality of scientific research institutions, 2004	14
1.09	Subsidies for firm-level R&D, 2004	20
1.10	Brain drain, 2004	17
1.11	Ease of access to loans, 2004	24
1.12	Administrative burden, 2004	89
1.13	Ease to start a new business, 2004	61
2.01	Effectiveness of lawmaking, 2004	53
2.02	Laws relating to ICT, 2004	30
2.03	Effectiveness of judiciary, 2004	21
2.04	Intellectual property protection, 2004	18
3.01	Telephone mainlines, 2002	22
3.02	Secure Internet servers, 2003	23
3.03	Internet hosts, 2003	29

Readiness Component Index		26
Individual Readiness		6
Business Readiness		11
Government Readiness		54
4.01	Quality of math and science education, 2004	4
4.02	Quality of educational system, 2004	4
4.03	Quality of public schools, 2004	3
4.04	Internet access in schools, 2004	27

Readiness Component Index (continued)		
4.05	Buyer sophistication, 2004	13
4.06	Buyer dynamism, 2004	24
4.07	Residential telephone connection charge, 2002	12
4.08	Affordability of Internet access, 2003	23
5.01	Investment in training, 2004	9
5.02	Availability of training services, 2004	10
5.03	Quality of business schools, 2004	15
5.04	Business investment in R&D, 2004	16
5.05	Business monthly telephone subscription, 2002–3	14
5.06	Business telephone connection charge, 2002–3	13
6.01	Government prioritization of ICT, 2004	62
6.02	Government procurement of ICT, 2004	38

Usage Component Index		23
Individual Usage		18
Business Usage		18
Government Usage		42
7.01	Cellular mobile subscribers, 2003	19
7.02	Telephone subscribers, 2002	16
7.03	Public payphones, 2002	66
7.04	Telephone lines, 2002	20
7.05	Television sets, 2002	10
7.06	Broadband-DSL Internet subscribers, 2002–3	7
7.07	Broadband-cable modem, 2002–3	12
7.08	Internet users per 100 inhabitants, 2002	23
8.01	Prevalence of foreign technology licensing, 2004	22
8.02	Firm-level technology absorption, 2004	30
8.03	Capacity for innovation, 2004	12
8.04	Availability of new telephone lines, 2004	24
8.05	Availability of cellular phones, 2004	19
9.01	Government success in ICT promotion, 2004	63
9.02	Government online services, 2003	24

Bolivia

Key Indicators

Population (mn), 2003	8.8
GDP per capita (PPP US$), 2003	2,546
Internet users per 100 inhabitants, 2002	3.2

RANK/104

Environment Component Index	103
Market Environment	103
Political and Regulatory Environment	103
Infrastructure Environment	79

1.01	Availability of scientists and engineers, 2004	95
1.02	Venture capital availability, 2004	97
1.03	Sophistication of financial markets, 2004	87
1.04	Technological sophistication, 2004	94
1.05	State of cluster development, 2004	82
1.06	Collaboration in clusters, 2004	100
1.07	University-industry collaboration, 2004	94
1.08	Quality of scientific research institutions, 2004	98
1.09	Subsidies for firm-level R&D, 2004	103
1.10	Brain drain, 2004	94
1.11	Ease of access to loans, 2004	101
1.12	Administrative burden, 2004	90
1.13	Ease to start a new business, 2004	100
2.01	Effectiveness of lawmaking, 2004	97
2.02	Laws relating to ICT, 2004	103
2.03	Effectiveness of judiciary, 2004	93
2.04	Intellectual property protection, 2004	103
3.01	Telephone mainlines, 2002	77
3.02	Secure Internet servers, 2003	75
3.03	Internet hosts, 2003	72

Readiness Component Index	97
Individual Readiness	89
Business Readiness	94
Government Readiness	102

4.01	Quality of math and science education, 2004	94
4.02	Quality of educational system, 2004	94
4.03	Quality of public schools, 2004	95
4.04	Internet access in schools, 2004	83

Readiness Component Index (continued)	
4.05	Buyer sophistication, 2004	102
4.06	Buyer dynamism, 2004	104
4.07	Residential telephone connection charge, 2002	81
4.08	Affordability of Internet access, 2003	80
5.01	Investment in training, 2004	99
5.02	Availability of training services, 2004	99
5.03	Quality of business schools, 2004	92
5.04	Business investment in R&D, 2004	99
5.05	Business monthly telephone subscription, 2002–3	83
5.06	Business telephone connection charge, 2002–3	84
6.01	Government prioritization of ICT, 2004	95
6.02	Government procurement of ICT, 2004	104

Usage Component Index	98
Individual Usage	76
Business Usage	93
Government Usage	102

7.01	Cellular mobile subscribers, 2003	70
7.02	Telephone subscribers, 2002	78
7.03	Public payphones, 2002	38
7.04	Telephone lines, 2002	74
7.05	Television sets, 2002	78
7.06	Broadband-DSL Internet subscribers, 2002–3	65
7.07	Broadband-cable modem, 2002–3	60
7.08	Internet users per 100 inhabitants, 2002	74
8.01	Prevalence of foreign technology licensing, 2004	103
8.02	Firm-level technology absorption, 2004	102
8.03	Capacity for innovation, 2004	96
8.04	Availability of new telephone lines, 2004	69
8.05	Availability of cellular phones, 2004	67
9.01	Government success in ICT promotion, 2004	101
9.02	Government online services, 2003	96

Bosnia and Herzegovina

Key Indicators

Population (mn), 2003	4.2
GDP per capita (PPP US$), 2003	6,029
Internet users per 100 inhabitants, 2002	2.6

Networked Readiness Index Rank

Year (number of countries) Rank

2004 (104) **89**

2003 (102) –

2002 (82) –

RANK/104

Environment Component Index		92
Market Environment		97
Political and Regulatory Environment		93
Infrastructure Environment		53
1.01	Availability of scientists and engineers, 2004	86
1.02	Venture capital availability, 2004	87
1.03	Sophistication of financial markets, 2004	98
1.04	Technological sophistication, 2004	102
1.05	State of cluster development, 2004	99
1.06	Collaboration in clusters, 2004	85
1.07	University-industry collaboration, 2004	73
1.08	Quality of scientific research institutions, 2004	97
1.09	Subsidies for firm-level R&D, 2004	88
1.10	Brain drain, 2004	95
1.11	Ease of access to loans, 2004	62
1.12	Administrative burden, 2004	101
1.13	Ease to start a new business, 2004	89
2.01	Effectiveness of lawmaking, 2004	92
2.02	Laws relating to ICT, 2004	91
2.03	Effectiveness of judiciary, 2004	78
2.04	Intellectual property protection, 2004	101
3.01	Telephone mainlines, 2002	50
3.02	Secure Internet servers, 2003	76
3.03	Internet hosts, 2003	62

Readiness Component Index		86
Individual Readiness		77
Business Readiness		83
Government Readiness		87
4.01	Quality of math and science education, 2004	47
4.02	Quality of educational system, 2004	59
4.03	Quality of public schools, 2004	65
4.04	Internet access in schools, 2004	76

Readiness Component Index (continued)		
4.05	Buyer sophistication, 2004	93
4.06	Buyer dynamism, 2004	103
4.07	Residential telephone connection charge, 2002	87
4.08	Affordability of Internet access, 2003	49
5.01	Investment in training, 2004	95
5.02	Availability of training services, 2004	80
5.03	Quality of business schools, 2004	88
5.04	Business investment in R&D, 2004	97
5.05	Business monthly telephone subscription, 2002–3	48
5.06	Business telephone connection charge, 2002–3	77
6.01	Government prioritization of ICT, 2004	80
6.02	Government procurement of ICT, 2004	95

Usage Component Index		92
Individual Usage		60
Business Usage		97
Government Usage		91
7.01	Cellular mobile subscribers, 2003	54
7.02	Telephone subscribers, 2002	52
7.03	Public payphones, 2002	55
7.04	Telephone lines, 2002	48
7.05	Television sets, 2002	63
7.06	Broadband-DSL Internet subscribers, 2002–3	65
7.07	Broadband-cable modem, 2002–3	60
7.08	Internet users per 100 inhabitants, 2002	77
8.01	Prevalence of foreign technology licensing, 2004	90
8.02	Firm-level technology absorption, 2004	104
8.03	Capacity for innovation, 2004	79
8.04	Availability of new telephone lines, 2004	73
8.05	Availability of cellular phones, 2004	96
9.01	Government success in ICT promotion, 2004	89
9.02	Government online services, 2003	71

Botswana

Key Indicators

Population (mn), 2003	1.8
GDP per capita (PPP US$), 2003	8,359
Internet users per 100 inhabitants, 2002	3.5

Networked Readiness Index Rank

Year (number of countries)	Rank
2004 (104)	**50**
2003 (102)	55
2002 (82)	44

RANK/104

Environment Component Index	48
Market Environment	43
Political and Regulatory Environment	44
Infrastructure Environment	74

1.01	Availability of scientists and engineers, 2004	91
1.02	Venture capital availability, 2004	43
1.03	Sophistication of financial markets, 2004	74
1.04	Technological sophistication, 2004	58
1.05	State of cluster development, 2004	58
1.06	Collaboration in clusters, 2004	69
1.07	University-industry collaboration, 2004	48
1.08	Quality of scientific research institutions, 2004	47
1.09	Subsidies for firm-level R&D, 2004	34
1.10	Brain drain, 2004	28
1.11	Ease of access to loans, 2004	34
1.12	Administrative burden, 2004	23
1.13	Ease to start a new business, 2004	54
2.01	Effectiveness of lawmaking, 2004	27
2.02	Laws relating to ICT, 2004	66
2.03	Effectiveness of judiciary, 2004	28
2.04	Intellectual property protection, 2004	58
3.01	Telephone mainlines, 2002	75
3.02	Secure Internet servers, 2003	46
3.03	Internet hosts, 2003	70

Readiness Component Index	38
Individual Readiness	54
Business Readiness	55
Government Readiness	30

4.01	Quality of math and science education, 2004	59
4.02	Quality of educational system, 2004	44
4.03	Quality of public schools, 2004	44
4.04	Internet access in schools, 2004	65

Readiness Component Index (continued)	
4.05	Buyer sophistication, 2004	66
4.06	Buyer dynamism, 2004	73
4.07	Residential telephone connection charge, 2002	48
4.08	Affordability of Internet access, 2003	56
5.01	Investment in training, 2004	70
5.02	Availability of training services, 2004	69
5.03	Quality of business schools, 2004	81
5.04	Business investment in R&D, 2004	44
5.05	Business monthly telephone subscription, 2002–3	29
5.06	Business telephone connection charge, 2002–3	46
6.01	Government prioritization of ICT, 2004	38
6.02	Government procurement of ICT, 2004	21

Usage Component Index	66
Individual Usage	80
Business Usage	65
Government Usage	60

7.01	Cellular mobile subscribers, 2003	51
7.02	Telephone subscribers, 2002	63
7.03	Public payphones, 2002	76
7.04	Telephone lines, 2002	73
7.05	Television sets, 2002	94
7.06	Broadband-DSL Internet subscribers, 2002–3	65
7.07	Broadband-cable modem, 2002–3	60
7.08	Internet users per 100 inhabitants, 2002	72
8.01	Prevalence of foreign technology licensing, 2004	54
8.02	Firm-level technology absorption, 2004	70
8.03	Capacity for innovation, 2004	68
8.04	Availability of new telephone lines, 2004	77
8.05	Availability of cellular phones, 2004	65
9.01	Government success in ICT promotion, 2004	44
9.02	Government online services, 2003	79

Brazil

Key Indicators

Population (mn), 2003	179
GDP per capita (PPP US$), 2003	7,767
Internet users per 100 inhabitants, 2002	8.2

Networked Readiness Index Rank

Year (number of countries)	Rank
2004 (104)	**46**
2003 (102)	39
2002 (82)	29

RANK/104

Environment Component Index		47
Market Environment		46
Political and Regulatory Environment		50
Infrastructure Environment		48
1.01	Availability of scientists and engineers, 2004	58
1.02	Venture capital availability, 2004	56
1.03	Sophistication of financial markets, 2004	26
1.04	Technological sophistication, 2004	36
1.05	State of cluster development, 2004	26
1.06	Collaboration in clusters, 2004	18
1.07	University-industry collaboration, 2004	28
1.08	Quality of scientific research institutions, 2004	37
1.09	Subsidies for firm-level R&D, 2004	40
1.10	Brain drain, 2004	29
1.11	Ease of access to loans, 2004	49
1.12	Administrative burden, 2004	96
1.13	Ease to start a new business, 2004	96
2.01	Effectiveness of lawmaking, 2004	61
2.02	Laws relating to ICT, 2004	41
2.03	Effectiveness of judiciary, 2004	51
2.04	Intellectual property protection, 2004	51
3.01	Telephone mainlines, 2002	53
3.02	Secure Internet servers, 2003	48
3.03	Internet hosts, 2003	37

Readiness Component Index		43
Individual Readiness		63
Business Readiness		32
Government Readiness		50
4.01	Quality of math and science education, 2004	79
4.02	Quality of educational system, 2004	85
4.03	Quality of public schools, 2004	81
4.04	Internet access in schools, 2004	50

Readiness Component Index (continued)		
4.05	Buyer sophistication, 2004	41
4.06	Buyer dynamism, 2004	31
4.07	Residential telephone connection charge, 2002	63
4.08	Affordability of Internet access, 2003	59
5.01	Investment in training, 2004	27
5.02	Availability of training services, 2004	17
5.03	Quality of business schools, 2004	28
5.04	Business investment in R&D, 2004	31
5.05	Business monthly telephone subscription, 2002–3	58
5.06	Business telephone connection charge, 2002–3	64
6.01	Government prioritization of ICT, 2004	54
6.02	Government procurement of ICT, 2004	41

Usage Component Index		43
Individual Usage		44
Business Usage		28
Government Usage		56
7.01	Cellular mobile subscribers, 2003	57
7.02	Telephone subscribers, 2002	54
7.03	Public payphones, 2002	4
7.04	Telephone lines, 2002	52
7.05	Television sets, 2002	56
7.06	Broadband-DSL Internet subscribers, 2002–3	38
7.07	Broadband-cable modem, 2002–3	42
7.08	Internet users per 100 inhabitants, 2002	53
8.01	Prevalence of foreign technology licensing, 2004	24
8.02	Firm-level technology absorption, 2004	33
8.03	Capacity for innovation, 2004	37
8.04	Availability of new telephone lines, 2004	35
8.05	Availability of cellular phones, 2004	32
9.01	Government success in ICT promotion, 2004	52
9.02	Government online services, 2003	61

Bulgaria

Key Indicators

Population (mn), 2003	7.9
GDP per capita (PPP US$), 2003	7,807
Internet users per 100 inhabitants, 2002	8.1

RANK/104

Environment Component Index		76
Market Environment		86
Political and Regulatory Environment		81
Infrastructure Environment		39
1.01	Availability of scientists and engineers, 2004	29
1.02	Venture capital availability, 2004	57
1.03	Sophistication of financial markets, 2004	100
1.04	Technological sophistication, 2004	87
1.05	State of cluster development, 2004	91
1.06	Collaboration in clusters, 2004	72
1.07	University-industry collaboration, 2004	89
1.08	Quality of scientific research institutions, 2004	61
1.09	Subsidies for firm-level R&D, 2004	77
1.10	Brain drain, 2004	98
1.11	Ease of access to loans, 2004	60
1.12	Administrative burden, 2004	65
1.13	Ease to start a new business, 2004	102
2.01	Effectiveness of lawmaking, 2004	82
2.02	Laws relating to ICT, 2004	61
2.03	Effectiveness of judiciary, 2004	81
2.04	Intellectual property protection, 2004	81
3.01	Telephone mainlines, 2002	32
3.02	Secure Internet servers, 2003	59
3.03	Internet hosts, 2003	46

Readiness Component Index		69
Individual Readiness		45
Business Readiness		64
Government Readiness		83
4.01	Quality of math and science education, 2004	17
4.02	Quality of educational system, 2004	40
4.03	Quality of public schools, 2004	42
4.04	Internet access in schools, 2004	66

Readiness Component Index (continued)		
4.05	Buyer sophistication, 2004	78
4.06	Buyer dynamism, 2004	76
4.07	Residential telephone connection charge, 2002	59
4.08	Affordability of Internet access, 2003	52
5.01	Investment in training, 2004	86
5.02	Availability of training services, 2004	62
5.03	Quality of business schools, 2004	78
5.04	Business investment in R&D, 2004	90
5.05	Business monthly telephone subscription, 2002–3	57
5.06	Business telephone connection charge, 2002–3	55
6.01	Government prioritization of ICT, 2004	85
6.02	Government procurement of ICT, 2004	79

Usage Component Index		74
Individual Usage		45
Business Usage		87
Government Usage		81
7.01	Cellular mobile subscribers, 2003	46
7.02	Telephone subscribers, 2002	41
7.03	Public payphones, 2002	44
7.04	Telephone lines, 2002	33
7.05	Television sets, 2002	54
7.06	Broadband-DSL Internet subscribers, 2002–3	65
7.07	Broadband-cable modem, 2002–3	29
7.08	Internet users per 100 inhabitants, 2002	54
8.01	Prevalence of foreign technology licensing, 2004	89
8.02	Firm-level technology absorption, 2004	94
8.03	Capacity for innovation, 2004	90
8.04	Availability of new telephone lines, 2004	68
8.05	Availability of cellular phones, 2004	74
9.01	Government success in ICT promotion, 2004	87
9.02	Government online services, 2003	57

Canada

Key Indicators

Population (mn), 2003	31.5
GDP per capita (PPP US$), 2003	30,463
Internet users per 100 inhabitants, 2002	51.3

Networked Readiness Index Rank

Year (number of countries)　　Rank

2004 (104)　**10**

2003 (102)　6

2002　(82)　6

RANK/104

Environment Component Index — 10

	Market Environment	12
	Political and Regulatory Environment	15
	Infrastructure Environment	4
1.01	Availability of scientists and engineers, 2004	13
1.02	Venture capital availability, 2004	16
1.03	Sophistication of financial markets, 2004	6
1.04	Technological sophistication, 2004	14
1.05	State of cluster development, 2004	15
1.06	Collaboration in clusters, 2004	13
1.07	University-industry collaboration, 2004	13
1.08	Quality of scientific research institutions, 2004	12
1.09	Subsidies for firm-level R&D, 2004	4
1.10	Brain drain, 2004	24
1.11	Ease of access to loans, 2004	22
1.12	Administrative burden, 2004	32
1.13	Ease to start a new business, 2004	15
2.01	Effectiveness of lawmaking, 2004	19
2.02	Laws relating to ICT, 2004	19
2.03	Effectiveness of judiciary, 2004	13
2.04	Intellectual property protection, 2004	16
3.01	Telephone mainlines, 2002	10
3.02	Secure Internet servers, 2003	3
3.03	Internet hosts, 2003	10

Readiness Component Index — 22

	Individual Readiness	12
	Business Readiness	12
	Government Readiness	41
4.01	Quality of math and science education, 2004	22
4.02	Quality of educational system, 2004	12
4.03	Quality of public schools, 2004	11
4.04	Internet access in schools, 2004	9

Readiness Component Index (continued)

4.05	Buyer sophistication, 2004	9
4.06	Buyer dynamism, 2004	19
4.07	Residential telephone connection charge, 2002	16
4.08	Affordability of Internet access, 2003	4
5.01	Investment in training, 2004	21
5.02	Availability of training services, 2004	14
5.03	Quality of business schools, 2004	5
5.04	Business investment in R&D, 2004	18
5.05	Business monthly telephone subscription, 2002–3	15
5.06	Business telephone connection charge, 2002–3	18
6.01	Government prioritization of ICT, 2004	42
6.02	Government procurement of ICT, 2004	35

Usage Component Index — 8

	Individual Usage	8
	Business Usage	21
	Government Usage	6
7.01	Cellular mobile subscribers, 2003	43
7.02	Telephone subscribers, 2002	33
7.03	Public payphones, 2002	8
7.04	Telephone lines, 2002	9
7.05	Television sets, 2002	23
7.06	Broadband-DSL Internet subscribers, 2002–3	6
7.07	Broadband-cable modem, 2002–3	3
7.08	Internet users per 100 inhabitants, 2002	5
8.01	Prevalence of foreign technology licensing, 2004	20
8.02	Firm-level technology absorption, 2004	27
8.03	Capacity for innovation, 2004	19
8.04	Availability of new telephone lines, 2004	16
8.05	Availability of cellular phones, 2004	31
9.01	Government success in ICT promotion, 2004	49
9.02	Government online services, 2003	4

Chad

Key Indicators

Population (mn), 2003	8.6
GDP per capita (PPP US$), 2003	1,206
Internet users per 100 inhabitants, 2002	0.2

Networked Readiness Index Rank

Year (number of countries)　Rank

2004 (104)　**104**

2003 (102)　102

2002　(82)　—

RANK/104

Environment Component Index		102
Market Environment		101
Political and Regulatory Environment		104
Infrastructure Environment		104
1.01	Availability of scientists and engineers, 2004	103
1.02	Venture capital availability, 2004	101
1.03	Sophistication of financial markets, 2004	103
1.04	Technological sophistication, 2004	104
1.05	State of cluster development, 2004	88
1.06	Collaboration in clusters, 2004	88
1.07	University-industry collaboration, 2004	103
1.08	Quality of scientific research institutions, 2004	104
1.09	Subsidies for firm-level R&D, 2004	89
1.10	Brain drain, 2004	99
1.11	Ease of access to loans, 2004	103
1.12	Administrative burden, 2004	53
1.13	Ease to start a new business, 2004	55
2.01	Effectiveness of lawmaking, 2004	91
2.02	Laws relating to ICT, 2004	104
2.03	Effectiveness of judiciary, 2004	100
2.04	Intellectual property protection, 2004	102
3.01	Telephone mainlines, 2002	104
3.02	Secure Internet servers, 2003	100
3.03	Internet hosts, 2003	101

Readiness Component Index		103
Individual Readiness		103
Business Readiness		100
Government Readiness		98
4.01	Quality of math and science education, 2004	100
4.02	Quality of educational system, 2004	101
4.03	Quality of public schools, 2004	100
4.04	Internet access in schools, 2004	104

Readiness Component Index (continued)		
4.05	Buyer sophistication, 2004	97
4.06	Buyer dynamism, 2004	87
4.07	Residential telephone connection charge, 2002	99
4.08	Affordability of Internet access, 2003	101
5.01	Investment in training, 2004	103
5.02	Availability of training services, 2004	102
5.03	Quality of business schools, 2004	103
5.04	Business investment in R&D, 2004	103
5.05	Business monthly telephone subscription, 2002–3	98
5.06	Business telephone connection charge, 2002–3	99
6.01	Government prioritization of ICT, 2004	97
6.02	Government procurement of ICT, 2004	99

Usage Component Index		104
Individual Usage		102
Business Usage		104
Government Usage		104
7.01	Cellular mobile subscribers, 2003	102
7.02	Telephone subscribers, 2002	104
7.03	Public payphones, 2002	87
7.04	Telephone lines, 2002	104
7.05	Television sets, 2002	104
7.06	Broadband-DSL Internet subscribers, 2002–3	65
7.07	Broadband-cable modem, 2002–3	60
7.08	Internet users per 100 inhabitants, 2002	102
8.01	Prevalence of foreign technology licensing, 2004	104
8.02	Firm-level technology absorption, 2004	99
8.03	Capacity for innovation, 2004	101
8.04	Availability of new telephone lines, 2004	93
8.05	Availability of cellular phones, 2004	103
9.01	Government success in ICT promotion, 2004	99
9.02	Government online services, 2003	102

Chile

Key Indicators

Population (mn), 2003	15.8
GDP per capita (PPP US$), 2003	10,206
Internet users per 100 inhabitants, 2002	23.8

Networked Readiness Index Rank

Year (number of countries)	Rank
2004 (104)	**35**
2003 (102)	32
2002 (82)	35

RANK/104

Environment Component Index	34
Market Environment	32
Political and Regulatory Environment	35
Infrastructure Environment	47

1.01	Availability of scientists and engineers, 2004	57
1.02	Venture capital availability, 2004	39
1.03	Sophistication of financial markets, 2004	22
1.04	Technological sophistication, 2004	23
1.05	State of cluster development, 2004	60
1.06	Collaboration in clusters, 2004	65
1.07	University-industry collaboration, 2004	44
1.08	Quality of scientific research institutions, 2004	59
1.09	Subsidies for firm-level R&D, 2004	73
1.10	Brain drain, 2004	7
1.11	Ease of access to loans, 2004	26
1.12	Administrative burden, 2004	21
1.13	Ease to start a new business, 2004	52
2.01	Effectiveness of lawmaking, 2004	42
2.02	Laws relating to ICT, 2004	28
2.03	Effectiveness of judiciary, 2004	37
2.04	Intellectual property protection, 2004	42
3.01	Telephone mainlines, 2002	52
3.02	Secure Internet servers, 2003	38
3.03	Internet hosts, 2003	39

Readiness Component Index	40
Individual Readiness	51
Business Readiness	33
Government Readiness	44

4.01	Quality of math and science education, 2004	87
4.02	Quality of educational system, 2004	71
4.03	Quality of public schools, 2004	75
4.04	Internet access in schools, 2004	33

Readiness Component Index (continued)	
4.05	Buyer sophistication, 2004	36
4.06	Buyer dynamism, 2004	40
4.07	Residential telephone connection charge, 2002	41
4.08	Affordability of Internet access, 2003	47
5.01	Investment in training, 2004	40
5.02	Availability of training services, 2004	39
5.03	Quality of business schools, 2004	17
5.04	Business investment in R&D, 2004	46
5.05	Business monthly telephone subscription, 2002–3	45
5.06	Business telephone connection charge, 2002–3	39
6.01	Government prioritization of ICT, 2004	39
6.02	Government procurement of ICT, 2004	50

Usage Component Index	36
Individual Usage	38
Business Usage	32
Government Usage	38

7.01	Cellular mobile subscribers, 2003	42
7.02	Telephone subscribers, 2002	44
7.03	Public payphones, 2002	15
7.04	Telephone lines, 2002	53
7.05	Television sets, 2002	28
7.06	Broadband-DSL Internet subscribers, 2002-3	35
7.07	Broadband-cable modem, 2002-3	27
7.08	Internet users per 100 inhabitants, 2002	34
8.01	Prevalence of foreign technology licensing, 2004	37
8.02	Firm-level technology absorption, 2004	31
8.03	Capacity for innovation, 2004	56
8.04	Availability of new telephone lines, 2004	15
8.05	Availability of cellular phones, 2004	20
9.01	Government success in ICT promotion, 2004	48
9.02	Government online services, 2003	34

China

Key Indicators

Population (mn), 2003	1,304
GDP per capita (PPP US$), 2003	4,995
Internet users per 100 inhabitants, 2002	4.6

Networked Readiness Index Rank

Year (number of countries)	Rank
2004 (104)	**41**
2003 (102)	51
2002 (82)	43

RANK/104

Environment Component Index	46
Market Environment	37
Political and Regulatory Environment	49
Infrastructure Environment	62

1.01	Availability of scientists and engineers, 2004	67
1.02	Venture capital availability, 2004	59
1.03	Sophistication of financial markets, 2004	77
1.04	Technological sophistication, 2004	60
1.05	State of cluster development, 2004	32
1.06	Collaboration in clusters, 2004	15
1.07	University-industry collaboration, 2004	22
1.08	Quality of scientific research institutions, 2004	40
1.09	Subsidies for firm-level R&D, 2004	24
1.10	Brain drain, 2004	41
1.11	Ease of access to loans, 2004	79
1.12	Administrative burden, 2004	18
1.13	Ease to start a new business, 2004	32
2.01	Effectiveness of lawmaking, 2004	24
2.02	Laws relating to ICT, 2004	51
2.03	Effectiveness of judiciary, 2004	61
2.04	Intellectual property protection, 2004	54
3.01	Telephone mainlines, 2002	60
3.02	Secure Internet servers, 2003	88
3.03	Internet hosts, 2003	86

Readiness Component Index	39
Individual Readiness	61
Business Readiness	60
Government Readiness	17

4.01	Quality of math and science education, 2004	55
4.02	Quality of educational system, 2004	65
4.03	Quality of public schools, 2004	60
4.04	Internet access in schools, 2004	49

Readiness Component Index (continued)	

4.05	Buyer sophistication, 2004	42
4.06	Buyer dynamism, 2004	49
4.07	Residential telephone connection charge, 2002	80
4.08	Affordability of Internet access, 2003	62
5.01	Investment in training, 2004	52
5.02	Availability of training services, 2004	34
5.03	Quality of business schools, 2004	71
5.04	Business investment in R&D, 2004	27
5.05	Business monthly telephone subscription, 2002–3	65
5.06	Business telephone connection charge, 2002–3	84
6.01	Government prioritization of ICT, 2004	43
6.02	Government procurement of ICT, 2004	12

Usage Component Index	38
Individual Usage	47
Business Usage	52
Government Usage	22

7.01	Cellular mobile subscribers, 2003	62
7.02	Telephone subscribers, 2002	65
7.03	Public payphones, 2002	3
7.04	Telephone lines, 2002	56
7.05	Television sets, 2002	58
7.06	Broadband-DSL Internet subscribers, 2002–3	43
7.07	Broadband-cable modem, 2002–3	34
7.08	Internet users per 100 inhabitants, 2002	66
8.01	Prevalence of foreign technology licensing, 2004	59
8.02	Firm-level technology absorption, 2004	34
8.03	Capacity for innovation, 2004	31
8.04	Availability of new telephone lines, 2004	60
8.05	Availability of cellular phones, 2004	89
9.01	Government success in ICT promotion, 2004	23
9.02	Government online services, 2003	18

Colombia

Key Indicators

Population (mn), 2003	44.2
GDP per capita (PPP US$), 2003	6,784
Internet users per 100 inhabitants, 2002	4.6

Networked Readiness Index Rank

Year (number of countries)	Rank
2004 (104)	**66**
2003 (102)	60
2002 (82)	59

RANK/104

Environment Component Index			72
Market Environment			71
Political and Regulatory Environment			69
Infrastructure Environment			59
1.01	Availability of scientists and engineers, 2004		82
1.02	Venture capital availability, 2004		75
1.03	Sophistication of financial markets, 2004		46
1.04	Technological sophistication, 2004		66
1.05	State of cluster development, 2004		69
1.06	Collaboration in clusters, 2004		78
1.07	University-industry collaboration, 2004		46
1.08	Quality of scientific research institutions, 2004		66
1.09	Subsidies for firm-level R&D, 2004		66
1.10	Brain drain, 2004		73
1.11	Ease of access to loans, 2004		52
1.12	Administrative burden, 2004		61
1.13	Ease to start a new business, 2004		85
2.01	Effectiveness of lawmaking, 2004		75
2.02	Laws relating to ICT, 2004		54
2.03	Effectiveness of judiciary, 2004		73
2.04	Intellectual property protection, 2004		65
3.01	Telephone mainlines, 2002		58
3.02	Secure Internet servers, 2003		64
3.03	Internet hosts, 2003		54

Readiness Component Index			66
Individual Readiness			71
Business Readiness			62
Government Readiness			69
4.01	Quality of math and science education, 2004		78
4.02	Quality of educational system, 2004		69
4.03	Quality of public schools, 2004		76
4.04	Internet access in schools, 2004		77

Readiness Component Index (continued)			
4.05	Buyer sophistication, 2004		67
4.06	Buyer dynamism, 2004		51
4.07	Residential telephone connection charge, 2002		74
4.08	Affordability of Internet access, 2003		60
5.01	Investment in training, 2004		62
5.02	Availability of training services, 2004		82
5.03	Quality of business schools, 2004		42
5.04	Business investment in R&D, 2004		58
5.05	Business monthly telephone subscription, 2002–3		61
5.06	Business telephone connection charge, 2002–3		72
6.01	Government prioritization of ICT, 2004		71
6.02	Government procurement of ICT, 2004		66

Usage Component Index			65
Individual Usage			61
Business Usage			73
Government Usage			72
7.01	Cellular mobile subscribers, 2003		71
7.02	Telephone subscribers, 2002		69
7.03	Public payphones, 2002		27
7.04	Telephone lines, 2002		58
7.05	Television sets, 2002		45
7.06	Broadband-DSL Internet subscribers, 2002–3		58
7.07	Broadband-cable modem, 2002–3		40
7.08	Internet users per 100 inhabitants, 2002		65
8.01	Prevalence of foreign technology licensing, 2004		82
8.02	Firm-level technology absorption, 2004		80
8.03	Capacity for innovation, 2004		51
8.04	Availability of new telephone lines, 2004		62
8.05	Availability of cellular phones, 2004		72
9.01	Government success in ICT promotion, 2004		62
9.02	Government online services, 2003		79

Costa Rica

Key Indicators

Population (mn), 2003	4.2
GDP per capita (PPP US$), 2003	9,490
Internet users per 100 inhabitants, 2002	19.3

Networked Readiness Index Rank

Year (number of countries)	Rank
2004 (104)	**61**
2003 (102)	49
2002 (82)	49

RANK/104

Environment Component Index		52
Market Environment		50
Political and Regulatory Environment		58
Infrastructure Environment		43
1.01	Availability of scientists and engineers, 2004	45
1.02	Venture capital availability, 2004	60
1.03	Sophistication of financial markets, 2004	47
1.04	Technological sophistication, 2004	37
1.05	State of cluster development, 2004	64
1.06	Collaboration in clusters, 2004	71
1.07	University-industry collaboration, 2004	52
1.08	Quality of scientific research institutions, 2004	38
1.09	Subsidies for firm-level R&D, 2004	85
1.10	Brain drain, 2004	20
1.11	Ease of access to loans, 2004	58
1.12	Administrative burden, 2004	47
1.13	Ease to start a new business, 2004	70
2.01	Effectiveness of lawmaking, 2004	88
2.02	Laws relating to ICT, 2004	71
2.03	Effectiveness of judiciary, 2004	33
2.04	Intellectual property protection, 2004	48
3.01	Telephone mainlines, 2002	47
3.02	Secure Internet servers, 2003	27
3.03	Internet hosts, 2003	55

Readiness Component Index		51
Individual Readiness		42
Business Readiness		31
Government Readiness		78
4.01	Quality of math and science education, 2004	52
4.02	Quality of educational system, 2004	24
4.03	Quality of public schools, 2004	49
4.04	Internet access in schools, 2004	60

Readiness Component Index (continued)		
4.05	Buyer sophistication, 2004	46
4.06	Buyer dynamism, 2004	59
4.07	Residential telephone connection charge, 2002	47
4.08	Affordability of Internet access, 2003	51
5.01	Investment in training, 2004	31
5.02	Availability of training services, 2004	36
5.03	Quality of business schools, 2004	24
5.04	Business investment in R&D, 2004	33
5.05	Business monthly telephone subscription, 2002–3	36
5.06	Business telephone connection charge, 2002–3	45
6.01	Government prioritization of ICT, 2004	75
6.02	Government procurement of ICT, 2004	76

Usage Component Index		77
Individual Usage		50
Business Usage		85
Government Usage		85
7.01	Cellular mobile subscribers, 2003	75
7.02	Telephone subscribers, 2002	61
7.03	Public payphones, 2002	7
7.04	Telephone lines, 2002	46
7.05	Television sets, 2002	74
7.06	Broadband-DSL Internet subscribers, 2002–3	56
7.07	Broadband-cable modem, 2002–3	60
7.08	Internet users per 100 inhabitants, 2002	39
8.01	Prevalence of foreign technology licensing, 2004	68
8.02	Firm-level technology absorption, 2004	47
8.03	Capacity for innovation, 2004	38
8.04	Availability of new telephone lines, 2004	98
8.05	Availability of cellular phones, 2004	99
9.01	Government success in ICT promotion, 2004	77
9.02	Government online services, 2003	86

Croatia

Key Indicators

Population (mn), 2003	4.4
GDP per capita (PPP US$), 2003	11,139
Internet users per 100 inhabitants, 2002	18.0

Networked Readiness Index Rank

Year (number of countries)	Rank
2004 (104)	**58**
2003 (102)	48
2002 (82)	48

RANK/104

Environment Component Index	64
Market Environment	75
Political and Regulatory Environment	76
Infrastructure Environment	34
1.01 Availability of scientists and engineers, 2004	63
1.02 Venture capital availability, 2004	91
1.03 Sophistication of financial markets, 2004	84
1.04 Technological sophistication, 2004	89
1.05 State of cluster development, 2004	80
1.06 Collaboration in clusters, 2004	26
1.07 University-industry collaboration, 2004	51
1.08 Quality of scientific research institutions, 2004	46
1.09 Subsidies for firm-level R&D, 2004	56
1.10 Brain drain, 2004	80
1.11 Ease of access to loans, 2004	67
1.12 Administrative burden, 2004	98
1.13 Ease to start a new business, 2004	59
2.01 Effectiveness of lawmaking, 2004	69
2.02 Laws relating to ICT, 2004	58
2.03 Effectiveness of judiciary, 2004	88
2.04 Intellectual property protection, 2004	76
3.01 Telephone mainlines, 2002	31
3.02 Secure Internet servers, 2003	33
3.03 Internet hosts, 2003	44

Readiness Component Index	64
Individual Readiness	50
Business Readiness	58
Government Readiness	79
4.01 Quality of math and science education, 2004	37
4.02 Quality of educational system, 2004	62
4.03 Quality of public schools, 2004	41
4.04 Internet access in schools, 2004	47

Readiness Component Index (continued)	
4.05 Buyer sophistication, 2004	89
4.06 Buyer dynamism, 2004	72
4.07 Residential telephone connection charge, 2002	49
4.08 Affordability of Internet access, 2003	40
5.01 Investment in training, 2004	88
5.02 Availability of training services, 2004	42
5.03 Quality of business schools, 2004	79
5.04 Business investment in R&D, 2004	54
5.05 Business monthly telephone subscription, 2002–3	43
5.06 Business telephone connection charge, 2002–3	47
6.01 Government prioritization of ICT, 2004	67
6.02 Government procurement of ICT, 2004	82

Usage Component Index	52
Individual Usage	36
Business Usage	59
Government Usage	70
7.01 Cellular mobile subscribers, 2003	36
7.02 Telephone subscribers, 2002	35
7.03 Public payphones, 2002	37
7.04 Telephone lines, 2002	30
7.05 Television sets, 2002	43
7.06 Broadband-DSL Internet subscribers, 2002–3	41
7.07 Broadband-cable modem, 2002–3	60
7.08 Internet users per 100 inhabitants, 2002	40
8.01 Prevalence of foreign technology licensing, 2004	43
8.02 Firm-level technology absorption, 2004	89
8.03 Capacity for innovation, 2004	60
8.04 Availability of new telephone lines, 2004	51
8.05 Availability of cellular phones, 2004	50
9.01 Government success in ICT promotion, 2004	64
9.02 Government online services, 2003	66

Cyprus

Key Indicators

Population (mn), 2003	0.802
GDP per capita (PPP US$), 2003	19,155
Internet users per 100 inhabitants, 2002	29.4

RANK/104

Environment Component Index		27
Market Environment		34
Political and Regulatory Environment		40
Infrastructure Environment		20
1.01	Availability of scientists and engineers, 2004	30
1.02	Venture capital availability, 2004	35
1.03	Sophistication of financial markets, 2004	42
1.04	Technological sophistication, 2004	59
1.05	State of cluster development, 2004	44
1.06	Collaboration in clusters, 2004	47
1.07	University-industry collaboration, 2004	86
1.08	Quality of scientific research institutions, 2004	89
1.09	Subsidies for firm-level R&D, 2004	59
1.10	Brain drain, 2004	43
1.11	Ease of access to loans, 2004	35
1.12	Administrative burden, 2004	19
1.13	Ease to start a new business, 2004	20
2.01	Effectiveness of lawmaking, 2004	37
2.02	Laws relating to ICT, 2004	62
2.03	Effectiveness of judiciary, 2004	27
2.04	Intellectual property protection, 2004	41
3.01	Telephone mainlines, 2002	6
3.02	Secure Internet servers, 2003	19
3.03	Internet hosts, 2003	43

Readiness Component Index		55
Individual Readiness		29
Business Readiness		49
Government Readiness		81
4.01	Quality of math and science education, 2004	31
4.02	Quality of educational system, 2004	26
4.03	Quality of public schools, 2004	30
4.04	Internet access in schools, 2004	36

Readiness Component Index (continued)		
4.05	Buyer sophistication, 2004	37
4.06	Buyer dynamism, 2004	36
4.07	Residential telephone connection charge, 2002	23
4.08	Affordability of Internet access, 2003	24
5.01	Investment in training, 2004	58
5.02	Availability of training services, 2004	63
5.03	Quality of business schools, 2004	62
5.04	Business investment in R&D, 2004	85
5.05	Business monthly telephone subscription, 2002–3	19
5.06	Business telephone connection charge, 2002–3	24
6.01	Government prioritization of ICT, 2004	77
6.02	Government procurement of ICT, 2004	80

Usage Component Index		35
Individual Usage		23
Business Usage		47
Government Usage		49
7.01	Cellular mobile subscribers, 2003	24
7.02	Telephone subscribers, 2002	17
7.03	Public payphones, 2002	28
7.04	Telephone lines, 2002	5
7.05	Television sets, 2002	18
7.06	Broadband-DSL Internet subscribers, 2002–3	29
7.07	Broadband-cable modem, 2002–3	60
7.08	Internet users per 100 inhabitants, 2002	29
8.01	Prevalence of foreign technology licensing, 2004	64
8.02	Firm-level technology absorption, 2004	59
8.03	Capacity for innovation, 2004	73
8.04	Availability of new telephone lines, 2004	21
8.05	Availability of cellular phones, 2004	28
9.01	Government success in ICT promotion, 2004	67
9.02	Government online services, 2003	29

127

Part 2 Country Profiles

Czech Republic

Key Indicators

Population (mn), 2003	10.2
GDP per capita (PPP US$), 2003	16,448
Internet users per 100 inhabitants, 2002	25.6

Networked Readiness Index Rank

Year (number of countries)	Rank
2004 (104)	**40**
2003 (102)	33
2002 (82)	28

RANK/104

Environment Component Index	41
Market Environment	47
Political and Regulatory Environment	48
Infrastructure Environment	33

1.01	Availability of scientists and engineers, 2004	35
1.02	Venture capital availability, 2004	45
1.03	Sophistication of financial markets, 2004	63
1.04	Technological sophistication, 2004	41
1.05	State of cluster development, 2004	59
1.06	Collaboration in clusters, 2004	41
1.07	University-industry collaboration, 2004	29
1.08	Quality of scientific research institutions, 2004	20
1.09	Subsidies for firm-level R&D, 2004	29
1.10	Brain drain, 2004	38
1.11	Ease of access to loans, 2004	50
1.12	Administrative burden, 2004	69
1.13	Ease to start a new business, 2004	78
2.01	Effectiveness of lawmaking, 2004	65
2.02	Laws relating to ICT, 2004	38
2.03	Effectiveness of judiciary, 2004	45
2.04	Intellectual property protection, 2004	43
3.01	Telephone mainlines, 2002	33
3.02	Secure Internet servers, 2003	35
3.03	Internet hosts, 2003	23

Readiness Component Index	41
Individual Readiness	30
Business Readiness	35
Government Readiness	61

4.01	Quality of math and science education, 2004	10
4.02	Quality of educational system, 2004	37
4.03	Quality of public schools, 2004	18
4.04	Internet access in schools, 2004	26

Readiness Component Index (continued)	

4.05	Buyer sophistication, 2004	55
4.06	Buyer dynamism, 2004	62
4.07	Residential telephone connection charge, 2002	51
4.08	Affordability of Internet access, 2003	41
5.01	Investment in training, 2004	46
5.02	Availability of training services, 2004	32
5.03	Quality of business schools, 2004	44
5.04	Business investment in R&D, 2004	35
5.05	Business monthly telephone subscription, 2002–3	39
5.06	Business telephone connection charge, 2002–3	49
6.01	Government prioritization of ICT, 2004	59
6.02	Government procurement of ICT, 2004	55

Usage Component Index	37
Individual Usage	33
Business Usage	26
Government Usage	69

7.01	Cellular mobile subscribers, 2003	5
7.02	Telephone subscribers, 2002	25
7.03	Public payphones, 2002	33
7.04	Telephone lines, 2002	34
7.05	Television sets, 2002	2
7.06	Broadband-DSL Internet subscribers, 2002–3	65
7.07	Broadband-cable modem, 2002–3	36
7.08	Internet users per 100 inhabitants, 2002	32
8.01	Prevalence of foreign technology licensing, 2004	21
8.02	Firm-level technology absorption, 2004	42
8.03	Capacity for innovation, 2004	28
8.04	Availability of new telephone lines, 2004	40
8.05	Availability of cellular phones, 2004	10
9.01	Government success in ICT promotion, 2004	75
9.02	Government online services, 2003	50

Denmark

Key Indicators

Population (mn), 2003	5.4
GDP per capita (PPP US$), 2003	30,588
Internet users per 100 inhabitants, 2002	51.3

Networked Readiness Index Rank

Year (number of countries)	Rank
2004 (104)	**4**
2003 (102)	5
2002 (82)	8

RANK/104

Environment Component Index		3
Market Environment		11
Political and Regulatory Environment		1
Infrastructure Environment		2
1.01	Availability of scientists and engineers, 2004	9
1.02	Venture capital availability, 2004	12
1.03	Sophistication of financial markets, 2004	13
1.04	Technological sophistication, 2004	5
1.05	State of cluster development, 2004	9
1.06	Collaboration in clusters, 2004	5
1.07	University-industry collaboration, 2004	11
1.08	Quality of scientific research institutions, 2004	10
1.09	Subsidies for firm-level R&D, 2004	23
1.10	Brain drain, 2004	23
1.11	Ease of access to loans, 2004	5
1.12	Administrative burden, 2004	45
1.13	Ease to start a new business, 2004	26
2.01	Effectiveness of lawmaking, 2004	3
2.02	Laws relating to ICT, 2004	6
2.03	Effectiveness of judiciary, 2004	1
2.04	Intellectual property protection, 2004	2
3.01	Telephone mainlines, 2002	5
3.02	Secure Internet servers, 2003	10
3.03	Internet hosts, 2003	3

Readiness Component Index		6
Individual Readiness		5
Business Readiness		9
Government Readiness		10
4.01	Quality of math and science education, 2004	6
4.02	Quality of educational system, 2004	5
4.03	Quality of public schools, 2004	6
4.04	Internet access in schools, 2004	7

Readiness Component Index (continued)		
4.05	Buyer sophistication, 2004	20
4.06	Buyer dynamism, 2004	17
4.07	Residential telephone connection charge, 2002	36
4.08	Affordability of Internet access, 2003	4
5.01	Investment in training, 2004	5
5.02	Availability of training services, 2004	9
5.03	Quality of business schools, 2004	14
5.04	Business investment in R&D, 2004	8
5.05	Business monthly telephone subscription, 2002–3	50
5.06	Business telephone connection charge, 2002–3	29
6.01	Government prioritization of ICT, 2004	5
6.02	Government procurement of ICT, 2004	22

Usage Component Index		4
Individual Usage		4
Business Usage		7
Government Usage		4
7.01	Cellular mobile subscribers, 2003	12
7.02	Telephone subscribers, 2002	6
7.03	Public payphones, 2002	40
7.04	Telephone lines, 2002	6
7.05	Television sets, 2002	15
7.06	Broadband-DSL Internet subscribers, 2002–3	4
7.07	Broadband-cable modem, 2002–3	7
7.08	Internet users per 100 inhabitants, 2002	6
8.01	Prevalence of foreign technology licensing, 2004	19
8.02	Firm-level technology absorption, 2004	12
8.03	Capacity for innovation, 2004	9
8.04	Availability of new telephone lines, 2004	3
8.05	Availability of cellular phones, 2004	5
9.01	Government success in ICT promotion, 2004	10
9.02	Government online services, 2003	5

Dominican Republic

Key Indicators

Population (mn), 2003	8.7
GDP per capita (PPP US$), 2003	6,703
Internet users per 100 inhabitants, 2002	6.1

Networked Readiness Index Rank

Year (number of countries)	Rank
2004 (104)	**78**
2003 (102)	57
2002 (82)	57

RANK/104

Environment Component Index		78
Market Environment		78
Political and Regulatory Environment		72
Infrastructure Environment		64
1.01	Availability of scientists and engineers, 2004	90
1.02	Venture capital availability, 2004	93
1.03	Sophistication of financial markets, 2004	69
1.04	Technological sophistication, 2004	42
1.05	State of cluster development, 2004	81
1.06	Collaboration in clusters, 2004	84
1.07	University-industry collaboration, 2004	93
1.08	Quality of scientific research institutions, 2004	90
1.09	Subsidies for firm-level R&D, 2004	100
1.10	Brain drain, 2004	44
1.11	Ease of access to loans, 2004	65
1.12	Administrative burden, 2004	58
1.13	Ease to start a new business, 2004	67
2.01	Effectiveness of lawmaking, 2004	95
2.02	Laws relating to ICT, 2004	60
2.03	Effectiveness of judiciary, 2004	67
2.04	Intellectual property protection, 2004	60
3.01	Telephone mainlines, 2002	69
3.02	Secure Internet servers, 2003	63
3.03	Internet hosts, 2003	42

Readiness Component Index		85
Individual Readiness		79
Business Readiness		78
Government Readiness		91
4.01	Quality of math and science education, 2004	97
4.02	Quality of educational system, 2004	91
4.03	Quality of public schools, 2004	90
4.04	Internet access in schools, 2004	72

Readiness Component Index (continued)		
4.05	Buyer sophistication, 2004	80
4.06	Buyer dynamism, 2004	85
4.07	Residential telephone connection charge, 2002	68
4.08	Affordability of Internet access, 2003	65
5.01	Investment in training, 2004	74
5.02	Availability of training services, 2004	75
5.03	Quality of business schools, 2004	65
5.04	Business investment in R&D, 2004	93
5.05	Business monthly telephone subscription, 2002–3	89
5.06	Business telephone connection charge, 2002–3	58
6.01	Government prioritization of ICT, 2004	92
6.02	Government procurement of ICT, 2004	85

Usage Component Index		75
Individual Usage		68
Business Usage		51
Government Usage		98
7.01	Cellular mobile subscribers, 2003	55
7.02	Telephone subscribers, 2002	66
7.03	Public payphones, 2002	71
7.04	Telephone lines, 2002	68
7.05	Television sets, 2002	68
7.06	Broadband-DSL Internet subscribers, 2002–3	65
7.07	Broadband-cable modem, 2002–3	60
7.08	Internet users per 100 inhabitants, 2002	58
8.01	Prevalence of foreign technology licensing, 2004	78
8.02	Firm-level technology absorption, 2004	67
8.03	Capacity for innovation, 2004	74
8.04	Availability of new telephone lines, 2004	25
8.05	Availability of cellular phones, 2004	9
9.01	Government success in ICT promotion, 2004	91
9.02	Government online services, 2003	96

Ecuador

Key Indicators

Population (mn), 2003	13
GDP per capita (PPP US$), 2003	3,684
Internet users per 100 inhabitants, 2002	4.2

Year (number of countries)	Rank
2004 (104)	**95**
2003 (102)	89
2002 (82)	75

RANK/104

Environment Component Index		98
Market Environment		98
Political and Regulatory Environment		99
Infrastructure Environment		72
1.01	Availability of scientists and engineers, 2004	92
1.02	Venture capital availability, 2004	92
1.03	Sophistication of financial markets, 2004	79
1.04	Technological sophistication, 2004	75
1.05	State of cluster development, 2004	83
1.06	Collaboration in clusters, 2004	99
1.07	University-industry collaboration, 2004	90
1.08	Quality of scientific research institutions, 2004	96
1.09	Subsidies for firm-level R&D, 2004	99
1.10	Brain drain, 2004	69
1.11	Ease of access to loans, 2004	95
1.12	Administrative burden, 2004	86
1.13	Ease to start a new business, 2004	98
2.01	Effectiveness of lawmaking, 2004	101
2.02	Laws relating to ICT, 2004	86
2.03	Effectiveness of judiciary, 2004	102
2.04	Intellectual property protection, 2004	95
3.01	Telephone mainlines, 2002	71
3.02	Secure Internet servers, 2003	67
3.03	Internet hosts, 2003	82

Readiness Component Index		93
Individual Readiness		82
Business Readiness		84
Government Readiness		101
4.01	Quality of math and science education, 2004	93
4.02	Quality of educational system, 2004	98
4.03	Quality of public schools, 2004	98
4.04	Internet access in schools, 2004	87

Readiness Component Index (continued)		
4.05	Buyer sophistication, 2004	95
4.06	Buyer dynamism, 2004	94
4.07	Residential telephone connection charge, 2002	66
4.08	Affordability of Internet access, 2003	78
5.01	Investment in training, 2004	94
5.02	Availability of training services, 2004	86
5.03	Quality of business schools, 2004	76
5.04	Business investment in R&D, 2004	91
5.05	Business monthly telephone subscription, 2002–3	79
5.06	Business telephone connection charge, 2002–3	82
6.01	Government prioritization of ICT, 2004	101
6.02	Government procurement of ICT, 2004	101

Usage Component Index		94
Individual Usage		74
Business Usage		95
Government Usage		95
7.01	Cellular mobile subscribers, 2003	66
7.02	Telephone subscribers, 2002	74
7.03	Public payphones, 2002	88
7.04	Telephone lines, 2002	65
7.05	Television sets, 2002	60
7.06	Broadband-DSL Internet subscribers, 2002–3	65
7.07	Broadband-cable modem, 2002–3	55
7.08	Internet users per 100 inhabitants, 2002	69
8.01	Prevalence of foreign technology licensing, 2004	92
8.02	Firm-level technology absorption, 2004	91
8.03	Capacity for innovation, 2004	94
8.04	Availability of new telephone lines, 2004	99
8.05	Availability of cellular phones, 2004	82
9.01	Government success in ICT promotion, 2004	102
9.02	Government online services, 2003	52

Egypt

Key Indicators

Population (mn), 2003	71.9
GDP per capita (PPP US$), 2003	3,950
Internet users per 100 inhabitants, 2002	2.8

Networked Readiness Index Rank

Year (number of countries)	Rank
2004 (104)	**57**
2003 (102)	65
2002 (82)	65

RANK/104

Environment Component Index		67
Market Environment		65
Political and Regulatory Environment		64
Infrastructure Environment		75
1.01	Availability of scientists and engineers, 2004	37
1.02	Venture capital availability, 2004	68
1.03	Sophistication of financial markets, 2004	83
1.04	Technological sophistication, 2004	52
1.05	State of cluster development, 2004	12
1.06	Collaboration in clusters, 2004	29
1.07	University-industry collaboration, 2004	58
1.08	Quality of scientific research institutions, 2004	56
1.09	Subsidies for firm-level R&D, 2004	35
1.10	Brain drain, 2004	89
1.11	Ease of access to loans, 2004	87
1.12	Administrative burden, 2004	85
1.13	Ease to start a new business, 2004	69
2.01	Effectiveness of lawmaking, 2004	47
2.02	Laws relating to ICT, 2004	63
2.03	Effectiveness of judiciary, 2004	98
2.04	Intellectual property protection, 2004	38
3.01	Telephone mainlines, 2002	70
3.02	Secure Internet servers, 2003	86
3.03	Internet hosts, 2003	92

Readiness Component Index		63
Individual Readiness		78
Business Readiness		81
Government Readiness		45
4.01	Quality of math and science education, 2004	73
4.02	Quality of educational system, 2004	79
4.03	Quality of public schools, 2004	73
4.04	Internet access in schools, 2004	63

Readiness Component Index (continued)		
4.05	Buyer sophistication, 2004	88
4.06	Buyer dynamism, 2004	41
4.07	Residential telephone connection charge, 2002	90
4.08	Affordability of Internet access, 2003	41
5.01	Investment in training, 2004	49
5.02	Availability of training services, 2004	49
5.03	Quality of business schools, 2004	70
5.04	Business investment in R&D, 2004	72
5.05	Business monthly telephone subscription, 2002–3	41
5.06	Business telephone connection charge, 2002–3	98
6.01	Government prioritization of ICT, 2004	13
6.02	Government procurement of ICT, 2004	75

Usage Component Index		51
Individual Usage		75
Business Usage		56
Government Usage		26
7.01	Cellular mobile subscribers, 2003	79
7.02	Telephone subscribers, 2002	76
7.03	Public payphones, 2002	85
7.04	Telephone lines, 2002	64
7.05	Television sets, 2002	29
7.06	Broadband-DSL Internet subscribers, 2002–3	63
7.07	Broadband-cable modem, 2002–3	60
7.08	Internet users per 100 inhabitants, 2002	75
8.01	Prevalence of foreign technology licensing, 2004	58
8.02	Firm-level technology absorption, 2004	37
8.03	Capacity for innovation, 2004	53
8.04	Availability of new telephone lines, 2004	56
8.05	Availability of cellular phones, 2004	87
9.01	Government success in ICT promotion, 2004	16
9.02	Government online services, 2003	47

El Salvador

Key Indicators

Population (mn), 2003	6.5
GDP per capita (PPP US$), 2003	4,994
Internet users per 100 inhabitants, 2002	4.6

RANK/104

Environment Component Index		77
Market Environment		69
Political and Regulatory Environment		77
Infrastructure Environment		73
1.01	Availability of scientists and engineers, 2004	98
1.02	Venture capital availability, 2004	50
1.03	Sophistication of financial markets, 2004	31
1.04	Technological sophistication, 2004	49
1.05	State of cluster development, 2004	83
1.06	Collaboration in clusters, 2004	90
1.07	University-industry collaboration, 2004	102
1.08	Quality of scientific research institutions, 2004	101
1.09	Subsidies for firm-level R&D, 2004	102
1.10	Brain drain, 2004	60
1.11	Ease of access to loans, 2004	40
1.12	Administrative burden, 2004	31
1.13	Ease to start a new business, 2004	73
2.01	Effectiveness of lawmaking, 2004	93
2.02	Laws relating to ICT, 2004	80
2.03	Effectiveness of judiciary, 2004	71
2.04	Intellectual property protection, 2004	52
3.01	Telephone mainlines, 2002	74
3.02	Secure Internet servers, 2003	58
3.03	Internet hosts, 2003	73

Readiness Component Index		74
Individual Readiness		75
Business Readiness		70
Government Readiness		76
4.01	Quality of math and science education, 2004	90
4.02	Quality of educational system, 2004	87
4.03	Quality of public schools, 2004	79
4.04	Internet access in schools, 2004	70

Readiness Component Index (continued)		
4.05	Buyer sophistication, 2004	63
4.06	Buyer dynamism, 2004	66
4.07	Residential telephone connection charge, 2002	77
4.08	Affordability of Internet access, 2003	79
5.01	Investment in training, 2004	60
5.02	Availability of training services, 2004	73
5.03	Quality of business schools, 2004	59
5.04	Business investment in R&D, 2004	92
5.05	Business monthly telephone subscription, 2002–3	71
5.06	Business telephone connection charge, 2002–3	70
6.01	Government prioritization of ICT, 2004	81
6.02	Government procurement of ICT, 2004	72

Usage Component Index		60
Individual Usage		64
Business Usage		46
Government Usage		84
7.01	Cellular mobile subscribers, 2003	69
7.02	Telephone subscribers, 2002	71
7.03	Public payphones, 2002	42
7.04	Telephone lines, 2002	67
7.05	Television sets, 2002	57
7.06	Broadband-DSL Internet subscribers, 2002–3	65
7.07	Broadband-cable modem, 2002–3	20
7.08	Internet users per 100 inhabitants, 2002	64
8.01	Prevalence of foreign technology licensing, 2004	56
8.02	Firm-level technology absorption, 2004	63
8.03	Capacity for innovation, 2004	72
8.04	Availability of new telephone lines, 2004	17
8.05	Availability of cellular phones, 2004	27
9.01	Government success in ICT promotion, 2004	82
9.02	Government online services, 2003	79

Estonia

Key Indicators

Population (mn), 2003	1.3
GDP per capita (PPP US$), 2003	13,348
Internet users per 100 inhabitants, 2002	32.8

Networked Readiness Index Rank

Year (number of countries)	Rank
2004 (104)	**25**
2003 (102)	25
2002 (82)	24

RANK/104

Environment Component Index — 21

Market Environment		24
Political and Regulatory Environment		16
Infrastructure Environment		27
1.01	Availability of scientists and engineers, 2004	54
1.02	Venture capital availability, 2004	28
1.03	Sophistication of financial markets, 2004	24
1.04	Technological sophistication, 2004	30
1.05	State of cluster development, 2004	78
1.06	Collaboration in clusters, 2004	59
1.07	University-industry collaboration, 2004	37
1.08	Quality of scientific research institutions, 2004	36
1.09	Subsidies for firm-level R&D, 2004	51
1.10	Brain drain, 2004	27
1.11	Ease of access to loans, 2004	17
1.12	Administrative burden, 2004	9
1.13	Ease to start a new business, 2004	8
2.01	Effectiveness of lawmaking, 2004	18
2.02	Laws relating to ICT, 2004	3
2.03	Effectiveness of judiciary, 2004	19
2.04	Intellectual property protection, 2004	29
3.01	Telephone mainlines, 2002	35
3.02	Secure Internet servers, 2003	22
3.03	Internet hosts, 2003	18

Readiness Component Index — 27

Individual Readiness		23
Business Readiness		29
Government Readiness		23
4.01	Quality of math and science education, 2004	20
4.02	Quality of educational system, 2004	28
4.03	Quality of public schools, 2004	16
4.04	Internet access in schools, 2004	10

Readiness Component Index (continued)

4.05	Buyer sophistication, 2004	30
4.06	Buyer dynamism, 2004	28
4.07	Residential telephone connection charge, 2002	39
4.08	Affordability of Internet access, 2003	34
5.01	Investment in training, 2004	28
5.02	Availability of training services, 2004	30
5.03	Quality of business schools, 2004	27
5.04	Business investment in R&D, 2004	39
5.05	Business monthly telephone subscription, 2002–3	31
5.06	Business telephone connection charge, 2002–3	38
6.01	Government prioritization of ICT, 2004	14
6.02	Government procurement of ICT, 2004	32

Usage Component Index — 22

Individual Usage		34
Business Usage		27
Government Usage		9
7.01	Cellular mobile subscribers, 2003	33
7.02	Telephone subscribers, 2002	34
7.03	Public payphones, 2002	57
7.04	Telephone lines, 2002	35
7.05	Television sets, 2002	48
7.06	Broadband-DSL Internet subscribers, 2002–3	16
7.07	Broadband-cable modem, 2002–3	24
7.08	Internet users per 100 inhabitants, 2002	24
8.01	Prevalence of foreign technology licensing, 2004	49
8.02	Firm-level technology absorption, 2004	22
8.03	Capacity for innovation, 2004	39
8.04	Availability of new telephone lines, 2004	32
8.05	Availability of cellular phones, 2004	13
9.01	Government success in ICT promotion, 2004	6
9.02	Government online services, 2003	16

Ethiopia

Key Indicators

Population (mn), 2003	70.7
GDP per capita (PPP US$), 2003	716
Internet users per 100 inhabitants, 2002	0.1

RANK/104

Environment Component Index	100
Market Environment	100
Political and Regulatory Environment	97
Infrastructure Environment	99
1.01 Availability of scientists and engineers, 2004	93
1.02 Venture capital availability, 2004	103
1.03 Sophistication of financial markets, 2004	102
1.04 Technological sophistication, 2004	103
1.05 State of cluster development, 2004	92
1.06 Collaboration in clusters, 2004	98
1.07 University-industry collaboration, 2004	101
1.08 Quality of scientific research institutions, 2004	76
1.09 Subsidies for firm-level R&D, 2004	87
1.10 Brain drain, 2004	96
1.11 Ease of access to loans, 2004	102
1.12 Administrative burden, 2004	54
1.13 Ease to start a new business, 2004	64
2.01 Effectiveness of lawmaking, 2004	87
2.02 Laws relating to ICT, 2004	99
2.03 Effectiveness of judiciary, 2004	86
2.04 Intellectual property protection, 2004	99
3.01 Telephone mainlines, 2002	98
3.02 Secure Internet servers, 2003	97
3.03 Internet hosts, 2003	103

Readiness Component Index	102
Individual Readiness	104
Business Readiness	103
Government Readiness	75
4.01 Quality of math and science education, 2004	89
4.02 Quality of educational system, 2004	80
4.03 Quality of public schools, 2004	89
4.04 Internet access in schools, 2004	102

Readiness Component Index (continued)	
4.05 Buyer sophistication, 2004	96
4.06 Buyer dynamism, 2004	98
4.07 Residential telephone connection charge, 2002	102
4.08 Affordability of Internet access, 2003	98
5.01 Investment in training, 2004	102
5.02 Availability of training services, 2004	96
5.03 Quality of business schools, 2004	97
5.04 Business investment in R&D, 2004	104
5.05 Business monthly telephone subscription, 2002–3	99
5.06 Business telephone connection charge, 2002–3	102
6.01 Government prioritization of ICT, 2004	83
6.02 Government procurement of ICT, 2004	70

Usage Component Index	103
Individual Usage	104
Business Usage	103
Government Usage	92
7.01 Cellular mobile subscribers, 2003	103
7.02 Telephone subscribers, 2002	103
7.03 Public payphones, 2002	101
7.04 Telephone lines, 2002	97
7.05 Television sets, 2002	102
7.06 Broadband-DSL Internet subscribers, 2002–3	65
7.07 Broadband-cable modem, 2002–3	60
7.08 Internet users per 100 inhabitants, 2002	104
8.01 Prevalence of foreign technology licensing, 2004	101
8.02 Firm-level technology absorption, 2004	96
8.03 Capacity for innovation, 2004	97
8.04 Availability of new telephone lines, 2004	96
8.05 Availability of cellular phones, 2004	104
9.01 Government success in ICT promotion, 2004	85
9.02 Government online services, 2003	90

Finland

Key Indicators

Population (mn), 2003	5.2
GDP per capita (PPP US$), 2003	27,252
Internet users per 100 inhabitants, 2002	50.9

Networked Readiness Index Rank

Year (number of countries)	Rank
2004 (104)	**3**
2003 (102)	3
2002 (82)	1

RANK/104

Environment Component Index	2
Market Environment	1
Political and Regulatory Environment	4
Infrastructure Environment	3
1.01 Availability of scientists and engineers, 2004	2
1.02 Venture capital availability, 2004	5
1.03 Sophistication of financial markets, 2004	9
1.04 Technological sophistication, 2004	3
1.05 State of cluster development, 2004	3
1.06 Collaboration in clusters, 2004	2
1.07 University-industry collaboration, 2004	1
1.08 Quality of scientific research institutions, 2004	4
1.09 Subsidies for firm-level R&D, 2004	5
1.10 Brain drain, 2004	11
1.11 Ease of access to loans, 2004	3
1.12 Administrative burden, 2004	5
1.13 Ease to start a new business, 2004	5
2.01 Effectiveness of lawmaking, 2004	14
2.02 Laws relating to ICT, 2004	4
2.03 Effectiveness of judiciary, 2004	2
2.04 Intellectual property protection, 2004	5
3.01 Telephone mainlines, 2002	17
3.02 Secure Internet servers, 2003	11
3.03 Internet hosts, 2003	2

Readiness Component Index	2
Individual Readiness	1
Business Readiness	4
Government Readiness	5
4.01 Quality of math and science education, 2004	2
4.02 Quality of educational system, 2004	1
4.03 Quality of public schools, 2004	1
4.04 Internet access in schools, 2004	2

Readiness Component Index (continued)	
4.05 Buyer sophistication, 2004	12
4.06 Buyer dynamism, 2004	6
4.07 Residential telephone connection charge, 2002	20
4.08 Affordability of Internet access, 2003	20
5.01 Investment in training, 2004	7
5.02 Availability of training services, 2004	4
5.03 Quality of business schools, 2004	13
5.04 Business investment in R&D, 2004	6
5.05 Business monthly telephone subscription, 2002–3	6
5.06 Business telephone connection charge, 2002–3	21
6.01 Government prioritization of ICT, 2004	8
6.02 Government procurement of ICT, 2004	7

Usage Component Index	9
Individual Usage	14
Business Usage	5
Government Usage	12
7.01 Cellular mobile subscribers, 2003	10
7.02 Telephone subscribers, 2002	11
7.03 Public payphones, 2002	75
7.04 Telephone lines, 2002	21
7.05 Television sets, 2002	51
7.06 Broadband-DSL Internet subscribers, 2002–3	9
7.07 Broadband-cable modem, 2002–3	17
7.08 Internet users per 100 inhabitants, 2002	7
8.01 Prevalence of foreign technology licensing, 2004	23
8.02 Firm-level technology absorption, 2004	7
8.03 Capacity for innovation, 2004	4
8.04 Availability of new telephone lines, 2004	5
8.05 Availability of cellular phones, 2004	7
9.01 Government success in ICT promotion, 2004	7
9.02 Government online services, 2003	20

France

Key Indicators

Population (mn), 2003	60.1
GDP per capita (PPP US$), 2003	27,327
Internet users per 100 inhabitants, 2002	31.4

Networked Readiness Index Rank

Year (number of countries)	Rank
2004 (104)	**20**
2003 (102)	19
2002 (82)	19

RANK/104

Environment Component Index		22
Market Environment		28
Political and Regulatory Environment		17
Infrastructure Environment		23
1.01	Availability of scientists and engineers, 2004	5
1.02	Venture capital availability, 2004	17
1.03	Sophistication of financial markets, 2004	17
1.04	Technological sophistication, 2004	19
1.05	State of cluster development, 2004	29
1.06	Collaboration in clusters, 2004	14
1.07	University-industry collaboration, 2004	25
1.08	Quality of scientific research institutions, 2004	15
1.09	Subsidies for firm-level R&D, 2004	9
1.10	Brain drain, 2004	46
1.11	Ease of access to loans, 2004	20
1.12	Administrative burden, 2004	100
1.13	Ease to start a new business, 2004	63
2.01	Effectiveness of lawmaking, 2004	13
2.02	Laws relating to ICT, 2004	15
2.03	Effectiveness of judiciary, 2004	34
2.04	Intellectual property protection, 2004	14
3.01	Telephone mainlines, 2002	13
3.02	Secure Internet servers, 2003	26
3.03	Internet hosts, 2003	19

Readiness Component Index		10
Individual Readiness		9
Business Readiness		8
Government Readiness		16
4.01	Quality of math and science education, 2004	3
4.02	Quality of educational system, 2004	14
4.03	Quality of public schools, 2004	9
4.04	Internet access in schools, 2004	31

Readiness Component Index (continued)		
4.05	Buyer sophistication, 2004	7
4.06	Buyer dynamism, 2004	16
4.07	Residential telephone connection charge, 2002	8
4.08	Affordability of Internet access, 2003	9
5.01	Investment in training, 2004	19
5.02	Availability of training services, 2004	15
5.03	Quality of business schools, 2004	2
5.04	Business investment in R&D, 2004	15
5.05	Business monthly telephone subscription, 2002–3	13
5.06	Business telephone connection charge, 2002–3	6
6.01	Government prioritization of ICT, 2004	45
6.02	Government procurement of ICT, 2004	10

Usage Component Index		17
Individual Usage		22
Business Usage		17
Government Usage		15
7.01	Cellular mobile subscribers, 2003	28
7.02	Telephone subscribers, 2002	24
7.03	Public payphones, 2002	32
7.04	Telephone lines, 2002	14
7.05	Television sets, 2002	27
7.06	Broadband-DSL Internet subscribers, 2002–3	20
7.07	Broadband-cable modem, 2002–3	26
7.08	Internet users per 100 inhabitants, 2002	26
8.01	Prevalence of foreign technology licensing, 2004	61
8.02	Firm-level technology absorption, 2004	35
8.03	Capacity for innovation, 2004	5
8.04	Availability of new telephone lines, 2004	8
8.05	Availability of cellular phones, 2004	11
9.01	Government success in ICT promotion, 2004	26
9.02	Government online services, 2003	14

Gambia

Key Indicators

Population (mn), 2003	1.4
GDP per capita (PPP US$), 2003	1,714
Internet users per 100 inhabitants, 2002	1.9

Networked Readiness Index Rank

Year (number of countries)	Rank
2004 (104)	**74**
2003 (102)	82
2002 (82)	–

RANK/104

Environment Component Index	63
Market Environment	59
Political and Regulatory Environment	59
Infrastructure Environment	90

1.01	Availability of scientists and engineers, 2004	96
1.02	Venture capital availability, 2004	62
1.03	Sophistication of financial markets, 2004	89
1.04	Technological sophistication, 2004	82
1.05	State of cluster development, 2004	57
1.06	Collaboration in clusters, 2004	79
1.07	University-industry collaboration, 2004	79
1.08	Quality of scientific research institutions, 2004	74
1.09	Subsidies for firm-level R&D, 2004	61
1.10	Brain drain, 2004	92
1.11	Ease of access to loans, 2004	61
1.12	Administrative burden, 2004	7
1.13	Ease to start a new business, 2004	31
2.01	Effectiveness of lawmaking, 2004	54
2.02	Laws relating to ICT, 2004	76
2.03	Effectiveness of judiciary, 2004	52
2.04	Intellectual property protection, 2004	67
3.01	Telephone mainlines, 2002	90
3.02	Secure Internet servers, 2003	100
3.03	Internet hosts, 2003	76

Readiness Component Index	70
Individual Readiness	91
Business Readiness	91
Government Readiness	39

4.01	Quality of math and science education, 2004	83
4.02	Quality of educational system, 2004	43
4.03	Quality of public schools, 2004	58
4.04	Internet access in schools, 2004	79

Readiness Component Index (continued)	
4.05	Buyer sophistication, 2004	83
4.06	Buyer dynamism, 2004	84
4.07	Residential telephone connection charge, 2002	96
4.08	Affordability of Internet access, 2003	89
5.01	Investment in training, 2004	71
5.02	Availability of training services, 2004	93
5.03	Quality of business schools, 2004	74
5.04	Business investment in R&D, 2004	73
5.05	Business monthly telephone subscription, 2002–3	84
5.06	Business telephone connection charge, 2002–3	95
6.01	Government prioritization of ICT, 2004	18
6.02	Government procurement of ICT, 2004	54

Usage Component Index	83
Individual Usage	90
Business Usage	82
Government Usage	62

7.01	Cellular mobile subscribers, 2003	81
7.02	Telephone subscribers, 2002	81
7.03	Public payphones, 2002	84
7.04	Telephone lines, 2002	88
7.05	Television sets, 2002	97
7.06	Broadband-DSL Internet subscribers, 2002–3	65
7.07	Broadband-cable modem, 2002–3	60
7.08	Internet users per 100 inhabitants, 2002	81
8.01	Prevalence of foreign technology licensing, 2004	83
8.02	Firm-level technology absorption, 2004	86
8.03	Capacity for innovation, 2004	58
8.04	Availability of new telephone lines, 2004	81
8.05	Availability of cellular phones, 2004	84
9.01	Government success in ICT promotion, 2004	25
9.02	Government online services, 2003	99

Georgia

Key Indicators

Population (mn), 2003	5.1
GDP per capita (PPP US$), 2003	2,569
Internet users per 100 inhabitants, 2002	1.5

RANK/104

Environment Component Index	91
Market Environment	89
Political and Regulatory Environment	95
Infrastructure Environment	65

1.01	Availability of scientists and engineers, 2004	31
1.02	Venture capital availability, 2004	86
1.03	Sophistication of financial markets, 2004	96
1.04	Technological sophistication, 2004	98
1.05	State of cluster development, 2004	93
1.06	Collaboration in clusters, 2004	83
1.07	University-industry collaboration, 2004	96
1.08	Quality of scientific research institutions, 2004	65
1.09	Subsidies for firm-level R&D, 2004	97
1.10	Brain drain, 2004	84
1.11	Ease of access to loans, 2004	76
1.12	Administrative burden, 2004	71
1.13	Ease to start a new business, 2004	83
2.01	Effectiveness of lawmaking, 2004	79
2.02	Laws relating to ICT, 2004	100
2.03	Effectiveness of judiciary, 2004	87
2.04	Intellectual property protection, 2004	94
3.01	Telephone mainlines, 2002	62
3.02	Secure Internet servers, 2003	77
3.03	Internet hosts, 2003	71

Readiness Component Index	89
Individual Readiness	87
Business Readiness	86
Government Readiness	94

4.01	Quality of math and science education, 2004	54
4.02	Quality of educational system, 2004	76
4.03	Quality of public schools, 2004	80
4.04	Internet access in schools, 2004	91

Readiness Component Index (continued)	
4.05	Buyer sophistication, 2004	92
4.06	Buyer dynamism, 2004	86
4.07	Residential telephone connection charge, 2002	92
4.08	Affordability of Internet access, 2003	84
5.01	Investment in training, 2004	89
5.02	Availability of training services, 2004	89
5.03	Quality of business schools, 2004	98
5.04	Business investment in R&D, 2004	83
5.05	Business monthly telephone subscription, 2002–3	64
5.06	Business telephone connection charge, 2002–3	90
6.01	Government prioritization of ICT, 2004	94
6.02	Government procurement of ICT, 2004	97

Usage Component Index	91
Individual Usage	77
Business Usage	77
Government Usage	96

7.01	Cellular mobile subscribers, 2003	76
7.02	Telephone subscribers, 2002	72
7.03	Public payphones, 2002	80
7.04	Telephone lines, 2002	62
7.05	Television sets, 2002	76
7.06	Broadband-DSL Internet subscribers, 2002–3	60
7.07	Broadband-cable modem, 2002–3	54
7.08	Internet users per 100 inhabitants, 2002	88
8.01	Prevalence of foreign technology licensing, 2004	95
8.02	Firm-level technology absorption, 2004	92
8.03	Capacity for innovation, 2004	75
8.04	Availability of new telephone lines, 2004	66
8.05	Availability of cellular phones, 2004	44
9.01	Government success in ICT promotion, 2004	96
9.02	Government online services, 2003	77

Germany

Key Indicators

Population (mn), 2003	82.5
GDP per capita (PPP US$), 2003	27,609
Internet users per 100 inhabitants, 2002	43.6

Networked Readiness Index Rank

Year (number of countries)	Rank
2004 (104)	**14**
2003 (102)	11
2002 (82)	10

RANK/104

Environment Component Index		18
Market Environment		21
Political and Regulatory Environment		11
Infrastructure Environment		19
1.01	Availability of scientists and engineers, 2004	16
1.02	Venture capital availability, 2004	29
1.03	Sophistication of financial markets, 2004	14
1.04	Technological sophistication, 2004	16
1.05	State of cluster development, 2004	17
1.06	Collaboration in clusters, 2004	7
1.07	University-industry collaboration, 2004	4
1.08	Quality of scientific research institutions, 2004	8
1.09	Subsidies for firm-level R&D, 2004	14
1.10	Brain drain, 2004	31
1.11	Ease of access to loans, 2004	45
1.12	Administrative burden, 2004	17
1.13	Ease to start a new business, 2004	45
2.01	Effectiveness of lawmaking, 2004	44
2.02	Laws relating to ICT, 2004	7
2.03	Effectiveness of judiciary, 2004	3
2.04	Intellectual property protection, 2004	4
3.01	Telephone mainlines, 2002	8
3.02	Secure Internet servers, 2003	18
3.03	Internet hosts, 2003	22

Readiness Component Index		8
Individual Readiness		24
Business Readiness		3
Government Readiness		13
4.01	Quality of math and science education, 2004	49
4.02	Quality of educational system, 2004	46
4.03	Quality of public schools, 2004	28
4.04	Internet access in schools, 2004	25

Readiness Component Index (continued)		
4.05	Buyer sophistication, 2004	17
4.06	Buyer dynamism, 2004	13
4.07	Residential telephone connection charge, 2002	5
4.08	Affordability of Internet access, 2003	4
5.01	Investment in training, 2004	2
5.02	Availability of training services, 2004	2
5.03	Quality of business schools, 2004	18
5.04	Business investment in R&D, 2004	3
5.05	Business monthly telephone subscription, 2002–3	9
5.06	Business telephone connection charge, 2002–3	5
6.01	Government prioritization of ICT, 2004	31
6.02	Government procurement of ICT, 2004	9

Usage Component Index		10
Individual Usage		15
Business Usage		2
Government Usage		18
7.01	Cellular mobile subscribers, 2003	20
7.02	Telephone subscribers, 2002	12
7.03	Public payphones, 2002	72
7.04	Telephone lines, 2002	8
7.05	Television sets, 2002	32
7.06	Broadband-DSL Internet subscribers, 2002-3	10
7.07	Broadband-cable modem, 2002-3	43
7.08	Internet users per 100 inhabitants, 2002	14
8.01	Prevalence of foreign technology licensing, 2004	10
8.02	Firm-level technology absorption, 2004	11
8.03	Capacity for innovation, 2004	1
8.04	Availability of new telephone lines, 2004	13
8.05	Availability of cellular phones, 2004	8
9.01	Government success in ICT promotion, 2004	50
9.02	Government online services, 2003	6

Ghana

Key Indicators

Population (mn), 2003	20.9
GDP per capita (PPP US$), 2003	2,234
Internet users per 100 inhabitants, 2002	0.8

Networked Readiness Index Rank
Year (number of countries) Rank

2004 (104) **65**

2003 (102) 74

2002 (82) —

RANK/104

Environment Component Index — 56

	Market Environment	70
	Political and Regulatory Environment	36
	Infrastructure Environment	92
1.01	Availability of scientists and engineers, 2004	62
1.02	Venture capital availability, 2004	79
1.03	Sophistication of financial markets, 2004	66
1.04	Technological sophistication, 2004	72
1.05	State of cluster development, 2004	89
1.06	Collaboration in clusters, 2004	76
1.07	University-industry collaboration, 2004	85
1.08	Quality of scientific research institutions, 2004	34
1.09	Subsidies for firm-level R&D, 2004	69
1.10	Brain drain, 2004	83
1.11	Ease of access to loans, 2004	69
1.12	Administrative burden, 2004	34
1.13	Ease to start a new business, 2004	74
2.01	Effectiveness of lawmaking, 2004	10
2.02	Laws relating to ICT, 2004	65
2.03	Effectiveness of judiciary, 2004	40
2.04	Intellectual property protection, 2004	44
3.01	Telephone mainlines, 2002	92
3.02	Secure Internet servers, 2003	85
3.03	Internet hosts, 2003	96

Readiness Component Index — 62

	Individual Readiness	92
	Business Readiness	76
	Government Readiness	26
4.01	Quality of math and science education, 2004	65
4.02	Quality of educational system, 2004	54
4.03	Quality of public schools, 2004	77
4.04	Internet access in schools, 2004	89

Readiness Component Index (continued)

4.05	Buyer sophistication, 2004	56
4.06	Buyer dynamism, 2004	50
4.07	Residential telephone connection charge, 2002	93
4.08	Affordability of Internet access, 2003	95
5.01	Investment in training, 2004	72
5.02	Availability of training services, 2004	65
5.03	Quality of business schools, 2004	52
5.04	Business investment in R&D, 2004	67
5.05	Business monthly telephone subscription, 2002–3	63
5.06	Business telephone connection charge, 2002–3	92
6.01	Government prioritization of ICT, 2004	10
6.02	Government procurement of ICT, 2004	52

Usage Component Index — 79

	Individual Usage	92
	Business Usage	81
	Government Usage	55
7.01	Cellular mobile subscribers, 2003	87
7.02	Telephone subscribers, 2002	92
7.03	Public payphones, 2002	92
7.04	Telephone lines, 2002	91
7.05	Television sets, 2002	92
7.06	Broadband-DSL Internet subscribers, 2002–3	65
7.07	Broadband-cable modem, 2002–3	60
7.08	Internet users per 100 inhabitants, 2002	92
8.01	Prevalence of foreign technology licensing, 2004	62
8.02	Firm-level technology absorption, 2004	60
8.03	Capacity for innovation, 2004	84
8.04	Availability of new telephone lines, 2004	100
8.05	Availability of cellular phones, 2004	85
9.01	Government success in ICT promotion, 2004	18
9.02	Government online services, 2003	99

Greece

Key Indicators

Population (mn), 2003	11
GDP per capita (PPP US$), 2003	19,973
Internet users per 100 inhabitants, 2002	13.5

Networked Readiness Index Rank

Year (number of countries)	Rank
2004 (104)	**42**
2003 (102)	34
2002 (82)	42

RANK/104

Environment Component Index		39
Market Environment		48
Political and Regulatory Environment		43
Infrastructure Environment		28
1.01	Availability of scientists and engineers, 2004	15
1.02	Venture capital availability, 2004	44
1.03	Sophistication of financial markets, 2004	38
1.04	Technological sophistication, 2004	70
1.05	State of cluster development, 2004	73
1.06	Collaboration in clusters, 2004	64
1.07	University-industry collaboration, 2004	42
1.08	Quality of scientific research institutions, 2004	52
1.09	Subsidies for firm-level R&D, 2004	31
1.10	Brain drain, 2004	37
1.11	Ease of access to loans, 2004	29
1.12	Administrative burden, 2004	67
1.13	Ease to start a new business, 2004	84
2.01	Effectiveness of lawmaking, 2004	39
2.02	Laws relating to ICT, 2004	64
2.03	Effectiveness of judiciary, 2004	42
2.04	Intellectual property protection, 2004	35
3.01	Telephone mainlines, 2002	23
3.02	Secure Internet servers, 2003	37
3.03	Internet hosts, 2003	36

Readiness Component Index		49
Individual Readiness		38
Business Readiness		40
Government Readiness		66
4.01	Quality of math and science education, 2004	30
4.02	Quality of educational system, 2004	50
4.03	Quality of public schools, 2004	46
4.04	Internet access in schools, 2004	43

Readiness Component Index (continued)		
4.05	Buyer sophistication, 2004	43
4.06	Buyer dynamism, 2004	44
4.07	Residential telephone connection charge, 2002	9
4.08	Affordability of Internet access, 2003	34
5.01	Investment in training, 2004	45
5.02	Availability of training services, 2004	61
5.03	Quality of business schools, 2004	56
5.04	Business investment in R&D, 2004	56
5.05	Business monthly telephone subscription, 2002–3	21
5.06	Business telephone connection charge, 2002–3	7
6.01	Government prioritization of ICT, 2004	60
6.02	Government procurement of ICT, 2004	68

Usage Component Index		39
Individual Usage		32
Business Usage		45
Government Usage		52
7.01	Cellular mobile subscribers, 2003	21
7.02	Telephone subscribers, 2002	14
7.03	Public payphones, 2002	5
7.04	Telephone lines, 2002	25
7.05	Television sets, 2002	35
7.06	Broadband-DSL Internet subscribers, 2002–3	65
7.07	Broadband-cable modem, 2002–3	60
7.08	Internet users per 100 inhabitants, 2002	44
8.01	Prevalence of foreign technology licensing, 2004	41
8.02	Firm-level technology absorption, 2004	65
8.03	Capacity for innovation, 2004	55
8.04	Availability of new telephone lines, 2004	33
8.05	Availability of cellular phones, 2004	26
9.01	Government success in ICT promotion, 2004	54
9.02	Government online services, 2003	52

Guatemala

Key Indicators

Population (mn), 2003	12.3
GDP per capita (PPP US$), 2003	4,122
Internet users per 100 inhabitants, 2002	3.3

RANK/104

Environment Component Index	90
Market Environment	83
Political and Regulatory Environment	91
Infrastructure Environment	76

1.01	Availability of scientists and engineers, 2004	102
1.02	Venture capital availability, 2004	82
1.03	Sophistication of financial markets, 2004	65
1.04	Technological sophistication, 2004	65
1.05	State of cluster development, 2004	77
1.06	Collaboration in clusters, 2004	93
1.07	University-industry collaboration, 2004	84
1.08	Quality of scientific research institutions, 2004	91
1.09	Subsidies for firm-level R&D, 2004	96
1.10	Brain drain, 2004	54
1.11	Ease of access to loans, 2004	72
1.12	Administrative burden, 2004	78
1.13	Ease to start a new business, 2004	66
2.01	Effectiveness of lawmaking, 2004	98
2.02	Laws relating to ICT, 2004	96
2.03	Effectiveness of judiciary, 2004	79
2.04	Intellectual property protection, 2004	78
3.01	Telephone mainlines, 2002	76
3.02	Secure Internet servers, 2003	60
3.03	Internet hosts, 2003	64

Readiness Component Index	88
Individual Readiness	86
Business Readiness	71
Government Readiness	93

4.01	Quality of math and science education, 2004	103
4.02	Quality of educational system, 2004	103
4.03	Quality of public schools, 2004	103
4.04	Internet access in schools, 2004	81

Readiness Component Index (continued)	

4.05	Buyer sophistication, 2004	91
4.06	Buyer dynamism, 2004	91
4.07	Residential telephone connection charge, 2002	76
4.08	Affordability of Internet access, 2003	72
5.01	Investment in training, 2004	79
5.02	Availability of training services, 2004	71
5.03	Quality of business schools, 2004	60
5.04	Business investment in R&D, 2004	70
5.05	Business monthly telephone subscription, 2002-3	75
5.06	Business telephone connection charge, 2002-3	71
6.01	Government prioritization of ICT, 2004	93
6.02	Government procurement of ICT, 2004	90

Usage Component Index	76
Individual Usage	72
Business Usage	68
Government Usage	90

7.01	Cellular mobile subscribers, 2003	72
7.02	Telephone subscribers, 2002	75
7.03	Public payphones, 2002	26
7.04	Telephone lines, 2002	75
7.05	Television sets, 2002	70
7.06	Broadband-DSL Internet subscribers, 2002-3	65
7.07	Broadband-cable modem, 2002-3	60
7.08	Internet users per 100 inhabitants, 2002	73
8.01	Prevalence of foreign technology licensing, 2004	79
8.02	Firm-level technology absorption, 2004	76
8.03	Capacity for innovation, 2004	87
8.04	Availability of new telephone lines, 2004	54
8.05	Availability of cellular phones, 2004	41
9.01	Government success in ICT promotion, 2004	97
9.02	Government online services, 2003	52

Honduras

Key Indicators

Population (mn), 2003	6.9
GDP per capita (PPP US$), 2003	2,658
Internet users per 100 inhabitants, 2002	2.5

Networked Readiness Index Rank

Year (number of countries)	Rank
2004 (104)	**97**
2003 (102)	98
2002 (82)	81

RANK/104

Environment Component Index		94
Market Environment		96
Political and Regulatory Environment		89
Infrastructure Environment		82
1.01	Availability of scientists and engineers, 2004	100
1.02	Venture capital availability, 2004	90
1.03	Sophistication of financial markets, 2004	80
1.04	Technological sophistication, 2004	91
1.05	State of cluster development, 2004	79
1.06	Collaboration in clusters, 2004	95
1.07	University-industry collaboration, 2004	100
1.08	Quality of scientific research institutions, 2004	102
1.09	Subsidies for firm-level R&D, 2004	95
1.10	Brain drain, 2004	49
1.11	Ease of access to loans, 2004	98
1.12	Administrative burden, 2004	73
1.13	Ease to start a new business, 2004	93
2.01	Effectiveness of lawmaking, 2004	78
2.02	Laws relating to ICT, 2004	95
2.03	Effectiveness of judiciary, 2004	92
2.04	Intellectual property protection, 2004	86
3.01	Telephone mainlines, 2002	82
3.02	Secure Internet servers, 2003	65
3.03	Internet hosts, 2003	80

Readiness Component Index		94
Individual Readiness		90
Business Readiness		89
Government Readiness		100
4.01	Quality of math and science education, 2004	102
4.02	Quality of educational system, 2004	102
4.03	Quality of public schools, 2004	99
4.04	Internet access in schools, 2004	94

Readiness Component Index (continued)		
4.05	Buyer sophistication, 2004	98
4.06	Buyer dynamism, 2004	100
4.07	Residential telephone connection charge, 2002	58
4.08	Affordability of Internet access, 2003	85
5.01	Investment in training, 2004	93
5.02	Availability of training services, 2004	101
5.03	Quality of business schools, 2004	101
5.04	Business investment in R&D, 2004	96
5.05	Business monthly telephone subscription, 2002–3	81
5.06	Business telephone connection charge, 2002–3	69
6.01	Government prioritization of ICT, 2004	100
6.02	Government procurement of ICT, 2004	100

Usage Component Index		100
Individual Usage		84
Business Usage		100
Government Usage		101
7.01	Cellular mobile subscribers, 2003	84
7.02	Telephone subscribers, 2002	82
7.03	Public payphones, 2002	89
7.04	Telephone lines, 2002	80
7.05	Television sets, 2002	82
7.06	Broadband-DSL Internet subscribers, 2002–3	65
7.07	Broadband-cable modem, 2002–3	60
7.08	Internet users per 100 inhabitants, 2002	78
8.01	Prevalence of foreign technology licensing, 2004	94
8.02	Firm-level technology absorption, 2004	93
8.03	Capacity for innovation, 2004	89
8.04	Availability of new telephone lines, 2004	104
8.05	Availability of cellular phones, 2004	70
9.01	Government success in ICT promotion, 2004	98
9.02	Government online services, 2003	93

Hong Kong

Key Indicators

Population (mn), 2003	7
GDP per capita (PPP US$), 2003	28,027
Internet users per 100 inhabitants, 2002	43.0

2004 (104)   **7**

2003 (102)   18

2002 (82)   18

RANK/104

Environment Component Index	15
Market Environment	4
Political and Regulatory Environment	21
Infrastructure Environment	16

1.01	Availability of scientists and engineers, 2004	46
1.02	Venture capital availability, 2004	4
1.03	Sophistication of financial markets, 2004	5
1.04	Technological sophistication, 2004	10
1.05	State of cluster development, 2004	11
1.06	Collaboration in clusters, 2004	22
1.07	University-industry collaboration, 2004	17
1.08	Quality of scientific research institutions, 2004	28
1.09	Subsidies for firm-level R&D, 2004	33
1.10	Brain drain, 2004	10
1.11	Ease of access to loans, 2004	2
1.12	Administrative burden, 2004	2
1.13	Ease to start a new business, 2004	1
2.01	Effectiveness of lawmaking, 2004	52
2.02	Laws relating to ICT, 2004	9
2.03	Effectiveness of judiciary, 2004	14
2.04	Intellectual property protection, 2004	19
3.01	Telephone mainlines, 2002	14
3.02	Secure Internet servers, 2003	17
3.03	Internet hosts, 2003	12

Readiness Component Index	12
Individual Readiness	10
Business Readiness	22
Government Readiness	12

4.01	Quality of math and science education, 2004	15
4.02	Quality of educational system, 2004	17
4.03	Quality of public schools, 2004	25
4.04	Internet access in schools, 2004	4

Readiness Component Index (continued)	
4.05	Buyer sophistication, 2004	6
4.06	Buyer dynamism, 2004	5
4.07	Residential telephone connection charge, 2002	17
4.08	Affordability of Internet access, 2003	1
5.01	Investment in training, 2004	23
5.02	Availability of training services, 2004	23
5.03	Quality of business schools, 2004	25
5.04	Business investment in R&D, 2004	29
5.05	Business monthly telephone subscription, 2002–3	25
5.06	Business telephone connection charge, 2002–3	16
6.01	Government prioritization of ICT, 2004	19
6.02	Government procurement of ICT, 2004	15

Usage Component Index	2
Individual Usage	5
Business Usage	12
Government Usage	2

7.01	Cellular mobile subscribers, 2003	2
7.02	Telephone subscribers, 2002	7
7.03	Public payphones, 2002	74
7.04	Telephone lines, 2002	16
7.05	Television sets, 2002	7
7.06	Broadband-DSL Internet subscribers, 2002–3	2
7.07	Broadband-cable modem, 2002–3	10
7.08	Internet users per 100 inhabitants, 2002	15
8.01	Prevalence of foreign technology licensing, 2004	3
8.02	Firm-level technology absorption, 2004	10
8.03	Capacity for innovation, 2004	26
8.04	Availability of new telephone lines, 2004	2
8.05	Availability of cellular phones, 2004	4
9.01	Government success in ICT promotion, 2004	13
9.02	Government online services, 2003	2

145

Part 2 Country Profiles

Hungary

Key Indicators

Population (mn), 2003	9.9
GDP per capita (PPP US$), 2003	14,572
Internet users per 100 inhabitants, 2002	15.8

Networked Readiness Index Rank

Year (number of countries)	Rank
2004 (104)	**38**
2003 (102)	36
2002 (82)	30

RANK/104

Environment Component Index		37
Market Environment		41
Political and Regulatory Environment		45
Infrastructure Environment		32
1.01	Availability of scientists and engineers, 2004	25
1.02	Venture capital availability, 2004	41
1.03	Sophistication of financial markets, 2004	36
1.04	Technological sophistication, 2004	47
1.05	State of cluster development, 2004	86
1.06	Collaboration in clusters, 2004	74
1.07	University-industry collaboration, 2004	70
1.08	Quality of scientific research institutions, 2004	24
1.09	Subsidies for firm-level R&D, 2004	26
1.10	Brain drain, 2004	42
1.11	Ease of access to loans, 2004	43
1.12	Administrative burden, 2004	63
1.13	Ease to start a new business, 2004	16
2.01	Effectiveness of lawmaking, 2004	58
2.02	Laws relating to ICT, 2004	48
2.03	Effectiveness of judiciary, 2004	41
2.04	Intellectual property protection, 2004	37
3.01	Telephone mainlines, 2002	34
3.02	Secure Internet servers, 2003	42
3.03	Internet hosts, 2003	21

Readiness Component Index		45
Individual Readiness		33
Business Readiness		54
Government Readiness		59
4.01	Quality of math and science education, 2004	16
4.02	Quality of educational system, 2004	36
4.03	Quality of public schools, 2004	22
4.04	Internet access in schools, 2004	29

Readiness Component Index (continued)		
4.05	Buyer sophistication, 2004	73
4.06	Buyer dynamism, 2004	55
4.07	Residential telephone connection charge, 2002	53
4.08	Affordability of Internet access, 2003	28
5.01	Investment in training, 2004	68
5.02	Availability of training services, 2004	66
5.03	Quality of business schools, 2004	45
5.04	Business investment in R&D, 2004	52
5.05	Business monthly telephone subscription, 2002–3	44
5.06	Business telephone connection charge, 2002–3	66
6.01	Government prioritization of ICT, 2004	55
6.02	Government procurement of ICT, 2004	59

Usage Component Index		31
Individual Usage		35
Business Usage		34
Government Usage		40
7.01	Cellular mobile subscribers, 2003	22
7.02	Telephone subscribers, 2002	32
7.03	Public payphones, 2002	19
7.04	Telephone lines, 2002	36
7.05	Television sets, 2002	30
7.06	Broadband-DSL Internet subscribers, 2002–3	36
7.07	Broadband-cable modem, 2002–3	25
7.08	Internet users per 100 inhabitants, 2002	42
8.01	Prevalence of foreign technology licensing, 2004	42
8.02	Firm-level technology absorption, 2004	32
8.03	Capacity for innovation, 2004	36
8.04	Availability of new telephone lines, 2004	30
8.05	Availability of cellular phones, 2004	33
9.01	Government success in ICT promotion, 2004	37
9.02	Government online services, 2003	43

Iceland

Key Indicators

Population (mn), 2003	0.29
GDP per capita (PPP US$), 2003	30,658
Internet users per 100 inhabitants, 2002	64.8

RANK/104

Environment Component Index	1
Market Environment	8
Political and Regulatory Environment	7
Infrastructure Environment	1

1.01	Availability of scientists and engineers, 2004	7
1.02	Venture capital availability, 2004	34
1.03	Sophistication of financial markets, 2004	30
1.04	Technological sophistication, 2004	6
1.05	State of cluster development, 2004	43
1.06	Collaboration in clusters, 2004	46
1.07	University-industry collaboration, 2004	20
1.08	Quality of scientific research institutions, 2004	26
1.09	Subsidies for firm-level R&D, 2004	30
1.10	Brain drain, 2004	5
1.11	Ease of access to loans, 2004	8
1.12	Administrative burden, 2004	3
1.13	Ease to start a new business, 2004	2
2.01	Effectiveness of lawmaking, 2004	4
2.02	Laws relating to ICT, 2004	13
2.03	Effectiveness of judiciary, 2004	6
2.04	Intellectual property protection, 2004	9
3.01	Telephone mainlines, 2002	7
3.02	Secure Internet servers, 2003	1
3.03	Internet hosts, 2003	1

Readiness Component Index	15
Individual Readiness	3
Business Readiness	19
Government Readiness	28

4.01	Quality of math and science education, 2004	29
4.02	Quality of educational system, 2004	3
4.03	Quality of public schools, 2004	2
4.04	Internet access in schools, 2004	1

Readiness Component Index (continued)	
4.05	Buyer sophistication, 2004	19
4.06	Buyer dynamism, 2004	1
4.07	Residential telephone connection charge, 2002	14
4.08	Affordability of Internet access, 2003	13
5.01	Investment in training, 2004	13
5.02	Availability of training services, 2004	28
5.03	Quality of business schools, 2004	26
5.04	Business investment in R&D, 2004	13
5.05	Business monthly telephone subscription, 2002–3	7
5.06	Business telephone connection charge, 2002–3	14
6.01	Government prioritization of ICT, 2004	21
6.02	Government procurement of ICT, 2004	30

Usage Component Index	12
Individual Usage	3
Business Usage	6
Government Usage	27

7.01	Cellular mobile subscribers, 2003	4
7.02	Telephone subscribers, 2002	4
7.03	Public payphones, 2002	49
7.04	Telephone lines, 2002	7
7.05	Television sets, 2002	20
7.06	Broadband-DSL Internet subscribers, 2002–3	3
7.07	Broadband-cable modem, 2002–3	60
7.08	Internet users per 100 inhabitants, 2002	1
8.01	Prevalence of foreign technology licensing, 2004	17
8.02	Firm-level technology absorption, 2004	4
8.03	Capacity for innovation, 2004	16
8.04	Availability of new telephone lines, 2004	1
8.05	Availability of cellular phones, 2004	3
9.01	Government success in ICT promotion, 2004	20
9.02	Government online services, 2003	42

Part 2 Country Profiles | 147

India

Key Indicators

Population (mn), 2003	1,066
GDP per capita (PPP US$), 2003	2,909
Internet users per 100 inhabitants, 2002	1.6

Networked Readiness Index Rank

Year (number of countries)   Rank

2004 (104)   **39**

2003 (102)   45

2002 (82)   37

RANK/104

Environment Component Index		38
Market Environment		27
Political and Regulatory Environment		33
Infrastructure Environment		86
1.01	Availability of scientists and engineers, 2004	1
1.02	Venture capital availability, 2004	27
1.03	Sophistication of financial markets, 2004	32
1.04	Technological sophistication, 2004	26
1.05	State of cluster development, 2004	7
1.06	Collaboration in clusters, 2004	24
1.07	University-industry collaboration, 2004	34
1.08	Quality of scientific research institutions, 2004	17
1.09	Subsidies for firm-level R&D, 2004	17
1.10	Brain drain, 2004	50
1.11	Ease of access to loans, 2004	33
1.12	Administrative burden, 2004	38
1.13	Ease to start a new business, 2004	48
2.01	Effectiveness of lawmaking, 2004	21
2.02	Laws relating to ICT, 2004	33
2.03	Effectiveness of judiciary, 2004	32
2.04	Intellectual property protection, 2004	59
3.01	Telephone mainlines, 2002	86
3.02	Secure Internet servers, 2003	84
3.03	Internet hosts, 2003	91

Readiness Component Index		34
Individual Readiness		44
Business Readiness		36
Government Readiness		34
4.01	Quality of math and science education, 2004	11
4.02	Quality of educational system, 2004	39
4.03	Quality of public schools, 2004	83
4.04	Internet access in schools, 2004	46

Readiness Component Index (continued)		
4.05	Buyer sophistication, 2004	34
4.06	Buyer dynamism, 2004	34
4.07	Residential telephone connection charge, 2002	69
4.08	Affordability of Internet access, 2003	74
5.01	Investment in training, 2004	29
5.02	Availability of training services, 2004	27
5.03	Quality of business schools, 2004	6
5.04	Business investment in R&D, 2004	26
5.05	Business monthly telephone subscription, 2002–3	93
5.06	Business telephone connection charge, 2002–3	63
6.01	Government prioritization of ICT, 2004	9
6.02	Government procurement of ICT, 2004	64

Usage Component Index		42
Individual Usage		85
Business Usage		23
Government Usage		23
7.01	Cellular mobile subscribers, 2003	93
7.02	Telephone subscribers, 2002	89
7.03	Public payphones, 2002	58
7.04	Telephone lines, 2002	82
7.05	Television sets, 2002	86
7.06	Broadband-DSL Internet subscribers, 2002–3	61
7.07	Broadband-cable modem, 2002–3	57
7.08	Internet users per 100 inhabitants, 2002	87
8.01	Prevalence of foreign technology licensing, 2004	8
8.02	Firm-level technology absorption, 2004	18
8.03	Capacity for innovation, 2004	33
8.04	Availability of new telephone lines, 2004	46
8.05	Availability of cellular phones, 2004	46
9.01	Government success in ICT promotion, 2004	11
9.02	Government online services, 2003	43

Indonesia

Key Indicators

Population (mn), 2003	220
GDP per capita (PPP US$), 2003	3,364
Internet users per 100 inhabitants, 2002	2.1

Networked Readiness Index Rank

Year (number of countries)	Rank
2004 (104)	**51**
2003 (102)	73
2002 (82)	64

RANK/104

Environment Component Index	42
Market Environment	26
Political and Regulatory Environment	42
Infrastructure Environment	89

1.01	Availability of scientists and engineers, 2004	71
1.02	Venture capital availability, 2004	20
1.03	Sophistication of financial markets, 2004	40
1.04	Technological sophistication, 2004	57
1.05	State of cluster development, 2004	20
1.06	Collaboration in clusters, 2004	32
1.07	University-industry collaboration, 2004	27
1.08	Quality of scientific research institutions, 2004	49
1.09	Subsidies for firm-level R&D, 2004	22
1.10	Brain drain, 2004	35
1.11	Ease of access to loans, 2004	27
1.12	Administrative burden, 2004	15
1.13	Ease to start a new business, 2004	33
2.01	Effectiveness of lawmaking, 2004	29
2.02	Laws relating to ICT, 2004	37
2.03	Effectiveness of judiciary, 2004	58
2.04	Intellectual property protection, 2004	47
3.01	Telephone mainlines, 2002	88
3.02	Secure Internet servers, 2003	83
3.03	Internet hosts, 2003	79

Readiness Component Index	46
Individual Readiness	53
Business Readiness	48
Government Readiness	48

4.01	Quality of math and science education, 2004	57
4.02	Quality of educational system, 2004	35
4.03	Quality of public schools, 2004	48
4.04	Internet access in schools, 2004	45

Readiness Component Index (continued)	
4.05	Buyer sophistication, 2004	39
4.06	Buyer dynamism, 2004	68
4.07	Residential telephone connection charge, 2002	65
4.08	Affordability of Internet access, 2003	82
5.01	Investment in training, 2004	39
5.02	Availability of training services, 2004	58
5.03	Quality of business schools, 2004	57
5.04	Business investment in R&D, 2004	28
5.05	Business monthly telephone subscription, 2002–3	73
5.06	Business telephone connection charge, 2002–3	67
6.01	Government prioritization of ICT, 2004	65
6.02	Government procurement of ICT, 2004	26

Usage Component Index	69
Individual Usage	81
Business Usage	84
Government Usage	43

7.01	Cellular mobile subscribers, 2003	78
7.02	Telephone subscribers, 2002	84
7.03	Public payphones, 2002	60
7.04	Telephone lines, 2002	86
7.05	Television sets, 2002	80
7.06	Broadband-DSL Internet subscribers, 2002–3	54
7.07	Broadband-cable modem, 2002–3	49
7.08	Internet users per 100 inhabitants, 2002	80
8.01	Prevalence of foreign technology licensing, 2004	71
8.02	Firm-level technology absorption, 2004	85
8.03	Capacity for innovation, 2004	28
8.04	Availability of new telephone lines, 2004	85
8.05	Availability of cellular phones, 2004	98
9.01	Government success in ICT promotion, 2004	33
9.02	Government online services, 2003	56

Ireland

Key Indicators

Population (mn), 2003	4
GDP per capita (PPP US$), 2003	36,775
Internet users per 100 inhabitants, 2002	28.0

RANK/104

Environment Component Index	17
Market Environment	14
Political and Regulatory Environment	22
Infrastructure Environment	17

1.01	Availability of scientists and engineers, 2004	26
1.02	Venture capital availability, 2004	7
1.03	Sophistication of financial markets, 2004	21
1.04	Technological sophistication, 2004	20
1.05	State of cluster development, 2004	13
1.06	Collaboration in clusters, 2004	27
1.07	University-industry collaboration, 2004	21
1.08	Quality of scientific research institutions, 2004	22
1.09	Subsidies for firm-level R&D, 2004	8
1.10	Brain drain, 2004	19
1.11	Ease of access to loans, 2004	9
1.12	Administrative burden, 2004	14
1.13	Ease to start a new business, 2004	28
2.01	Effectiveness of lawmaking, 2004	22
2.02	Laws relating to ICT, 2004	24
2.03	Effectiveness of judiciary, 2004	17
2.04	Intellectual property protection, 2004	21
3.01	Telephone mainlines, 2002	21
3.02	Secure Internet servers, 2003	9
3.03	Internet hosts, 2003	20

Readiness Component Index	24
Individual Readiness	19
Business Readiness	20
Government Readiness	31

4.01	Quality of math and science education, 2004	25
4.02	Quality of educational system, 2004	9
4.03	Quality of public schools, 2004	7
4.04	Internet access in schools, 2004	34

Readiness Component Index (continued)	
4.05	Buyer sophistication, 2004	22
4.06	Buyer dynamism, 2004	26
4.07	Residential telephone connection charge, 2002	21
4.08	Affordability of Internet access, 2003	22
5.01	Investment in training, 2004	20
5.02	Availability of training services, 2004	22
5.03	Quality of business schools, 2004	19
5.04	Business investment in R&D, 2004	25
5.05	Business monthly telephone subscription, 2002–3	18
5.06	Business telephone connection charge, 2002–3	22
6.01	Government prioritization of ICT, 2004	24
6.02	Government procurement of ICT, 2004	29

Usage Component Index	25
Individual Usage	31
Business Usage	31
Government Usage	21

7.01	Cellular mobile subscribers, 2003	16
7.02	Telephone subscribers, 2002	18
7.03	Public payphones, 2002	52
7.04	Telephone lines, 2002	22
7.05	Television sets, 2002	19
7.06	Broadband-DSL Internet subscribers, 2002–3	47
7.07	Broadband-cable modem, 2002–3	38
7.08	Internet users per 100 inhabitants, 2002	30
8.01	Prevalence of foreign technology licensing, 2004	35
8.02	Firm-level technology absorption, 2004	23
8.03	Capacity for innovation, 2004	23
8.04	Availability of new telephone lines, 2004	36
8.05	Availability of cellular phones, 2004	73
9.01	Government success in ICT promotion, 2004	19
9.02	Government online services, 2003	18

Israel

151

Key Indicators

Population (mn), 2003	6.4
GDP per capita (PPP US$), 2003	19,678
Internet users per 100 inhabitants, 2002	30.1

Networked Readiness Index Rank

Year (number of countries)	Rank
2004 (104)	**18**
2003 (102)	16
2002 (82)	12

RANK/104

Environment Component Index		20
Market Environment		18
Political and Regulatory Environment		25
Infrastructure Environment		22
1.01	Availability of scientists and engineers, 2004	3
1.02	Venture capital availability, 2004	2
1.03	Sophistication of financial markets, 2004	18
1.04	Technological sophistication, 2004	1
1.05	State of cluster development, 2004	16
1.06	Collaboration in clusters, 2004	19
1.07	University-industry collaboration, 2004	12
1.08	Quality of scientific research institutions, 2004	3
1.09	Subsidies for firm-level R&D, 2004	7
1.10	Brain drain, 2004	13
1.11	Ease of access to loans, 2004	63
1.12	Administrative burden, 2004	49
1.13	Ease to start a new business, 2004	22
2.01	Effectiveness of lawmaking, 2004	34
2.02	Laws relating to ICT, 2004	31
2.03	Effectiveness of judiciary, 2004	20
2.04	Intellectual property protection, 2004	23
3.01	Telephone mainlines, 2002	27
3.02	Secure Internet servers, 2003	21
3.03	Internet hosts, 2003	15

Readiness Component Index		19
Individual Readiness		26
Business Readiness		14
Government Readiness		21
4.01	Quality of math and science education, 2004	42
4.02	Quality of educational system, 2004	29
4.03	Quality of public schools, 2004	29
4.04	Internet access in schools, 2004	23

Readiness Component Index (continued)		
4.05	Buyer sophistication, 2004	26
4.06	Buyer dynamism, 2004	21
4.07	Residential telephone connection charge, 2002	22
4.08	Affordability of Internet access, 2003	27
5.01	Investment in training, 2004	25
5.02	Availability of training services, 2004	13
5.03	Quality of business schools, 2004	23
5.04	Business investment in R&D, 2004	7
5.05	Business monthly telephone subscription, 2002–3	12
5.06	Business telephone connection charge, 2002–3	23
6.01	Government prioritization of ICT, 2004	27
6.02	Government procurement of ICT, 2004	20

Usage Component Index		13
Individual Usage		19
Business Usage		8
Government Usage		10
7.01	Cellular mobile subscribers, 2003	6
7.02	Telephone subscribers, 2002	10
7.03	Public payphones, 2002	47
7.04	Telephone lines, 2002	26
7.05	Television sets, 2002	41
7.06	Broadband-DSL Internet subscribers, 2002–3	17
7.07	Broadband-cable modem, 2002–3	11
7.08	Internet users per 100 inhabitants, 2002	28
8.01	Prevalence of foreign technology licensing, 2004	40
8.02	Firm-level technology absorption, 2004	6
8.03	Capacity for innovation, 2004	8
8.04	Availability of new telephone lines, 2004	7
8.05	Availability of cellular phones, 2004	1
9.01	Government success in ICT promotion, 2004	21
9.02	Government online services, 2003	8

Italy

Key Indicators

Population (mn), 2003	57.4
GDP per capita (PPP US$), 2003	27,050
Internet users per 100 inhabitants, 2002	35.2

Networked Readiness Index Rank

Year (number of countries)	Rank
2004 (104)	**45**
2003 (102)	28
2002 (82)	26

RANK/104

Environment Component Index	51
Market Environment	60
Political and Regulatory Environment	61
Infrastructure Environment	30

1.01	Availability of scientists and engineers, 2004	53
1.02	Venture capital availability, 2004	65
1.03	Sophistication of financial markets, 2004	43
1.04	Technological sophistication, 2004	45
1.05	State of cluster development, 2004	5
1.06	Collaboration in clusters, 2004	16
1.07	University-industry collaboration, 2004	64
1.08	Quality of scientific research institutions, 2004	79
1.09	Subsidies for firm-level R&D, 2004	48
1.10	Brain drain, 2004	45
1.11	Ease of access to loans, 2004	74
1.12	Administrative burden, 2004	103
1.13	Ease to start a new business, 2004	76
2.01	Effectiveness of lawmaking, 2004	71
2.02	Laws relating to ICT, 2004	59
2.03	Effectiveness of judiciary, 2004	63
2.04	Intellectual property protection, 2004	45
3.01	Telephone mainlines, 2002	26
3.02	Secure Internet servers, 2003	32
3.03	Internet hosts, 2003	41

Readiness Component Index	53
Individual Readiness	37
Business Readiness	38
Government Readiness	74

4.01	Quality of math and science education, 2004	50
4.02	Quality of educational system, 2004	52
4.03	Quality of public schools, 2004	34
4.04	Internet access in schools, 2004	53

Readiness Component Index (continued)	
4.05	Buyer sophistication, 2004	31
4.06	Buyer dynamism, 2004	43
4.07	Residential telephone connection charge, 2002	15
4.08	Affordability of Internet access, 2003	15
5.01	Investment in training, 2004	61
5.02	Availability of training services, 2004	25
5.03	Quality of business schools, 2004	33
5.04	Business investment in R&D, 2004	78
5.05	Business monthly telephone subscription, 2002–3	53
5.06	Business telephone connection charge, 2002–3	15
6.01	Government prioritization of ICT, 2004	73
6.02	Government procurement of ICT, 2004	77

Usage Component Index	30
Individual Usage	17
Business Usage	48
Government Usage	45

7.01	Cellular mobile subscribers, 2003	3
7.02	Telephone subscribers, 2002	9
7.03	Public payphones, 2002	6
7.04	Telephone lines, 2002	23
7.05	Television sets, 2002	23
7.06	Broadband-DSL Internet subscribers, 2002–3	24
7.07	Broadband-cable modem, 2002–3	60
7.08	Internet users per 100 inhabitants, 2002	21
8.01	Prevalence of foreign technology licensing, 2004	50
8.02	Firm-level technology absorption, 2004	79
8.03	Capacity for innovation, 2004	21
8.04	Availability of new telephone lines, 2004	61
8.05	Availability of cellular phones, 2004	54
9.01	Government success in ICT promotion, 2004	66
9.02	Government online services, 2003	27

Jamaica

Key Indicators

Population (mn), 2003	2.7
GDP per capita (PPP US$), 2003	4,184
Internet users per 100 inhabitants, 2002	22.8

Networked Readiness Index Rank

Year (number of countries)   Rank

2004 (104)   **49**

2003 (102)   53

2002 (82)   60

Environment Component Index	RANK/104 58
Market Environment	64
Political and Regulatory Environment	51
Infrastructure Environment	61
1.01 Availability of scientists and engineers, 2004	80
1.02 Venture capital availability, 2004	69
1.03 Sophistication of financial markets, 2004	37
1.04 Technological sophistication, 2004	44
1.05 State of cluster development, 2004	53
1.06 Collaboration in clusters, 2004	56
1.07 University-industry collaboration, 2004	43
1.08 Quality of scientific research institutions, 2004	35
1.09 Subsidies for firm-level R&D, 2004	68
1.10 Brain drain, 2004	78
1.11 Ease of access to loans, 2004	77
1.12 Administrative burden, 2004	72
1.13 Ease to start a new business, 2004	57
2.01 Effectiveness of lawmaking, 2004	56
2.02 Laws relating to ICT, 2004	53
2.03 Effectiveness of judiciary, 2004	49
2.04 Intellectual property protection, 2004	56
3.01 Telephone mainlines, 2002	59
3.02 Secure Internet servers, 2003	54
3.03 Internet hosts, 2003	75

Readiness Component Index	48
Individual Readiness	62
Business Readiness	47
Government Readiness	47
4.01 Quality of math and science education, 2004	85
4.02 Quality of educational system, 2004	74
4.03 Quality of public schools, 2004	68
4.04 Internet access in schools, 2004	73

Readiness Component Index (continued)	
4.05 Buyer sophistication, 2004	51
4.06 Buyer dynamism, 2004	38
4.07 Residential telephone connection charge, 2002	34
4.08 Affordability of Internet access, 2003	68
5.01 Investment in training, 2004	53
5.02 Availability of training services, 2004	54
5.03 Quality of business schools, 2004	39
5.04 Business investment in R&D, 2004	40
5.05 Business monthly telephone subscription, 2002–3	78
5.06 Business telephone connection charge, 2002–3	37
6.01 Government prioritization of ICT, 2004	46
6.02 Government procurement of ICT, 2004	47

Usage Component Index	44
Individual Usage	43
Business Usage	54
Government Usage	32
7.01 Cellular mobile subscribers, 2003	38
7.02 Telephone subscribers, 2002	40
7.03 Public payphones, 2002	35
7.04 Telephone lines, 2002	60
7.05 Television sets, 2002	77
7.06 Broadband-DSL Internet subscribers, 2002–3	23
7.07 Broadband-cable modem, 2002–3	60
7.08 Internet users per 100 inhabitants, 2002	36
8.01 Prevalence of foreign technology licensing, 2004	38
8.02 Firm-level technology absorption, 2004	51
8.03 Capacity for innovation, 2004	66
8.04 Availability of new telephone lines, 2004	75
8.05 Availability of cellular phones, 2004	51
9.01 Government success in ICT promotion, 2004	46
9.02 Government online services, 2003	30

Japan

Key Indicators

Population (mn), 2003	127.7
GDP per capita (PPP US$), 2003	28,162
Internet users per 100 inhabitants, 2002	44.9

Networked Readiness Index Rank

Year (number of countries)	Rank
2004 (104)	**8**
2003 (102)	12
2002 (82)	20

RANK/104

Environment Component Index	16
Market Environment	17
Political and Regulatory Environment	23
Infrastructure Environment	15

1.01	Availability of scientists and engineers, 2004	4
1.02	Venture capital availability, 2004	23
1.03	Sophistication of financial markets, 2004	19
1.04	Technological sophistication, 2004	7
1.05	State of cluster development, 2004	1
1.06	Collaboration in clusters, 2004	1
1.07	University-industry collaboration, 2004	8
1.08	Quality of scientific research institutions, 2004	7
1.09	Subsidies for firm-level R&D, 2004	10
1.10	Brain drain, 2004	6
1.11	Ease of access to loans, 2004	56
1.12	Administrative burden, 2004	57
1.13	Ease to start a new business, 2004	37
2.01	Effectiveness of lawmaking, 2004	27
2.02	Laws relating to ICT, 2004	17
2.03	Effectiveness of judiciary, 2004	22
2.04	Intellectual property protection, 2004	20
3.01	Telephone mainlines, 2002	15
3.02	Secure Internet servers, 2003	20
3.03	Internet hosts, 2003	11

Readiness Component Index	3
Individual Readiness	15
Business Readiness	6
Government Readiness	2

4.01	Quality of math and science education, 2004	14
4.02	Quality of educational system, 2004	31
4.03	Quality of public schools, 2004	17
4.04	Internet access in schools, 2004	22

Readiness Component Index (continued)	
4.05	Buyer sophistication, 2004	2
4.06	Buyer dynamism, 2004	2
4.07	Residential telephone connection charge, 2002	55
4.08	Affordability of Internet access, 2003	9
5.01	Investment in training, 2004	4
5.02	Availability of training services, 2004	3
5.03	Quality of business schools, 2004	37
5.04	Business investment in R&D, 2004	1
5.05	Business monthly telephone subscription, 2002–3	22
5.06	Business telephone connection charge, 2002–3	53
6.01	Government prioritization of ICT, 2004	2
6.02	Government procurement of ICT, 2004	4

Usage Component Index	6
Individual Usage	10
Business Usage	1
Government Usage	8

7.01	Cellular mobile subscribers, 2003	31
7.02	Telephone subscribers, 2002	26
7.03	Public payphones, 2002	16
7.04	Telephone lines, 2002	15
7.05	Television sets, 2002	6
7.06	Broadband-DSL Internet subscribers, 2002–3	5
7.07	Broadband-cable modem, 2002–3	16
7.08	Internet users per 100 inhabitants, 2002	13
8.01	Prevalence of foreign technology licensing, 2004	7
8.02	Firm-level technology absorption, 2004	1
8.03	Capacity for innovation, 2004	3
8.04	Availability of new telephone lines, 2004	12
8.05	Availability of cellular phones, 2004	18
9.01	Government success in ICT promotion, 2004	17
9.02	Government online services, 2003	8

Jordan

Key Indicators

Population (mn), 2003	5.5
GDP per capita (PPP US$), 2003	4,319
Internet users per 100 inhabitants, 2002	5.8

Networked Readiness Index Rank

Year (number of countries)	Rank
2004 (104)	**44**
2003 (102)	46
2002 (82)	51

RANK/104

Environment Component Index	**43**
Market Environment	42
Political and Regulatory Environment	32
Infrastructure Environment	66

1.01	Availability of scientists and engineers, 2004	17
1.02	Venture capital availability, 2004	71
1.03	Sophistication of financial markets, 2004	48
1.04	Technological sophistication, 2004	38
1.05	State of cluster development, 2004	65
1.06	Collaboration in clusters, 2004	63
1.07	University-industry collaboration, 2004	47
1.08	Quality of scientific research institutions, 2004	43
1.09	Subsidies for firm-level R&D, 2004	49
1.10	Brain drain, 2004	64
1.11	Ease of access to loans, 2004	46
1.12	Administrative burden, 2004	16
1.13	Ease to start a new business, 2004	30
2.01	Effectiveness of lawmaking, 2004	40
2.02	Laws relating to ICT, 2004	35
2.03	Effectiveness of judiciary, 2004	43
2.04	Intellectual property protection, 2004	28
3.01	Telephone mainlines, 2002	63
3.02	Secure Internet servers, 2003	68
3.03	Internet hosts, 2003	74

Readiness Component Index	**44**
Individual Readiness	47
Business Readiness	75
Government Readiness	32

4.01	Quality of math and science education, 2004	28
4.02	Quality of educational system, 2004	41
4.03	Quality of public schools, 2004	51
4.04	Internet access in schools, 2004	40

Readiness Component Index (continued)	
4.05	Buyer sophistication, 2004	59
4.06	Buyer dynamism, 2004	65
4.07	Residential telephone connection charge, 2002	78
4.08	Affordability of Internet access, 2003	67
5.01	Investment in training, 2004	63
5.02	Availability of training services, 2004	53
5.03	Quality of business schools, 2004	67
5.04	Business investment in R&D, 2004	65
5.05	Business monthly telephone subscription, 2002–3	85
5.06	Business telephone connection charge, 2002–3	80
6.01	Government prioritization of ICT, 2004	17
6.02	Government procurement of ICT, 2004	39

Usage Component Index	**47**
Individual Usage	65
Business Usage	41
Government Usage	29

7.01	Cellular mobile subscribers, 2003	61
7.02	Telephone subscribers, 2002	62
7.03	Public payphones, 2002	69
7.04	Telephone lines, 2002	69
7.05	Television sets, 2002	36
7.06	Broadband-DSL Internet subscribers, 2002–3	49
7.07	Broadband-cable modem, 2002–3	60
7.08	Internet users per 100 inhabitants, 2002	60
8.01	Prevalence of foreign technology licensing, 2004	18
8.02	Firm-level technology absorption, 2004	46
8.03	Capacity for innovation, 2004	63
8.04	Availability of new telephone lines, 2004	37
8.05	Availability of cellular phones, 2004	43
9.01	Government success in ICT promotion, 2004	9
9.02	Government online services, 2003	62

Kenya

Key Indicators

Population (mn), 2003	32
GDP per capita (PPP US$), 2003	1,035
Internet users per 100 inhabitants, 2002	1.3

Networked Readiness Index Rank

Year (number of countries)	Rank
2004 (104)	**75**
2003 (102)	84
2002 (82)	—

RANK/104

Environment Component Index	66
Market Environment	55
Political and Regulatory Environment	66
Infrastructure Environment	93

1.01	Availability of scientists and engineers, 2004	55
1.02	Venture capital availability, 2004	54
1.03	Sophistication of financial markets, 2004	56
1.04	Technological sophistication, 2004	68
1.05	State of cluster development, 2004	33
1.06	Collaboration in clusters, 2004	52
1.07	University-industry collaboration, 2004	63
1.08	Quality of scientific research institutions, 2004	31
1.09	Subsidies for firm-level R&D, 2004	67
1.10	Brain drain, 2004	74
1.11	Ease of access to loans, 2004	30
1.12	Administrative burden, 2004	70
1.13	Ease to start a new business, 2004	58
2.01	Effectiveness of lawmaking, 2004	49
2.02	Laws relating to ICT, 2004	73
2.03	Effectiveness of judiciary, 2004	69
2.04	Intellectual property protection, 2004	71
3.01	Telephone mainlines, 2002	93
3.02	Secure Internet servers, 2003	90
3.03	Internet hosts, 2003	81

Readiness Component Index	79
Individual Readiness	84
Business Readiness	80
Government Readiness	72

4.01	Quality of math and science education, 2004	68
4.02	Quality of educational system, 2004	42
4.03	Quality of public schools, 2004	67
4.04	Internet access in schools, 2004	98

Readiness Component Index (continued)	
4.05	Buyer sophistication, 2004	74
4.06	Buyer dynamism, 2004	74
4.07	Residential telephone connection charge, 2002	86
4.08	Affordability of Internet access, 2003	93
5.01	Investment in training, 2004	56
5.02	Availability of training services, 2004	57
5.03	Quality of business schools, 2004	50
5.04	Business investment in R&D, 2004	32
5.05	Business monthly telephone subscription, 2002–3	97
5.06	Business telephone connection charge, 2002–3	75
6.01	Government prioritization of ICT, 2004	69
6.02	Government procurement of ICT, 2004	74

Usage Component Index	85
Individual Usage	91
Business Usage	80
Government Usage	68

7.01	Cellular mobile subscribers, 2003	82
7.02	Telephone subscribers, 2002	90
7.03	Public payphones, 2002	91
7.04	Telephone lines, 2002	92
7.05	Television sets, 2002	93
7.06	Broadband-DSL Internet subscribers, 2002–3	65
7.07	Broadband-cable modem, 2002–3	60
7.08	Internet users per 100 inhabitants, 2002	89
8.01	Prevalence of foreign technology licensing, 2004	34
8.02	Firm-level technology absorption, 2004	71
8.03	Capacity for innovation, 2004	67
8.04	Availability of new telephone lines, 2004	95
8.05	Availability of cellular phones, 2004	90
9.01	Government success in ICT promotion, 2004	65
9.02	Government online services, 2003	62

Korea

Key Indicators

Population (mn), 2003	47.7
GDP per capita (PPP US$), 2003	17,908
Internet users per 100 inhabitants, 2002	55.2

Networked Readiness Index Rank

Year (number of countries)	Rank
2004 (104)	**24**
2003 (102)	20
2002 (82)	14

RANK/104

Environment Component Index — 36

Market Environment	39
Political and Regulatory Environment	41
Infrastructure Environment	31

1.01	Availability of scientists and engineers, 2004	52
1.02	Venture capital availability, 2004	52
1.03	Sophistication of financial markets, 2004	49
1.04	Technological sophistication, 2004	24
1.05	State of cluster development, 2004	21
1.06	Collaboration in clusters, 2004	20
1.07	University-industry collaboration, 2004	24
1.08	Quality of scientific research institutions, 2004	32
1.09	Subsidies for firm-level R&D, 2004	21
1.10	Brain drain, 2004	40
1.11	Ease of access to loans, 2004	81
1.12	Administrative burden, 2004	66
1.13	Ease to start a new business, 2004	44
2.01	Effectiveness of lawmaking, 2004	81
2.02	Laws relating to ICT, 2004	14
2.03	Effectiveness of judiciary, 2004	48
2.04	Intellectual property protection, 2004	36
3.01	Telephone mainlines, 2002	25
3.02	Secure Internet servers, 2003	39
3.03	Internet hosts, 2003	49

Readiness Component Index — 25

Individual Readiness	40
Business Readiness	24
Government Readiness	11

4.01	Quality of math and science education, 2004	41
4.02	Quality of educational system, 2004	60
4.03	Quality of public schools, 2004	39
4.04	Internet access in schools, 2004	3

Readiness Component Index (continued)

4.05	Buyer sophistication, 2004	11
4.06	Buyer dynamism, 2004	10
4.07	Residential telephone connection charge, 2002	26
4.08	Affordability of Internet access, 2003	93
5.01	Investment in training, 2004	18
5.02	Availability of training services, 2004	26
5.03	Quality of business schools, 2004	58
5.04	Business investment in R&D, 2004	14
5.05	Business monthly telephone subscription, 2002–3	4
5.06	Business telephone connection charge, 2002–3	25
6.01	Government prioritization of ICT, 2004	15
6.02	Government procurement of ICT, 2004	17

Usage Component Index — 7

Individual Usage	2
Business Usage	22
Government Usage	19

7.01	Cellular mobile subscribers, 2003	29
7.02	Telephone subscribers, 2002	28
7.03	Public payphones, 2002	1
7.04	Telephone lines, 2002	18
7.05	Television sets, 2002	44
7.06	Broadband-DSL Internet subscribers, 2002–3	1
7.07	Broadband-cable modem, 2002–3	2
7.08	Internet users per 100 inhabitants, 2002	3
8.01	Prevalence of foreign technology licensing, 2004	63
8.02	Firm-level technology absorption, 2004	14
8.03	Capacity for innovation, 2004	15
8.04	Availability of new telephone lines, 2004	22
8.05	Availability of cellular phones, 2004	16
9.01	Government success in ICT promotion, 2004	14
9.02	Government online services, 2003	26

Latvia

Key Indicators

Population (mn), 2003	2.3
GDP per capita (PPP US$), 2003	9,981
Internet users per 100 inhabitants, 2002	13.3

Networked Readiness Index Rank

Year (number of countries)	Rank
2004 (104)	**56**
2003 (102)	35
2002 (82)	38

RANK/104

Environment Component Index — 55

Market Environment	57
Political and Regulatory Environment	68
Infrastructure Environment	36

1.01	Availability of scientists and engineers, 2004	75
1.02	Venture capital availability, 2004	55
1.03	Sophistication of financial markets, 2004	51
1.04	Technological sophistication, 2004	63
1.05	State of cluster development, 2004	72
1.06	Collaboration in clusters, 2004	53
1.07	University-industry collaboration, 2004	77
1.08	Quality of scientific research institutions, 2004	77
1.09	Subsidies for firm-level R&D, 2004	81
1.10	Brain drain, 2004	51
1.11	Ease of access to loans, 2004	48
1.12	Administrative burden, 2004	52
1.13	Ease to start a new business, 2004	39
2.01	Effectiveness of lawmaking, 2004	66
2.02	Laws relating to ICT, 2004	68
2.03	Effectiveness of judiciary, 2004	66
2.04	Intellectual property protection, 2004	72
3.01	Telephone mainlines, 2002	38
3.02	Secure Internet servers, 2003	34
3.03	Internet hosts, 2003	35

Readiness Component Index — 61

Individual Readiness	41
Business Readiness	44
Government Readiness	86

4.01	Quality of math and science education, 2004	32
4.02	Quality of educational system, 2004	38
4.03	Quality of public schools, 2004	36
4.04	Internet access in schools, 2004	39

Readiness Component Index (continued)

4.05	Buyer sophistication, 2004	52
4.06	Buyer dynamism, 2004	57
4.07	Residential telephone connection charge, 2002	61
4.08	Affordability of Internet access, 2003	70
5.01	Investment in training, 2004	51
5.02	Availability of training services, 2004	44
5.03	Quality of business schools, 2004	46
5.04	Business investment in R&D, 2004	60
5.05	Business monthly telephone subscription, 2002–3	47
5.06	Business telephone connection charge, 2002–3	57
6.01	Government prioritization of ICT, 2004	76
6.02	Government procurement of ICT, 2004	91

Usage Component Index — 58

Individual Usage	46
Business Usage	49
Government Usage	77

7.01	Cellular mobile subscribers, 2003	39
7.02	Telephone subscribers, 2002	42
7.03	Public payphones, 2002	64
7.04	Telephone lines, 2002	39
7.05	Television sets, 2002	71
7.06	Broadband-DSL Internet subscribers, 2002–3	37
7.07	Broadband-cable modem, 2002–3	60
7.08	Internet users per 100 inhabitants, 2002	45
8.01	Prevalence of foreign technology licensing, 2004	60
8.02	Firm-level technology absorption, 2004	53
8.03	Capacity for innovation, 2004	49
8.04	Availability of new telephone lines, 2004	48
8.05	Availability of cellular phones, 2004	49
9.01	Government success in ICT promotion, 2004	83
9.02	Government online services, 2003	62

Lithuania

Key Indicators

Population (mn), 2003	3.4
GDP per capita (PPP US$), 2003	11,250
Internet users per 100 inhabitants, 2002	14.4

RANK/104

Environment Component Index	45
Market Environment	38
Political and Regulatory Environment	53
Infrastructure Environment	42

1.01	Availability of scientists and engineers, 2004	27
1.02	Venture capital availability, 2004	18
1.03	Sophistication of financial markets, 2004	50
1.04	Technological sophistication, 2004	64
1.05	State of cluster development, 2004	52
1.06	Collaboration in clusters, 2004	42
1.07	University-industry collaboration, 2004	36
1.08	Quality of scientific research institutions, 2004	29
1.09	Subsidies for firm-level R&D, 2004	36
1.10	Brain drain, 2004	68
1.11	Ease of access to loans, 2004	6
1.12	Administrative burden, 2004	46
1.13	Ease to start a new business, 2004	72
2.01	Effectiveness of lawmaking, 2004	50
2.02	Laws relating to ICT, 2004	46
2.03	Effectiveness of judiciary, 2004	68
2.04	Intellectual property protection, 2004	61
3.01	Telephone mainlines, 2002	44
3.02	Secure Internet servers, 2003	49
3.03	Internet hosts, 2003	31

Readiness Component Index	36
Individual Readiness	36
Business Readiness	34
Government Readiness	46

4.01	Quality of math and science education, 2004	19
4.02	Quality of educational system, 2004	33
4.03	Quality of public schools, 2004	35
4.04	Internet access in schools, 2004	35

Readiness Component Index (continued)	
4.05	Buyer sophistication, 2004	71
4.06	Buyer dynamism, 2004	42
4.07	Residential telephone connection charge, 2002	50
4.08	Affordability of Internet access, 2003	57
5.01	Investment in training, 2004	48
5.02	Availability of training services, 2004	33
5.03	Quality of business schools, 2004	41
5.04	Business investment in R&D, 2004	34
5.05	Business monthly telephone subscription, 2002–3	38
5.06	Business telephone connection charge, 2002–3	48
6.01	Government prioritization of ICT, 2004	48
6.02	Government procurement of ICT, 2004	45

Usage Component Index	48
Individual Usage	42
Business Usage	36
Government Usage	61

7.01	Cellular mobile subscribers, 2003	32
7.02	Telephone subscribers, 2002	39
7.03	Public payphones, 2002	59
7.04	Telephone lines, 2002	45
7.05	Television sets, 2002	21
7.06	Broadband-DSL Internet subscribers, 2002–3	40
7.07	Broadband-cable modem, 2002–3	28
7.08	Internet users per 100 inhabitants, 2002	43
8.01	Prevalence of foreign technology licensing, 2004	44
8.02	Firm-level technology absorption, 2004	36
8.03	Capacity for innovation, 2004	44
8.04	Availability of new telephone lines, 2004	43
8.05	Availability of cellular phones, 2004	29
9.01	Government success in ICT promotion, 2004	55
9.02	Government online services, 2003	66

Luxembourg

Key Indicators

Population (mn), 2003	0.453
GDP per capita (PPP US$), 2003	62,844
Internet users per 100 inhabitants, 2002	37.0

Networked Readiness Index Rank

Year (number of countries)	Rank
2004 (104)	**17**
2003 (102)	14
2002 (82)	27

RANK/104

Environment Component Index	13
Market Environment	19
Political and Regulatory Environment	13
Infrastructure Environment	10

1.01	Availability of scientists and engineers, 2004	66
1.02	Venture capital availability, 2004	14
1.03	Sophistication of financial markets, 2004	3
1.04	Technological sophistication, 2004	25
1.05	State of cluster development, 2004	36
1.06	Collaboration in clusters, 2004	21
1.07	University-industry collaboration, 2004	56
1.08	Quality of scientific research institutions, 2004	48
1.09	Subsidies for firm-level R&D, 2004	2
1.10	Brain drain, 2004	14
1.11	Ease of access to loans, 2004	12
1.12	Administrative burden, 2004	25
1.13	Ease to start a new business, 2004	14
2.01	Effectiveness of lawmaking, 2004	7
2.02	Laws relating to ICT, 2004	26
2.03	Effectiveness of judiciary, 2004	11
2.04	Intellectual property protection, 2004	17
3.01	Telephone mainlines, 2002	1
3.02	Secure Internet servers, 2003	7
3.03	Internet hosts, 2003	16

Readiness Component Index	14
Individual Readiness	20
Business Readiness	27
Government Readiness	6

4.01	Quality of math and science education, 2004	44
4.02	Quality of educational system, 2004	19
4.03	Quality of public schools, 2004	21
4.04	Internet access in schools, 2004	19

Readiness Component Index (continued)	
4.05	Buyer sophistication, 2004	5
4.06	Buyer dynamism, 2004	23
4.07	Residential telephone connection charge, 2002	2
4.08	Affordability of Internet access, 2003	13
5.01	Investment in training, 2004	12
5.02	Availability of training services, 2004	48
5.03	Quality of business schools, 2004	91
5.04	Business investment in R&D, 2004	17
5.05	Business monthly telephone subscription, 2002–3	2
5.06	Business telephone connection charge, 2002–3	3
6.01	Government prioritization of ICT, 2004	35
6.02	Government procurement of ICT, 2004	2

Usage Component Index	27
Individual Usage	13
Business Usage	25
Government Usage	48

7.01	Cellular mobile subscribers, 2003	1
7.02	Telephone subscribers, 2002	1
7.03	Public payphones, 2002	78
7.04	Telephone lines, 2002	1
7.05	Television sets, 2002	33
7.06	Broadband-DSL Internet subscribers, 2002–3	26
7.07	Broadband-cable modem, 2002–3	51
7.08	Internet users per 100 inhabitants, 2002	20
8.01	Prevalence of foreign technology licensing, 2004	47
8.02	Firm-level technology absorption, 2004	25
8.03	Capacity for innovation, 2004	17
8.04	Availability of new telephone lines, 2004	28
8.05	Availability of cellular phones, 2004	34
9.01	Government success in ICT promotion, 2004	39
9.02	Government online services, 2003	57

Macedonia

Key Indicators

Population (mn), 2003	2.1
GDP per capita (PPP US$), 2003	6,762
Internet users per 100 inhabitants, 2002	4.8

Networked Readiness Index Rank

Year (number of countries)	Rank
2004 (104)	**85**
2003 (102)	75
2002 (82)	—

RANK/104

Environment Component Index		86
Market Environment		92
Political and Regulatory Environment		88
Infrastructure Environment		51
1.01	Availability of scientists and engineers, 2004	38
1.02	Venture capital availability, 2004	36
1.03	Sophistication of financial markets, 2004	94
1.04	Technological sophistication, 2004	100
1.05	State of cluster development, 2004	102
1.06	Collaboration in clusters, 2004	96
1.07	University-industry collaboration, 2004	74
1.08	Quality of scientific research institutions, 2004	84
1.09	Subsidies for firm-level R&D, 2004	80
1.10	Brain drain, 2004	102
1.11	Ease of access to loans, 2004	90
1.12	Administrative burden, 2004	68
1.13	Ease to start a new business, 2004	103
2.01	Effectiveness of lawmaking, 2004	80
2.02	Laws relating to ICT, 2004	89
2.03	Effectiveness of judiciary, 2004	91
2.04	Intellectual property protection, 2004	87
3.01	Telephone mainlines, 2002	42
3.02	Secure Internet servers, 2003	66
3.03	Internet hosts, 2003	63

Readiness Component Index		75
Individual Readiness		65
Business Readiness		72
Government Readiness		84
4.01	Quality of math and science education, 2004	43
4.02	Quality of educational system, 2004	45
4.03	Quality of public schools, 2004	54
4.04	Internet access in schools, 2004	74

Readiness Component Index (continued)		
4.05	Buyer sophistication, 2004	86
4.06	Buyer dynamism, 2004	90
4.07	Residential telephone connection charge, 2002	71
4.08	Affordability of Internet access, 2003	63
5.01	Investment in training, 2004	64
5.02	Availability of training services, 2004	88
5.03	Quality of business schools, 2004	87
5.04	Business investment in R&D, 2004	89
5.05	Business monthly telephone subscription, 2002–3	56
5.06	Business telephone connection charge, 2002–3	68
6.01	Government prioritization of ICT, 2004	87
6.02	Government procurement of ICT, 2004	78

Usage Component Index		86
Individual Usage		57
Business Usage		86
Government Usage		94
7.01	Cellular mobile subscribers, 2003	68
7.02	Telephone subscribers, 2002	51
7.03	Public payphones, 2002	51
7.04	Telephone lines, 2002	43
7.05	Television sets, 2002	54
7.06	Broadband-DSL Internet subscribers, 2002–3	65
7.07	Broadband-cable modem, 2002–3	60
7.08	Internet users per 100 inhabitants, 2002	63
8.01	Prevalence of foreign technology licensing, 2004	93
8.02	Firm-level technology absorption, 2004	101
8.03	Capacity for innovation, 2004	65
8.04	Availability of new telephone lines, 2004	57
8.05	Availability of cellular phones, 2004	81
9.01	Government success in ICT promotion, 2004	93
9.02	Government online services, 2003	72

Madagascar

Key Indicators

Population (mn), 2003	17.4
GDP per capita (PPP US$), 2003	808
Internet users per 100 inhabitants, 2002	0.3

Networked Readiness Index Rank

Year (number of countries)	Rank
2004 (104)	**87**
2003 (102)	92
2002 (82)	–

RANK/104

Environment Component Index		82
Market Environment		76
Political and Regulatory Environment		82
Infrastructure Environment		102
1.01	Availability of scientists and engineers, 2004	68
1.02	Venture capital availability, 2004	78
1.03	Sophistication of financial markets, 2004	78
1.04	Technological sophistication, 2004	90
1.05	State of cluster development, 2004	54
1.06	Collaboration in clusters, 2004	49
1.07	University-industry collaboration, 2004	56
1.08	Quality of scientific research institutions, 2004	82
1.09	Subsidies for firm-level R&D, 2004	69
1.10	Brain drain, 2004	75
1.11	Ease of access to loans, 2004	96
1.12	Administrative burden, 2004	81
1.13	Ease to start a new business, 2004	47
2.01	Effectiveness of lawmaking, 2004	68
2.02	Laws relating to ICT, 2004	90
2.03	Effectiveness of judiciary, 2004	80
2.04	Intellectual property protection, 2004	69
3.01	Telephone mainlines, 2002	102
3.02	Secure Internet servers, 2003	95
3.03	Internet hosts, 2003	93

Readiness Component Index		84
Individual Readiness		94
Business Readiness		88
Government Readiness		42
4.01	Quality of math and science education, 2004	64
4.02	Quality of educational system, 2004	92
4.03	Quality of public schools, 2004	93
4.04	Internet access in schools, 2004	95

Readiness Component Index (continued)		
4.05	Buyer sophistication, 2004	101
4.06	Buyer dynamism, 2004	58
4.07	Residential telephone connection charge, 2002	89
4.08	Affordability of Internet access, 2003	99
5.01	Investment in training, 2004	91
5.02	Availability of training services, 2004	76
5.03	Quality of business schools, 2004	64
5.04	Business investment in R&D, 2004	55
5.05	Business monthly telephone subscription, 2002–3	95
5.06	Business telephone connection charge, 2002–3	83
6.01	Government prioritization of ICT, 2004	44
6.02	Government procurement of ICT, 2004	37

Usage Component Index		88
Individual Usage		101
Business Usage		88
Government Usage		64
7.01	Cellular mobile subscribers, 2003	98
7.02	Telephone subscribers, 2002	100
7.03	Public payphones, 2002	99
7.04	Telephone lines, 2002	102
7.05	Television sets, 2002	99
7.06	Broadband-DSL Internet subscribers, 2002–3	65
7.07	Broadband-cable modem, 2002–3	60
7.08	Internet users per 100 inhabitants, 2002	96
8.01	Prevalence of foreign technology licensing, 2004	96
8.02	Firm-level technology absorption, 2004	48
8.03	Capacity for innovation, 2004	81
8.04	Availability of new telephone lines, 2004	91
8.05	Availability of cellular phones, 2004	92
9.01	Government success in ICT promotion, 2004	32
9.02	Government online services, 2003	96

Malawi

Key Indicators

Population (mn), 2003	12.1
GDP per capita (PPP US$), 2003	618
Internet users per 100 inhabitants, 2002	0.3

RANK/104

Environment Component Index	73
Market Environment	77
Political and Regulatory Environment	55
Infrastructure Environment	94

1.01	Availability of scientists and engineers, 2004	76
1.02	Venture capital availability, 2004	96
1.03	Sophistication of financial markets, 2004	99
1.04	Technological sophistication, 2004	95
1.05	State of cluster development, 2004	70
1.06	Collaboration in clusters, 2004	75
1.07	University-industry collaboration, 2004	81
1.08	Quality of scientific research institutions, 2004	54
1.09	Subsidies for firm-level R&D, 2004	81
1.10	Brain drain, 2004	78
1.11	Ease of access to loans, 2004	89
1.12	Administrative burden, 2004	29
1.13	Ease to start a new business, 2004	53
2.01	Effectiveness of lawmaking, 2004	57
2.02	Laws relating to ICT, 2004	87
2.03	Effectiveness of judiciary, 2004	36
2.04	Intellectual property protection, 2004	57
3.01	Telephone mainlines, 2002	94
3.02	Secure Internet servers, 2003	100
3.03	Internet hosts, 2003	100

Readiness Component Index	96
Individual Readiness	97
Business Readiness	90
Government Readiness	82

4.01	Quality of math and science education, 2004	81
4.02	Quality of educational system, 2004	75
4.03	Quality of public schools, 2004	101
4.04	Internet access in schools, 2004	100

Readiness Component Index (continued)	
4.05	Buyer sophistication, 2004	87
4.06	Buyer dynamism, 2004	96
4.07	Residential telephone connection charge, 2002	91
4.08	Affordability of Internet access, 2003	103
5.01	Investment in training, 2004	75
5.02	Availability of training services, 2004	85
5.03	Quality of business schools, 2004	94
5.04	Business investment in R&D, 2004	77
5.05	Business monthly telephone subscription, 2002–3	91
5.06	Business telephone connection charge, 2002–3	87
6.01	Government prioritization of ICT, 2004	88
6.02	Government procurement of ICT, 2004	71

Usage Component Index	96
Individual Usage	103
Business Usage	91
Government Usage	88

7.01	Cellular mobile subscribers, 2003	99
7.02	Telephone subscribers, 2002	99
7.03	Public payphones, 2002	102
7.04	Telephone lines, 2002	93
7.05	Television sets, 2002	103
7.06	Broadband-DSL Internet subscribers, 2002–3	65
7.07	Broadband-cable modem, 2002–3	59
7.08	Internet users per 100 inhabitants, 2002	99
8.01	Prevalence of foreign technology licensing, 2004	70
8.02	Firm-level technology absorption, 2004	88
8.03	Capacity for innovation, 2004	93
8.04	Availability of new telephone lines, 2004	97
8.05	Availability of cellular phones, 2004	91
9.01	Government success in ICT promotion, 2004	73
9.02	Government online services, 2003	93

Malaysia

Key Indicators

Population (mn), 2003	24.4
GDP per capita (PPP US$), 2003	9,696
Internet users per 100 inhabitants, 2002	32.0

Networked Readiness Index Rank

Year (number of countries)	Rank
2004 (104)	**27**
2003 (102)	26
2002 (82)	32

RANK/104

Environment Component Index — 25

Market Environment		16
Political and Regulatory Environment		20
Infrastructure Environment		56
1.01	Availability of scientists and engineers, 2004	43
1.02	Venture capital availability, 2004	19
1.03	Sophistication of financial markets, 2004	29
1.04	Technological sophistication, 2004	28
1.05	State of cluster development, 2004	18
1.06	Collaboration in clusters, 2004	23
1.07	University-industry collaboration, 2004	15
1.08	Quality of scientific research institutions, 2004	25
1.09	Subsidies for firm-level R&D, 2004	6
1.10	Brain drain, 2004	18
1.11	Ease of access to loans, 2004	23
1.12	Administrative burden, 2004	10
1.13	Ease to start a new business, 2004	17
2.01	Effectiveness of lawmaking, 2004	6
2.02	Laws relating to ICT, 2004	16
2.03	Effectiveness of judiciary, 2004	31
2.04	Intellectual property protection, 2004	25
3.01	Telephone mainlines, 2002	57
3.02	Secure Internet servers, 2003	51
3.03	Internet hosts, 2003	51

Readiness Component Index — 17

Individual Readiness		22
Business Readiness		28
Government Readiness		8
4.01	Quality of math and science education, 2004	27
4.02	Quality of educational system, 2004	18
4.03	Quality of public schools, 2004	19
4.04	Internet access in schools, 2004	28

Readiness Component Index (continued)

4.05	Buyer sophistication, 2004	24
4.06	Buyer dynamism, 2004	29
4.07	Residential telephone connection charge, 2002	19
4.08	Affordability of Internet access, 2003	32
5.01	Investment in training, 2004	26
5.02	Availability of training services, 2004	35
5.03	Quality of business schools, 2004	31
5.04	Business investment in R&D, 2004	23
5.05	Business monthly telephone subscription, 2002–3	54
5.06	Business telephone connection charge, 2002–3	20
6.01	Government prioritization of ICT, 2004	23
6.02	Government procurement of ICT, 2004	5

Usage Component Index — 29

Individual Usage		39
Business Usage		44
Government Usage		14
7.01	Cellular mobile subscribers, 2003	41
7.02	Telephone subscribers, 2002	46
7.03	Public payphones, 2002	13
7.04	Telephone lines, 2002	59
7.05	Television sets, 2002	59
7.06	Broadband-DSL Internet subscribers, 2002–3	45
7.07	Broadband-cable modem, 2002–3	60
7.08	Internet users per 100 inhabitants, 2002	25
8.01	Prevalence of foreign technology licensing, 2004	16
8.02	Firm-level technology absorption, 2004	41
8.03	Capacity for innovation, 2004	32
8.04	Availability of new telephone lines, 2004	63
8.05	Availability of cellular phones, 2004	83
9.01	Government success in ICT promotion, 2004	8
9.02	Government online services, 2003	20

Mali

Key Indicators

Population (mn), 2003	13
GDP per capita (PPP US$), 2003	994
Internet users per 100 inhabitants, 2002	0.2

RANK/104

Environment Component Index	80
Market Environment	80
Political and Regulatory Environment	70
Infrastructure Environment	97

1.01	Availability of scientists and engineers, 2004	85
1.02	Venture capital availability, 2004	94
1.03	Sophistication of financial markets, 2004	82
1.04	Technological sophistication, 2004	86
1.05	State of cluster development, 2004	103
1.06	Collaboration in clusters, 2004	50
1.07	University-industry collaboration, 2004	38
1.08	Quality of scientific research institutions, 2004	85
1.09	Subsidies for firm-level R&D, 2004	38
1.10	Brain drain, 2004	88
1.11	Ease of access to loans, 2004	91
1.12	Administrative burden, 2004	40
1.13	Ease to start a new business, 2004	68
2.01	Effectiveness of lawmaking, 2004	46
2.02	Laws relating to ICT, 2004	88
2.03	Effectiveness of judiciary, 2004	76
2.04	Intellectual property protection, 2004	66
3.01	Telephone mainlines, 2002	97
3.02	Secure Internet servers, 2003	93
3.03	Internet hosts, 2003	97

Readiness Component Index	95
Individual Readiness	98
Business Readiness	99
Government Readiness	43

4.01	Quality of math and science education, 2004	77
4.02	Quality of educational system, 2004	86
4.03	Quality of public schools, 2004	91
4.04	Internet access in schools, 2004	96

Readiness Component Index (continued)	
4.05	Buyer sophistication, 2004	94
4.06	Buyer dynamism, 2004	88
4.07	Residential telephone connection charge, 2002	101
4.08	Affordability of Internet access, 2003	96
5.01	Investment in training, 2004	100
5.02	Availability of training services, 2004	78
5.03	Quality of business schools, 2004	85
5.04	Business investment in R&D, 2004	82
5.05	Business monthly telephone subscription, 2002–3	101
5.06	Business telephone connection charge, 2002–3	101
6.01	Government prioritization of ICT, 2004	29
6.02	Government procurement of ICT, 2004	53

Usage Component Index	93
Individual Usage	96
Business Usage	98
Government Usage	67

7.01	Cellular mobile subscribers, 2003	94
7.02	Telephone subscribers, 2002	102
7.03	Public payphones, 2002	82
7.04	Telephone lines, 2002	99
7.05	Television sets, 2002	95
7.06	Broadband-DSL Internet subscribers, 2002–3	65
7.07	Broadband-cable modem, 2002–3	60
7.08	Internet users per 100 inhabitants, 2002	100
8.01	Prevalence of foreign technology licensing, 2004	97
8.02	Firm-level technology absorption, 2004	72
8.03	Capacity for innovation, 2004	83
8.04	Availability of new telephone lines, 2004	92
8.05	Availability of cellular phones, 2004	100
9.01	Government success in ICT promotion, 2004	24
9.02	Government online services, 2003	102

Part 2 Country Profiles

165

Malta

Key Indicators

Population (mn), 2003	0.394
GDP per capita (PPP US$), 2003	18,203
Internet users per 100 inhabitants, 2002	30.3

Networked Readiness Index Rank

Year (number of countries)	Rank
2004 (104)	**28**
2003 (102)	27
2002 (82)	—

RANK/104

Environment Component Index	28
Market Environment	58
Political and Regulatory Environment	29
Infrastructure Environment	21

1.01	Availability of scientists and engineers, 2004	79
1.02	Venture capital availability, 2004	73
1.03	Sophistication of financial markets, 2004	54
1.04	Technological sophistication, 2004	35
1.05	State of cluster development, 2004	95
1.06	Collaboration in clusters, 2004	94
1.07	University-industry collaboration, 2004	76
1.08	Quality of scientific research institutions, 2004	72
1.09	Subsidies for firm-level R&D, 2004	57
1.10	Brain drain, 2004	36
1.11	Ease of access to loans, 2004	37
1.12	Administrative burden, 2004	74
1.13	Ease to start a new business, 2004	46
2.01	Effectiveness of lawmaking, 2004	23
2.02	Laws relating to ICT, 2004	22
2.03	Effectiveness of judiciary, 2004	25
2.04	Intellectual property protection, 2004	50
3.01	Telephone mainlines, 2002	18
3.02	Secure Internet servers, 2003	15
3.03	Internet hosts, 2003	34

Readiness Component Index	35
Individual Readiness	31
Business Readiness	59
Government Readiness	36

4.01	Quality of math and science education, 2004	35
4.02	Quality of educational system, 2004	20
4.03	Quality of public schools, 2004	37
4.04	Internet access in schools, 2004	17

Readiness Component Index (continued)	
4.05	Buyer sophistication, 2004	53
4.06	Buyer dynamism, 2004	54
4.07	Residential telephone connection charge, 2002	35
4.08	Affordability of Internet access, 2003	28
5.01	Investment in training, 2004	47
5.02	Availability of training services, 2004	92
5.03	Quality of business schools, 2004	63
5.04	Business investment in R&D, 2004	84
5.05	Business monthly telephone subscription, 2002–3	32
5.06	Business telephone connection charge, 2002–3	40
6.01	Government prioritization of ICT, 2004	11
6.02	Government procurement of ICT, 2004	61

Usage Component Index	26
Individual Usage	21
Business Usage	50
Government Usage	11

7.01	Cellular mobile subscribers, 2003	26
7.02	Telephone subscribers, 2002	23
7.03	Public payphones, 2002	54
7.04	Telephone lines, 2002	19
7.05	Television sets, 2002	38
7.06	Broadband-DSL Internet subscribers, 2002–3	14
7.07	Broadband-cable modem, 2002–3	15
7.08	Internet users per 100 inhabitants, 2002	27
8.01	Prevalence of foreign technology licensing, 2004	52
8.02	Firm-level technology absorption, 2004	49
8.03	Capacity for innovation, 2004	78
8.04	Availability of new telephone lines, 2004	44
8.05	Availability of cellular phones, 2004	53
9.01	Government success in ICT promotion, 2004	5
9.02	Government online services, 2003	27

Mauritius

Key Indicators

Population (mn), 2003	1.2
GDP per capita (PPP US$), 2003	11,258
Internet users per 100 inhabitants, 2002	10.3

RANK/104

Environment Component Index	50
Market Environment	56
Political and Regulatory Environment	46
Infrastructure Environment	45

1.01	Availability of scientists and engineers, 2004	81
1.02	Venture capital availability, 2004	47
1.03	Sophistication of financial markets, 2004	58
1.04	Technological sophistication, 2004	55
1.05	State of cluster development, 2004	63
1.06	Collaboration in clusters, 2004	51
1.07	University-industry collaboration, 2004	59
1.08	Quality of scientific research institutions, 2004	57
1.09	Subsidies for firm-level R&D, 2004	54
1.10	Brain drain, 2004	56
1.11	Ease of access to loans, 2004	32
1.12	Administrative burden, 2004	84
1.13	Ease to start a new business, 2004	60
2.01	Effectiveness of lawmaking, 2004	32
2.02	Laws relating to ICT, 2004	42
2.03	Effectiveness of judiciary, 2004	55
2.04	Intellectual property protection, 2004	55
3.01	Telephone mainlines, 2002	43
3.02	Secure Internet servers, 2003	41
3.03	Internet hosts, 2003	53

Readiness Component Index	37
Individual Readiness	52
Business Readiness	53
Government Readiness	27

4.01	Quality of math and science education, 2004	66
4.02	Quality of educational system, 2004	57
4.03	Quality of public schools, 2004	57
4.04	Internet access in schools, 2004	56

Readiness Component Index (continued)	
4.05	Buyer sophistication, 2004	45
4.06	Buyer dynamism, 2004	71
4.07	Residential telephone connection charge, 2002	40
4.08	Affordability of Internet access, 2003	44
5.01	Investment in training, 2004	37
5.02	Availability of training services, 2004	77
5.03	Quality of business schools, 2004	80
5.04	Business investment in R&D, 2004	50
5.05	Business monthly telephone subscription, 2002–3	42
5.06	Business telephone connection charge, 2002–3	52
6.01	Government prioritization of ICT, 2004	4
6.02	Government procurement of ICT, 2004	58

Usage Component Index	50
Individual Usage	48
Business Usage	57
Government Usage	34

7.01	Cellular mobile subscribers, 2003	45
7.02	Telephone subscribers, 2002	47
7.03	Public payphones, 2002	48
7.04	Telephone lines, 2002	38
7.05	Television sets, 2002	39
7.06	Broadband-DSL Internet subscribers, 2002–3	52
7.07	Broadband-cable modem, 2002–3	60
7.08	Internet users per 100 inhabitants, 2002	49
8.01	Prevalence of foreign technology licensing, 2004	48
8.02	Firm-level technology absorption, 2004	55
8.03	Capacity for innovation, 2004	59
8.04	Availability of new telephone lines, 2004	65
8.05	Availability of cellular phones, 2004	66
9.01	Government success in ICT promotion, 2004	22
9.02	Government online services, 2003	47

Mexico

Key Indicators

Population (mn), 2003	104
GDP per capita (PPP US$), 2003	9,136
Internet users per 100 inhabitants, 2002	9.8

Networked Readiness Index Rank

Year (number of countries) Rank

2004 (104) **60**

2003 (102) 44

2002 (82) 47

RANK/104

Environment Component Index	74
Market Environment	73
Political and Regulatory Environment	73
Infrastructure Environment	57
1.01 Availability of scientists and engineers, 2004	89
1.02 Venture capital availability, 2004	85
1.03 Sophistication of financial markets, 2004	35
1.04 Technological sophistication, 2004	46
1.05 State of cluster development, 2004	49
1.06 Collaboration in clusters, 2004	58
1.07 University-industry collaboration, 2004	45
1.08 Quality of scientific research institutions, 2004	58
1.09 Subsidies for firm-level R&D, 2004	62
1.10 Brain drain, 2004	52
1.11 Ease of access to loans, 2004	83
1.12 Administrative burden, 2004	87
1.13 Ease to start a new business, 2004	91
2.01 Effectiveness of lawmaking, 2004	96
2.02 Laws relating to ICT, 2004	57
2.03 Effectiveness of judiciary, 2004	65
2.04 Intellectual property protection, 2004	62
3.01 Telephone mainlines, 2002	61
3.02 Secure Internet servers, 2003	56
3.03 Internet hosts, 2003	38

Readiness Component Index	57
Individual Readiness	68
Business Readiness	45
Government Readiness	64
4.01 Quality of math and science education, 2004	88
4.02 Quality of educational system, 2004	77
4.03 Quality of public schools, 2004	74
4.04 Internet access in schools, 2004	62

Readiness Component Index (continued)	
4.05 Buyer sophistication, 2004	70
4.06 Buyer dynamism, 2004	77
4.07 Residential telephone connection charge, 2002	42
4.08 Affordability of Internet access, 2003	43
5.01 Investment in training, 2004	67
5.02 Availability of training services, 2004	45
5.03 Quality of business schools, 2004	36
5.04 Business investment in R&D, 2004	57
5.05 Business monthly telephone subscription, 2002–3	67
5.06 Business telephone connection charge, 2002–3	1
6.01 Government prioritization of ICT, 2004	64
6.02 Government procurement of ICT, 2004	63

Usage Component Index	55
Individual Usage	55
Business Usage	61
Government Usage	50
7.01 Cellular mobile subscribers, 2003	52
7.02 Telephone subscribers, 2002	56
7.03 Public payphones, 2002	53
7.04 Telephone lines, 2002	61
7.05 Television sets, 2002	34
7.06 Broadband-DSL Internet subscribers, 2002–3	48
7.07 Broadband-cable modem, 2002–3	31
7.08 Internet users per 100 inhabitants, 2002	51
8.01 Prevalence of foreign technology licensing, 2004	39
8.02 Firm-level technology absorption, 2004	68
8.03 Capacity for innovation, 2004	54
8.04 Availability of new telephone lines, 2004	67
8.05 Availability of cellular phones, 2004	76
9.01 Government success in ICT promotion, 2004	73
9.02 Government online services, 2003	24

Morocco

Key Indicators

Population (mn), 2003	30.6
GDP per capita (PPP US$), 2003	4,012
Internet users per 100 inhabitants, 2002	2.4

RANK/104

Environment Component Index	53
Market Environment	52
Political and Regulatory Environment	47
Infrastructure Environment	88

1.01	Availability of scientists and engineers, 2004	56
1.02	Venture capital availability, 2004	40
1.03	Sophistication of financial markets, 2004	52
1.04	Technological sophistication, 2004	71
1.05	State of cluster development, 2004	28
1.06	Collaboration in clusters, 2004	36
1.07	University-industry collaboration, 2004	55
1.08	Quality of scientific research institutions, 2004	81
1.09	Subsidies for firm-level R&D, 2004	42
1.10	Brain drain, 2004	72
1.11	Ease of access to loans, 2004	53
1.12	Administrative burden, 2004	44
1.13	Ease to start a new business, 2004	50
2.01	Effectiveness of lawmaking, 2004	48
2.02	Laws relating to ICT, 2004	50
2.03	Effectiveness of judiciary, 2004	47
2.04	Intellectual property protection, 2004	40
3.01	Telephone mainlines, 2002	87
3.02	Secure Internet servers, 2003	82
3.03	Internet hosts, 2003	87

Readiness Component Index	54
Individual Readiness	55
Business Readiness	63
Government Readiness	57

4.01	Quality of math and science education, 2004	48
4.02	Quality of educational system, 2004	51
4.03	Quality of public schools, 2004	47
4.04	Internet access in schools, 2004	58

Readiness Component Index (continued)	
4.05	Buyer sophistication, 2004	47
4.06	Buyer dynamism, 2004	47
4.07	Residential telephone connection charge, 2002	73
4.08	Affordability of Internet access, 2003	76
5.01	Investment in training, 2004	41
5.02	Availability of training services, 2004	52
5.03	Quality of business schools, 2004	34
5.04	Business investment in R&D, 2004	66
5.05	Business monthly telephone subscription, 2002–3	88
5.06	Business telephone connection charge, 2002–3	76
6.01	Government prioritization of ICT, 2004	62
6.02	Government procurement of ICT, 2004	48

Usage Component Index	57
Individual Usage	73
Business Usage	55
Government Usage	46

7.01	Cellular mobile subscribers, 2003	60
7.02	Telephone subscribers, 2002	70
7.03	Public payphones, 2002	45
7.04	Telephone lines, 2002	85
7.05	Television sets, 2002	75
7.06	Broadband-DSL Internet subscribers, 2002–3	65
7.07	Broadband-cable modem, 2002–3	60
7.08	Internet users per 100 inhabitants, 2002	79
8.01	Prevalence of foreign technology licensing, 2004	77
8.02	Firm-level technology absorption, 2004	74
8.03	Capacity for innovation, 2004	40
8.04	Availability of new telephone lines, 2004	45
8.05	Availability of cellular phones, 2004	55
9.01	Government success in ICT promotion, 2004	36
9.02	Government online services, 2003	57

Mozambique

Key Indicators

Population (mn), 2003	18.9
GDP per capita (PPP US$), 2003	1,133
Internet users per 100 inhabitants, 2002	0.3

Networked Readiness Index Rank

Year (number of countries) Rank

2004 (104) **96**

2003 (102) 97

2002 (82) —

RANK/104

Environment Component Index		95
Market Environment		102
Political and Regulatory Environment		84
Infrastructure Environment		98
1.01	Availability of scientists and engineers, 2004	97
1.02	Venture capital availability, 2004	89
1.03	Sophistication of financial markets, 2004	95
1.04	Technological sophistication, 2004	99
1.05	State of cluster development, 2004	85
1.06	Collaboration in clusters, 2004	86
1.07	University-industry collaboration, 2004	75
1.08	Quality of scientific research institutions, 2004	80
1.09	Subsidies for firm-level R&D, 2004	84
1.10	Brain drain, 2004	65
1.11	Ease of access to loans, 2004	104
1.12	Administrative burden, 2004	88
1.13	Ease to start a new business, 2004	104
2.01	Effectiveness of lawmaking, 2004	73
2.02	Laws relating to ICT, 2004	82
2.03	Effectiveness of judiciary, 2004	83
2.04	Intellectual property protection, 2004	89
3.01	Telephone mainlines, 2002	101
3.02	Secure Internet servers, 2003	91
3.03	Internet hosts, 2003	84

Readiness Component Index		100
Individual Readiness		95
Business Readiness		102
Government Readiness		68
4.01	Quality of math and science education, 2004	91
4.02	Quality of educational system, 2004	83
4.03	Quality of public schools, 2004	88
4.04	Internet access in schools, 2004	97

Readiness Component Index (continued)		
4.05	Buyer sophistication, 2004	103
4.06	Buyer dynamism, 2004	101
4.07	Residential telephone connection charge, 2002	88
4.08	Affordability of Internet access, 2003	97
5.01	Investment in training, 2004	87
5.02	Availability of training services, 2004	100
5.03	Quality of business schools, 2004	99
5.04	Business investment in R&D, 2004	81
5.05	Business monthly telephone subscription, 2002–3	104
5.06	Business telephone connection charge, 2002–3	81
6.01	Government prioritization of ICT, 2004	51
6.02	Government procurement of ICT, 2004	81

Usage Component Index		90
Individual Usage		99
Business Usage		90
Government Usage		59
7.01	Cellular mobile subscribers, 2003	95
7.02	Telephone subscribers, 2002	96
7.03	Public payphones, 2002	93
7.04	Telephone lines, 2002	100
7.05	Television sets, 2002	100
7.06	Broadband-DSL Internet subscribers, 2002–3	65
7.07	Broadband-cable modem, 2002–3	60
7.08	Internet users per 100 inhabitants, 2002	98
8.01	Prevalence of foreign technology licensing, 2004	85
8.02	Firm-level technology absorption, 2004	97
8.03	Capacity for innovation, 2004	100
8.04	Availability of new telephone lines, 2004	78
8.05	Availability of cellular phones, 2004	86
9.01	Government success in ICT promotion, 2004	41
9.02	Government online services, 2003	86

Namibia

Key Indicators

Population (mn), 2003	2
GDP per capita (PPP US$), 2003	6,375
Internet users per 100 inhabitants, 2002	2.7

Part 2 Country Profiles

Networked Readiness Index Rank

Year (number of countries)	Rank
2004 (104)	**55**
2003 (102)	59
2002 (82)	53

RANK/104

Environment Component Index — 44

Market Environment		44
Political and Regulatory Environment		30
Infrastructure Environment		78
1.01	Availability of scientists and engineers, 2004	98
1.02	Venture capital availability, 2004	63
1.03	Sophistication of financial markets, 2004	45
1.04	Technological sophistication, 2004	40
1.05	State of cluster development, 2004	71
1.06	Collaboration in clusters, 2004	87
1.07	University-industry collaboration, 2004	66
1.08	Quality of scientific research institutions, 2004	71
1.09	Subsidies for firm-level R&D, 2004	47
1.10	Brain drain, 2004	57
1.11	Ease of access to loans, 2004	42
1.12	Administrative burden, 2004	11
1.13	Ease to start a new business, 2004	24
2.01	Effectiveness of lawmaking, 2004	15
2.02	Laws relating to ICT, 2004	52
2.03	Effectiveness of judiciary, 2004	23
2.04	Intellectual property protection, 2004	33
3.01	Telephone mainlines, 2002	79
3.02	Secure Internet servers, 2003	53
3.03	Internet hosts, 2003	66

Readiness Component Index — 56

Individual Readiness		60
Business Readiness		68
Government Readiness		55
4.01	Quality of math and science education, 2004	76
4.02	Quality of educational system, 2004	58
4.03	Quality of public schools, 2004	59
4.04	Internet access in schools, 2004	51

Readiness Component Index (continued)

4.05	Buyer sophistication, 2004	38
4.06	Buyer dynamism, 2004	70
4.07	Residential telephone connection charge, 2002	60
4.08	Affordability of Internet access, 2003	75
5.01	Investment in training, 2004	34
5.02	Availability of training services, 2004	98
5.03	Quality of business schools, 2004	102
5.04	Business investment in R&D, 2004	42
5.05	Business monthly telephone subscription, 2002–3	72
5.06	Business telephone connection charge, 2002–3	56
6.01	Government prioritization of ICT, 2004	57
6.02	Government procurement of ICT, 2004	49

Usage Component Index — 71

Individual Usage		83
Business Usage		53
Government Usage		76
7.01	Cellular mobile subscribers, 2003	74
7.02	Telephone subscribers, 2002	80
7.03	Public payphones, 2002	67
7.04	Telephone lines, 2002	78
7.05	Television sets, 2002	84
7.06	Broadband-DSL Internet subscribers, 2002–3	65
7.07	Broadband-cable modem, 2002–3	60
7.08	Internet users per 100 inhabitants, 2002	76
8.01	Prevalence of foreign technology licensing, 2004	36
8.02	Firm-level technology absorption, 2004	45
8.03	Capacity for innovation, 2004	88
8.04	Availability of new telephone lines, 2004	53
8.05	Availability of cellular phones, 2004	75
9.01	Government success in ICT promotion, 2004	61
9.02	Government online services, 2003	93

Netherlands

Key Indicators

Population (mn), 2003	16.1
GDP per capita (PPP US$), 2003	29,412
Internet users per 100 inhabitants, 2002	50.6

Networked Readiness Index Rank

Year (number of countries)	Rank
2004 (104)	**16**
2003 (102)	13
2002 (82)	11

RANK/104

Environment Component Index		14
Market Environment		13
Political and Regulatory Environment		12
Infrastructure Environment		12
1.01	Availability of scientists and engineers, 2004	36
1.02	Venture capital availability, 2004	6
1.03	Sophistication of financial markets, 2004	10
1.04	Technological sophistication, 2004	20
1.05	State of cluster development, 2004	30
1.06	Collaboration in clusters, 2004	10
1.07	University-industry collaboration, 2004	10
1.08	Quality of scientific research institutions, 2004	11
1.09	Subsidies for firm-level R&D, 2004	18
1.10	Brain drain, 2004	8
1.11	Ease of access to loans, 2004	10
1.12	Administrative burden, 2004	49
1.13	Ease to start a new business, 2004	23
2.01	Effectiveness of lawmaking, 2004	17
2.02	Laws relating to ICT, 2004	29
2.03	Effectiveness of judiciary, 2004	5
2.04	Intellectual property protection, 2004	10
3.01	Telephone mainlines, 2002	11
3.02	Secure Internet servers, 2003	57
3.03	Internet hosts, 2003	4

Readiness Component Index		16
Individual Readiness		13
Business Readiness		10
Government Readiness		33
4.01	Quality of math and science education, 2004	24
4.02	Quality of educational system, 2004	16
4.03	Quality of public schools, 2004	4
4.04	Internet access in schools, 2004	16

Readiness Component Index (continued)		
4.05	Buyer sophistication, 2004	16
4.06	Buyer dynamism, 2004	15
4.07	Residential telephone connection charge, 2002	25
4.08	Affordability of Internet access, 2003	20
5.01	Investment in training, 2004	6
5.02	Availability of training services, 2004	7
5.03	Quality of business schools, 2004	9
5.04	Business investment in R&D, 2004	10
5.05	Business monthly telephone subscription, 2002–3	52
5.06	Business telephone connection charge, 2002–3	26
6.01	Government prioritization of ICT, 2004	30
6.02	Government procurement of ICT, 2004	27

Usage Component Index		21
Individual Usage		9
Business Usage		14
Government Usage		44
7.01	Cellular mobile subscribers, 2003	23
7.02	Telephone subscribers, 2002	13
7.03	Public payphones, 2002	41
7.04	Telephone lines, 2002	11
7.05	Television sets, 2002	9
7.06	Broadband-DSL Internet subscribers, 2002–3	19
7.07	Broadband-cable modem, 2002–3	4
7.08	Internet users per 100 inhabitants, 2002	8
8.01	Prevalence of foreign technology licensing, 2004	13
8.02	Firm-level technology absorption, 2004	40
8.03	Capacity for innovation, 2004	11
8.04	Availability of new telephone lines, 2004	10
8.05	Availability of cellular phones, 2004	15
9.01	Government success in ICT promotion, 2004	59
9.02	Government online services, 2003	36

New Zealand

Key Indicators

Population (mn), 2003	3.9
GDP per capita (PPP US$), 2003	21,177
Internet users per 100 inhabitants, 2002	48.4

RANK/104

Environment Component Index	12
Market Environment	23
Political and Regulatory Environment	9
Infrastructure Environment	7

1.01	Availability of scientists and engineers, 2004	32
1.02	Venture capital availability, 2004	9
1.03	Sophistication of financial markets, 2004	11
1.04	Technological sophistication, 2004	22
1.05	State of cluster development, 2004	47
1.06	Collaboration in clusters, 2004	34
1.07	University-industry collaboration, 2004	26
1.08	Quality of scientific research institutions, 2004	17
1.09	Subsidies for firm-level R&D, 2004	74
1.10	Brain drain, 2004	47
1.11	Ease of access to loans, 2004	7
1.12	Administrative burden, 2004	49
1.13	Ease to start a new business, 2004	3
2.01	Effectiveness of lawmaking, 2004	11
2.02	Laws relating to ICT, 2004	10
2.03	Effectiveness of judiciary, 2004	12
2.04	Intellectual property protection, 2004	12
3.01	Telephone mainlines, 2002	29
3.02	Secure Internet servers, 2003	4
3.03	Internet hosts, 2003	7

Readiness Component Index	28
Individual Readiness	16
Business Readiness	21
Government Readiness	53

4.01	Quality of math and science education, 2004	36
4.02	Quality of educational system, 2004	21
4.03	Quality of public schools, 2004	13
4.04	Internet access in schools, 2004	15

Readiness Component Index (continued)	
4.05	Buyer sophistication, 2004	10
4.06	Buyer dynamism, 2004	7
4.07	Residential telephone connection charge, 2002	6
4.08	Affordability of Internet access, 2003	16
5.01	Investment in training, 2004	22
5.02	Availability of training services, 2004	19
5.03	Quality of business schools, 2004	22
5.04	Business investment in R&D, 2004	30
5.05	Business monthly telephone subscription, 2002–3	37
5.06	Business telephone connection charge, 2002–3	10
6.01	Government prioritization of ICT, 2004	53
6.02	Government procurement of ICT, 2004	46

Usage Component Index	24
Individual Usage	25
Business Usage	15
Government Usage	36

7.01	Cellular mobile subscribers, 2003	34
7.02	Telephone subscribers, 2002	31
7.03	Public payphones, 2002	14
7.04	Telephone lines, 2002	28
7.05	Television sets, 2002	12
7.06	Broadband-DSL Internet subscribers, 2002–3	27
7.07	Broadband-cable modem, 2002–3	39
7.08	Internet users per 100 inhabitants, 2002	11
8.01	Prevalence of foreign technology licensing, 2004	6
8.02	Firm-level technology absorption, 2004	17
8.03	Capacity for innovation, 2004	25
8.04	Availability of new telephone lines, 2004	14
8.05	Availability of cellular phones, 2004	23
9.01	Government success in ICT promotion, 2004	68
9.02	Government online services, 2003	15

Nicaragua

Key Indicators

Population (mn), 2003	5.5
GDP per capita (PPP US$), 2003	2,523
Internet users per 100 inhabitants, 2002	1.7

Networked Readiness Index Rank

Year (number of countries)	Rank
2004 (104)	**103**
2003 (102)	94
2002 (82)	79

RANK/104

Environment Component Index	101
Market Environment	99
Political and Regulatory Environment	102
Infrastructure Environment	87

1.01	Availability of scientists and engineers, 2004	94
1.02	Venture capital availability, 2004	81
1.03	Sophistication of financial markets, 2004	86
1.04	Technological sophistication, 2004	96
1.05	State of cluster development, 2004	97
1.06	Collaboration in clusters, 2004	102
1.07	University-industry collaboration, 2004	95
1.08	Quality of scientific research institutions, 2004	99
1.09	Subsidies for firm-level R&D, 2004	98
1.10	Brain drain, 2004	71
1.11	Ease of access to loans, 2004	79
1.12	Administrative burden, 2004	82
1.13	Ease to start a new business, 2004	88
2.01	Effectiveness of lawmaking, 2004	103
2.02	Laws relating to ICT, 2004	92
2.03	Effectiveness of judiciary, 2004	104
2.04	Intellectual property protection, 2004	91
3.01	Telephone mainlines, 2002	89
3.02	Secure Internet servers, 2003	70
3.03	Internet hosts, 2003	69

Readiness Component Index	104
Individual Readiness	101
Business Readiness	104
Government Readiness	92

4.01	Quality of math and science education, 2004	95
4.02	Quality of educational system, 2004	97
4.03	Quality of public schools, 2004	92
4.04	Internet access in schools, 2004	82

Readiness Component Index (continued)	
4.05	Buyer sophistication, 2004	100
4.06	Buyer dynamism, 2004	102
4.07	Residential telephone connection charge, 2002	103
4.08	Affordability of Internet access, 2003	91
5.01	Investment in training, 2004	96
5.02	Availability of training services, 2004	95
5.03	Quality of business schools, 2004	75
5.04	Business investment in R&D, 2004	95
5.05	Business monthly telephone subscription, 2002–3	103
5.06	Business telephone connection charge, 2002–3	104
6.01	Government prioritization of ICT, 2004	86
6.02	Government procurement of ICT, 2004	93

Usage Component Index	99
Individual Usage	88
Business Usage	99
Government Usage	99

7.01	Cellular mobile subscribers, 2003	86
7.02	Telephone subscribers, 2002	87
7.03	Public payphones, 2002	97
7.04	Telephone lines, 2002	87
7.05	Television sets, 2002	85
7.06	Broadband-DSL Internet subscribers, 2002–3	65
7.07	Broadband-cable modem, 2002–3	48
7.08	Internet users per 100 inhabitants, 2002	85
8.01	Prevalence of foreign technology licensing, 2004	102
8.02	Firm-level technology absorption, 2004	100
8.03	Capacity for innovation, 2004	103
8.04	Availability of new telephone lines, 2004	90
8.05	Availability of cellular phones, 2004	61
9.01	Government success in ICT promotion, 2004	90
9.02	Government online services, 2003	99

Nigeria

Key Indicators

Population (mn), 2003	124
GDP per capita (PPP US$), 2003	1,024
Internet users per 100 inhabitants, 2002	0.3

RANK/104

Environment Component Index	75
Market Environment	66
Political and Regulatory Environment	71
Infrastructure Environment	96

1.01	Availability of scientists and engineers, 2004	44
1.02	Venture capital availability, 2004	67
1.03	Sophistication of financial markets, 2004	75
1.04	Technological sophistication, 2004	83
1.05	State of cluster development, 2004	24
1.06	Collaboration in clusters, 2004	40
1.07	University-industry collaboration, 2004	53
1.08	Quality of scientific research institutions, 2004	55
1.09	Subsidies for firm-level R&D, 2004	52
1.10	Brain drain, 2004	76
1.11	Ease of access to loans, 2004	92
1.12	Administrative burden, 2004	39
1.13	Ease to start a new business, 2004	79
2.01	Effectiveness of lawmaking, 2004	62
2.02	Laws relating to ICT, 2004	70
2.03	Effectiveness of judiciary, 2004	70
2.04	Intellectual property protection, 2004	73
3.01	Telephone mainlines, 2002	96
3.02	Secure Internet servers, 2003	98
3.03	Internet hosts, 2003	98

Readiness Component Index	87
Individual Readiness	96
Business Readiness	82
Government Readiness	60

4.01	Quality of math and science education, 2004	72
4.02	Quality of educational system, 2004	68
4.03	Quality of public schools, 2004	94
4.04	Internet access in schools, 2004	90

Readiness Component Index (continued)	
4.05	Buyer sophistication, 2004	69
4.06	Buyer dynamism, 2004	63
4.07	Residential telephone connection charge, 2002	95
4.08	Affordability of Internet access, 2003	100
5.01	Investment in training, 2004	59
5.02	Availability of training services, 2004	37
5.03	Quality of business schools, 2004	55
5.04	Business investment in R&D, 2004	47
5.05	Business monthly telephone subscription, 2002–3	90
5.06	Business telephone connection charge, 2002–3	94
6.01	Government prioritization of ICT, 2004	74
6.02	Government procurement of ICT, 2004	34

Usage Component Index	87
Individual Usage	94
Business Usage	79
Government Usage	73

7.01	Cellular mobile subscribers, 2003	91
7.02	Telephone subscribers, 2002	95
7.03	Public payphones, 2002	103
7.04	Telephone lines, 2002	95
7.05	Television sets, 2002	91
7.06	Broadband-DSL Internet subscribers, 2002–3	65
7.07	Broadband-cable modem, 2002–3	60
7.08	Internet users per 100 inhabitants, 2002	95
8.01	Prevalence of foreign technology licensing, 2004	66
8.02	Firm-level technology absorption, 2004	75
8.03	Capacity for innovation, 2004	50
8.04	Availability of new telephone lines, 2004	80
8.05	Availability of cellular phones, 2004	95
9.01	Government success in ICT promotion, 2004	58
9.02	Government online services, 2003	90

Norway

Key Indicators

Population (mn), 2003	4.5
GDP per capita (PPP US$), 2003	37,063
Internet users per 100 inhabitants, 2002	50.3

Networked Readiness Index Rank

Year (number of countries)	Rank
2004 (104)	**13**
2003 (102)	8
2002 (82)	17

RANK/104

Environment Component Index	8
Market Environment	10
Political and Regulatory Environment	5
Infrastructure Environment	9

1.01	Availability of scientists and engineers, 2004	10
1.02	Venture capital availability, 2004	10
1.03	Sophistication of financial markets, 2004	20
1.04	Technological sophistication, 2004	15
1.05	State of cluster development, 2004	25
1.06	Collaboration in clusters, 2004	25
1.07	University-industry collaboration, 2004	23
1.08	Quality of scientific research institutions, 2004	16
1.09	Subsidies for firm-level R&D, 2004	19
1.10	Brain drain, 2004	2
1.11	Ease of access to loans, 2004	14
1.12	Administrative burden, 2004	28
1.13	Ease to start a new business, 2004	21
2.01	Effectiveness of lawmaking, 2004	12
2.02	Laws relating to ICT, 2004	2
2.03	Effectiveness of judiciary, 2004	4
2.04	Intellectual property protection, 2004	7
3.01	Telephone mainlines, 2002	4
3.02	Secure Internet servers, 2003	14
3.03	Internet hosts, 2003	6

Readiness Component Index	20
Individual Readiness	18
Business Readiness	17
Government Readiness	25

4.01	Quality of math and science education, 2004	45
4.02	Quality of educational system, 2004	13
4.03	Quality of public schools, 2004	14
4.04	Internet access in schools, 2004	18

Readiness Component Index (continued)	

4.05	Buyer sophistication, 2004	18
4.06	Buyer dynamism, 2004	18
4.07	Residential telephone connection charge, 2002	10
4.08	Affordability of Internet access, 2003	9
5.01	Investment in training, 2004	10
5.02	Availability of training services, 2004	18
5.03	Quality of business schools, 2004	20
5.04	Business investment in R&D, 2004	20
5.05	Business monthly telephone subscription, 2002–3	10
5.06	Business telephone connection charge, 2002–3	9
6.01	Government prioritization of ICT, 2004	22
6.02	Government procurement of ICT, 2004	28

Usage Component Index	18
Individual Usage	6
Business Usage	16
Government Usage	32

7.01	Cellular mobile subscribers, 2003	8
7.02	Telephone subscribers, 2002	3
7.03	Public payphones, 2002	30
7.04	Telephone lines, 2002	4
7.05	Television sets, 2002	4
7.06	Broadband-DSL Internet subscribers, 2002–3	12
7.07	Broadband-cable modem, 2002–3	19
7.08	Internet users per 100 inhabitants, 2002	10
8.01	Prevalence of foreign technology licensing, 2004	32
8.02	Firm-level technology absorption, 2004	16
8.03	Capacity for innovation, 2004	20
8.04	Availability of new telephone lines, 2004	11
8.05	Availability of cellular phones, 2004	1
9.01	Government success in ICT promotion, 2004	45
9.02	Government online services, 2003	30

Pakistan

Key Indicators

Population (mn), 2003	154
GDP per capita (PPP US$), 2003	1,971
Internet users per 100 inhabitants, 2002	1.0

Networked Readiness Index Rank

Year (number of countries)	Rank
2004 (104)	**63**
2003 (102)	76
2002 (82)	–

RANK/104

Environment Component Index		85
Market Environment		84
Political and Regulatory Environment		85
Infrastructure Environment		91
1.01	Availability of scientists and engineers, 2004	61
1.02	Venture capital availability, 2004	76
1.03	Sophistication of financial markets, 2004	64
1.04	Technological sophistication, 2004	84
1.05	State of cluster development, 2004	19
1.06	Collaboration in clusters, 2004	60
1.07	University-industry collaboration, 2004	98
1.08	Quality of scientific research institutions, 2004	94
1.09	Subsidies for firm-level R&D, 2004	93
1.10	Brain drain, 2004	58
1.11	Ease of access to loans, 2004	73
1.12	Administrative burden, 2004	99
1.13	Ease to start a new business, 2004	92
2.01	Effectiveness of lawmaking, 2004	90
2.02	Laws relating to ICT, 2004	49
2.03	Effectiveness of judiciary, 2004	89
2.04	Intellectual property protection, 2004	98
3.01	Telephone mainlines, 2002	91
3.02	Secure Internet servers, 2003	87
3.03	Internet hosts, 2003	90

Readiness Component Index		58
Individual Readiness		76
Business Readiness		95
Government Readiness		19
4.01	Quality of math and science education, 2004	99
4.02	Quality of educational system, 2004	99
4.03	Quality of public schools, 2004	87
4.04	Internet access in schools, 2004	54

Readiness Component Index (continued)		
4.05	Buyer sophistication, 2004	29
4.06	Buyer dynamism, 2004	27
4.07	Residential telephone connection charge, 2002	84
4.08	Affordability of Internet access, 2003	83
5.01	Investment in training, 2004	104
5.02	Availability of training services, 2004	90
5.03	Quality of business schools, 2004	84
5.04	Business investment in R&D, 2004	101
5.05	Business monthly telephone subscription, 2002–3	92
5.06	Business telephone connection charge, 2002–3	73
6.01	Government prioritization of ICT, 2004	7
6.02	Government procurement of ICT, 2004	43

Usage Component Index		56
Individual Usage		86
Business Usage		60
Government Usage		25
7.01	Cellular mobile subscribers, 2003	97
7.02	Telephone subscribers, 2002	91
7.03	Public payphones, 2002	83
7.04	Telephone lines, 2002	89
7.05	Television sets, 2002	83
7.06	Broadband-DSL Internet subscribers, 2002–3	65
7.07	Broadband-cable modem, 2002–3	60
7.08	Internet users per 100 inhabitants, 2002	91
8.01	Prevalence of foreign technology licensing, 2004	67
8.02	Firm-level technology absorption, 2004	44
8.03	Capacity for innovation, 2004	34
8.04	Availability of new telephone lines, 2004	83
8.05	Availability of cellular phones, 2004	64
9.01	Government success in ICT promotion, 2004	15
9.02	Government online services, 2003	39

Panama

Key Indicators

Population (mn), 2003	3.1
GDP per capita (PPP US$), 2003	6,475
Internet users per 100 inhabitants, 2002	4.1

Networked Readiness Index Rank

Year (number of countries)	Rank
2004 (104)	**69**
2003 (102)	58
2002 (82)	61

RANK/104

Environment Component Index	61
Market Environment	49
Political and Regulatory Environment	78
Infrastructure Environment	60

1.01	Availability of scientists and engineers, 2004	74
1.02	Venture capital availability, 2004	37
1.03	Sophistication of financial markets, 2004	25
1.04	Technological sophistication, 2004	27
1.05	State of cluster development, 2004	48
1.06	Collaboration in clusters, 2004	67
1.07	University-industry collaboration, 2004	83
1.08	Quality of scientific research institutions, 2004	87
1.09	Subsidies for firm-level R&D, 2004	86
1.10	Brain drain, 2004	32
1.11	Ease of access to loans, 2004	25
1.12	Administrative burden, 2004	79
1.13	Ease to start a new business, 2004	62
2.01	Effectiveness of lawmaking, 2004	99
2.02	Laws relating to ICT, 2004	47
2.03	Effectiveness of judiciary, 2004	96
2.04	Intellectual property protection, 2004	46
3.01	Telephone mainlines, 2002	64
3.02	Secure Internet servers, 2003	31
3.03	Internet hosts, 2003	57

Readiness Component Index	80
Individual Readiness	66
Business Readiness	61
Government Readiness	97

4.01	Quality of math and science education, 2004	74
4.02	Quality of educational system, 2004	90
4.03	Quality of public schools, 2004	71
4.04	Internet access in schools, 2004	69

Readiness Component Index (continued)	
4.05	Buyer sophistication, 2004	54
4.06	Buyer dynamism, 2004	48
4.07	Residential telephone connection charge, 2002	43
4.08	Affordability of Internet access, 2003	55
5.01	Investment in training, 2004	76
5.02	Availability of training services, 2004	55
5.03	Quality of business schools, 2004	61
5.04	Business investment in R&D, 2004	74
5.05	Business monthly telephone subscription, 2002–3	69
5.06	Business telephone connection charge, 2002–3	44
6.01	Government prioritization of ICT, 2004	96
6.02	Government procurement of ICT, 2004	98

Usage Component Index	61
Individual Usage	71
Business Usage	43
Government Usage	87

7.01	Cellular mobile subscribers, 2003	56
7.02	Telephone subscribers, 2002	67
7.03	Public payphones, 2002	68
7.04	Telephone lines, 2002	63
7.05	Television sets, 2002	72
7.06	Broadband-DSL Internet subscribers, 2002–3	65
7.07	Broadband-cable modem, 2002–3	60
7.08	Internet users per 100 inhabitants, 2002	70
8.01	Prevalence of foreign technology licensing, 2004	27
8.02	Firm-level technology absorption, 2004	39
8.03	Capacity for innovation, 2004	61
8.04	Availability of new telephone lines, 2004	55
8.05	Availability of cellular phones, 2004	45
9.01	Government success in ICT promotion, 2004	88
9.02	Government online services, 2003	66

Paraguay

Key Indicators

Population (mn), 2003	5.9
GDP per capita (PPP US$), 2003	4,724
Internet users per 100 inhabitants, 2002	1.7

RANK/104

Environment Component Index		99
Market Environment		94
Political and Regulatory Environment		101
Infrastructure Environment		81
1.01	Availability of scientists and engineers, 2004	101
1.02	Venture capital availability, 2004	99
1.03	Sophistication of financial markets, 2004	88
1.04	Technological sophistication, 2004	76
1.05	State of cluster development, 2004	96
1.06	Collaboration in clusters, 2004	101
1.07	University-industry collaboration, 2004	99
1.08	Quality of scientific research institutions, 2004	103
1.09	Subsidies for firm-level R&D, 2004	104
1.10	Brain drain, 2004	67
1.11	Ease of access to loans, 2004	71
1.12	Administrative burden, 2004	62
1.13	Ease to start a new business, 2004	82
2.01	Effectiveness of lawmaking, 2004	89
2.02	Laws relating to ICT, 2004	102
2.03	Effectiveness of judiciary, 2004	99
2.04	Intellectual property protection, 2004	97
3.01	Telephone mainlines, 2002	83
3.02	Secure Internet servers, 2003	78
3.03	Internet hosts, 2003	67

Readiness Component Index		98
Individual Readiness		88
Business Readiness		93
Government Readiness		103
4.01	Quality of math and science education, 2004	98
4.02	Quality of educational system, 2004	100
4.03	Quality of public schools, 2004	86
4.04	Internet access in schools, 2004	92

Readiness Component Index (continued)		
4.05	Buyer sophistication, 2004	99
4.06	Buyer dynamism, 2004	97
4.07	Residential telephone connection charge, 2002	81
4.08	Affordability of Internet access, 2003	81
5.01	Investment in training, 2004	97
5.02	Availability of training services, 2004	97
5.03	Quality of business schools, 2004	82
5.04	Business investment in R&D, 2004	100
5.05	Business monthly telephone subscription, 2002–3	82
5.06	Business telephone connection charge, 2002–3	84
6.01	Government prioritization of ICT, 2004	104
6.02	Government procurement of ICT, 2004	102

Usage Component Index		97
Individual Usage		78
Business Usage		94
Government Usage		100
7.01	Cellular mobile subscribers, 2003	50
7.02	Telephone subscribers, 2002	64
7.03	Public payphones, 2002	86
7.04	Telephone lines, 2002	83
7.05	Television sets, 2002	78
7.06	Broadband-DSL Internet subscribers, 2002–3	65
7.07	Broadband-cable modem, 2002–3	56
7.08	Internet users per 100 inhabitants, 2002	84
8.01	Prevalence of foreign technology licensing, 2004	100
8.02	Firm-level technology absorption, 2004	98
8.03	Capacity for innovation, 2004	98
8.04	Availability of new telephone lines, 2004	89
8.05	Availability of cellular phones, 2004	56
9.01	Government success in ICT promotion, 2004	103
9.02	Government online services, 2003	72

Peru

Key Indicators

Population (mn), 2003	27.2
GDP per capita (PPP US$), 2003	5,267
Internet users per 100 inhabitants, 2002	9.0

Networked Readiness Index Rank

Year (number of countries)	Rank
2004 (104)	**90**
2003 (102)	70
2002 (82)	67

RANK/104

Environment Component Index		93
Market Environment		91
Political and Regulatory Environment		90
Infrastructure Environment		77
1.01	Availability of scientists and engineers, 2004	87
1.02	Venture capital availability, 2004	80
1.03	Sophistication of financial markets, 2004	53
1.04	Technological sophistication, 2004	67
1.05	State of cluster development, 2004	90
1.06	Collaboration in clusters, 2004	92
1.07	University-industry collaboration, 2004	91
1.08	Quality of scientific research institutions, 2004	95
1.09	Subsidies for firm-level R&D, 2004	101
1.10	Brain drain, 2004	90
1.11	Ease of access to loans, 2004	64
1.12	Administrative burden, 2004	94
1.13	Ease to start a new business, 2004	94
2.01	Effectiveness of lawmaking, 2004	100
2.02	Laws relating to ICT, 2004	67
2.03	Effectiveness of judiciary, 2004	94
2.04	Intellectual property protection, 2004	90
3.01	Telephone mainlines, 2002	78
3.02	Secure Internet servers, 2003	62
3.03	Internet hosts, 2003	56

Readiness Component Index		92
Individual Readiness		83
Business Readiness		77
Government Readiness		104
4.01	Quality of math and science education, 2004	101
4.02	Quality of educational system, 2004	96
4.03	Quality of public schools, 2004	102
4.04	Internet access in schools, 2004	59

Readiness Component Index (continued)		
4.05	Buyer sophistication, 2004	90
4.06	Buyer dynamism, 2004	82
4.07	Residential telephone connection charge, 2002	85
4.08	Affordability of Internet access, 2003	69
5.01	Investment in training, 2004	82
5.02	Availability of training services, 2004	74
5.03	Quality of business schools, 2004	40
5.04	Business investment in R&D, 2004	87
5.05	Business monthly telephone subscription, 2002–3	87
5.06	Business telephone connection charge, 2002–3	74
6.01	Government prioritization of ICT, 2004	102
6.02	Government procurement of ICT, 2004	103

Usage Component Index		78
Individual Usage		66
Business Usage		66
Government Usage		97
7.01	Cellular mobile subscribers, 2003	77
7.02	Telephone subscribers, 2002	79
7.03	Public payphones, 2002	24
7.04	Telephone lines, 2002	77
7.05	Television sets, 2002	50
7.06	Broadband-DSL Internet subscribers, 2002–3	46
7.07	Broadband-cable modem, 2002–3	47
7.08	Internet users per 100 inhabitants, 2002	52
8.01	Prevalence of foreign technology licensing, 2004	84
8.02	Firm-level technology absorption, 2004	84
8.03	Capacity for innovation, 2004	82
8.04	Availability of new telephone lines, 2004	39
8.05	Availability of cellular phones, 2004	40
9.01	Government success in ICT promotion, 2004	100
9.02	Government online services, 2003	66

Philippines

Key Indicators

Population (mn), 2003	80
GDP per capita (PPP US$), 2003	4,321
Internet users per 100 inhabitants, 2002	4.4

Networked Readiness Index Rank

Year (number of countries)	Rank
2004 (104)	**67**
2003 (102)	69
2002 (82)	62

RANK/104

Environment Component Index		81
Market Environment		81
Political and Regulatory Environment		75
Infrastructure Environment		85
1.01	Availability of scientists and engineers, 2004	78
1.02	Venture capital availability, 2004	77
1.03	Sophistication of financial markets, 2004	67
1.04	Technological sophistication, 2004	77
1.05	State of cluster development, 2004	50
1.06	Collaboration in clusters, 2004	70
1.07	University-industry collaboration, 2004	68
1.08	Quality of scientific research institutions, 2004	75
1.09	Subsidies for firm-level R&D, 2004	64
1.10	Brain drain, 2004	97
1.11	Ease of access to loans, 2004	82
1.12	Administrative burden, 2004	95
1.13	Ease to start a new business, 2004	51
2.01	Effectiveness of lawmaking, 2004	86
2.02	Laws relating to ICT, 2004	44
2.03	Effectiveness of judiciary, 2004	74
2.04	Intellectual property protection, 2004	82
3.01	Telephone mainlines, 2002	85
3.02	Secure Internet servers, 2003	73
3.03	Internet hosts, 2003	77

Readiness Component Index		76
Individual Readiness		70
Business Readiness		92
Government Readiness		70
4.01	Quality of math and science education, 2004	84
4.02	Quality of educational system, 2004	67
4.03	Quality of public schools, 2004	85
4.04	Internet access in schools, 2004	57

Readiness Component Index (continued)		
4.05	Buyer sophistication, 2004	49
4.06	Buyer dynamism, 2004	78
4.07	Residential telephone connection charge, 2002	56
4.08	Affordability of Internet access, 2003	71
5.01	Investment in training, 2004	36
5.02	Availability of training services, 2004	81
5.03	Quality of business schools, 2004	30
5.04	Business investment in R&D, 2004	49
5.05	Business monthly telephone subscription, 2002–3	102
5.06	Business telephone connection charge, 2002–3	60
6.01	Government prioritization of ICT, 2004	50
6.02	Government procurement of ICT, 2004	88

Usage Component Index		54
Individual Usage		70
Business Usage		58
Government Usage		35
7.01	Cellular mobile subscribers, 2003	65
7.02	Telephone subscribers, 2002	73
7.03	Public payphones, 2002	21
7.04	Telephone lines, 2002	84
7.05	Television sets, 2002	73
7.06	Broadband-DSL Internet subscribers, 2002–3	50
7.07	Broadband-cable modem, 2002–3	52
7.08	Internet users per 100 inhabitants, 2002	67
8.01	Prevalence of foreign technology licensing, 2004	46
8.02	Firm-level technology absorption, 2004	58
8.03	Capacity for innovation, 2004	80
8.04	Availability of new telephone lines, 2004	71
8.05	Availability of cellular phones, 2004	39
9.01	Government success in ICT promotion, 2004	60
9.02	Government online services, 2003	20

Poland

Key Indicators

Population (mn), 2003	38.6
GDP per capita (PPP US$), 2003	11,623
Internet users per 100 inhabitants, 2002	23.0

Networked Readiness Index Rank

Year (number of countries) Rank

2004 (104) **72**

2003 (102) 47

2002 (82) 39

RANK/104

Environment Component Index — 70

Market Environment	72
Political and Regulatory Environment	79
Infrastructure Environment	40

1.01	Availability of scientists and engineers, 2004	39
1.02	Venture capital availability, 2004	74
1.03	Sophistication of financial markets, 2004	59
1.04	Technological sophistication, 2004	61
1.05	State of cluster development, 2004	66
1.06	Collaboration in clusters, 2004	62
1.07	University-industry collaboration, 2004	49
1.08	Quality of scientific research institutions, 2004	39
1.09	Subsidies for firm-level R&D, 2004	55
1.10	Brain drain, 2004	87
1.11	Ease of access to loans, 2004	66
1.12	Administrative burden, 2004	97
1.13	Ease to start a new business, 2004	81
2.01	Effectiveness of lawmaking, 2004	84
2.02	Laws relating to ICT, 2004	56
2.03	Effectiveness of judiciary, 2004	72
2.04	Intellectual property protection, 2004	79
3.01	Telephone mainlines, 2002	39
3.02	Secure Internet servers, 2003	44
3.03	Internet hosts, 2003	30

Readiness Component Index — 73

Individual Readiness	49
Business Readiness	50
Government Readiness	95

4.01	Quality of math and science education, 2004	39
4.02	Quality of educational system, 2004	56
4.03	Quality of public schools, 2004	43
4.04	Internet access in schools, 2004	61

Readiness Component Index (continued)

4.05	Buyer sophistication, 2004	75
4.06	Buyer dynamism, 2004	83
4.07	Residential telephone connection charge, 2002	45
4.08	Affordability of Internet access, 2003	37
5.01	Investment in training, 2004	69
5.02	Availability of training services, 2004	40
5.03	Quality of business schools, 2004	54
5.04	Business investment in R&D, 2004	58
5.05	Business monthly telephone subscription, 2002–3	60
5.06	Business telephone connection charge, 2002–3	42
6.01	Government prioritization of ICT, 2004	99
6.02	Government procurement of ICT, 2004	86

Usage Component Index — 72

Individual Usage	40
Business Usage	74
Government Usage	93

7.01	Cellular mobile subscribers, 2003	40
7.02	Telephone subscribers, 2002	43
7.03	Public payphones, 2002	46
7.04	Telephone lines, 2002	37
7.05	Television sets, 2002	42
7.06	Broadband-DSL Internet subscribers, 2002–3	39
7.07	Broadband-cable modem, 2002–3	30
7.08	Internet users per 100 inhabitants, 2002	35
8.01	Prevalence of foreign technology licensing, 2004	73
8.02	Firm-level technology absorption, 2004	82
8.03	Capacity for innovation, 2004	46
8.04	Availability of new telephone lines, 2004	86
8.05	Availability of cellular phones, 2004	71
9.01	Government success in ICT promotion, 2004	94
9.02	Government online services, 2003	66

Portugal

Part 2 Country Profiles | 183

Key Indicators

Population (mn), 2003	10.1
GDP per capita (PPP US$), 2003	18,444
Internet users per 100 inhabitants, 2002	19.4

Networked Readiness Index Rank

Year (number of countries)	Rank
2004 (104)	**30**
2003 (102)	31
2002 (82)	31

RANK/104

Environment Component Index		29
Market Environment		33
Political and Regulatory Environment		31
Infrastructure Environment		29
1.01	Availability of scientists and engineers, 2004	47
1.02	Venture capital availability, 2004	22
1.03	Sophistication of financial markets, 2004	28
1.04	Technological sophistication, 2004	51
1.05	State of cluster development, 2004	34
1.06	Collaboration in clusters, 2004	38
1.07	University-industry collaboration, 2004	41
1.08	Quality of scientific research institutions, 2004	41
1.09	Subsidies for firm-level R&D, 2004	27
1.10	Brain drain, 2004	34
1.11	Ease of access to loans, 2004	21
1.12	Administrative burden, 2004	55
1.13	Ease to start a new business, 2004	65
2.01	Effectiveness of lawmaking, 2004	45
2.02	Laws relating to ICT, 2004	36
2.03	Effectiveness of judiciary, 2004	18
2.04	Intellectual property protection, 2004	30
3.01	Telephone mainlines, 2002	30
3.02	Secure Internet servers, 2003	28
3.03	Internet hosts, 2003	25

Readiness Component Index		42
Individual Readiness		43
Business Readiness		39
Government Readiness		49
4.01	Quality of math and science education, 2004	86
4.02	Quality of educational system, 2004	73
4.03	Quality of public schools, 2004	33
4.04	Internet access in schools, 2004	32

Readiness Component Index (continued)		
4.05	Buyer sophistication, 2004	40
4.06	Buyer dynamism, 2004	39
4.07	Residential telephone connection charge, 2002	33
4.08	Affordability of Internet access, 2003	28
5.01	Investment in training, 2004	55
5.02	Availability of training services, 2004	43
5.03	Quality of business schools, 2004	43
5.04	Business investment in R&D, 2004	63
5.05	Business monthly telephone subscription, 2002–3	26
5.06	Business telephone connection charge, 2002–3	34
6.01	Government prioritization of ICT, 2004	37
6.02	Government procurement of ICT, 2004	60

Usage Component Index		28
Individual Usage		26
Business Usage		38
Government Usage		31
7.01	Cellular mobile subscribers, 2003	9
7.02	Telephone subscribers, 2002	21
7.03	Public payphones, 2002	18
7.04	Telephone lines, 2002	31
7.05	Television sets, 2002	37
7.06	Broadband-DSL Internet subscribers, 2002–3	34
7.07	Broadband-cable modem, 2002–3	13
7.08	Internet users per 100 inhabitants, 2002	37
8.01	Prevalence of foreign technology licensing, 2004	25
8.02	Firm-level technology absorption, 2004	62
8.03	Capacity for innovation, 2004	43
8.04	Availability of new telephone lines, 2004	31
8.05	Availability of cellular phones, 2004	22
9.01	Government success in ICT promotion, 2004	27
9.02	Government online services, 2003	39

Romania

Key Indicators

Population (mn), 2003	22.3
GDP per capita (PPP US$), 2003	7,222
Internet users per 100 inhabitants, 2002	10.1

Networked Readiness Index Rank

Year (number of countries)	Rank
2004 (104)	**53**
2003 (102)	61
2002 (82)	72

RANK/104

Environment Component Index	65
Market Environment	68
Political and Regulatory Environment	65
Infrastructure Environment	58

1.01	Availability of scientists and engineers, 2004	18
1.02	Venture capital availability, 2004	83
1.03	Sophistication of financial markets, 2004	70
1.04	Technological sophistication, 2004	74
1.05	State of cluster development, 2004	76
1.06	Collaboration in clusters, 2004	61
1.07	University-industry collaboration, 2004	71
1.08	Quality of scientific research institutions, 2004	69
1.09	Subsidies for firm-level R&D, 2004	53
1.10	Brain drain, 2004	101
1.11	Ease of access to loans, 2004	55
1.12	Administrative burden, 2004	37
1.13	Ease to start a new business, 2004	75
2.01	Effectiveness of lawmaking, 2004	70
2.02	Laws relating to ICT, 2004	40
2.03	Effectiveness of judiciary, 2004	75
2.04	Intellectual property protection, 2004	64
3.01	Telephone mainlines, 2002	56
3.02	Secure Internet servers, 2003	71
3.03	Internet hosts, 2003	58

Readiness Component Index	47
Individual Readiness	39
Business Readiness	51
Government Readiness	58

4.01	Quality of math and science education, 2004	5
4.02	Quality of educational system, 2004	32
4.03	Quality of public schools, 2004	32
4.04	Internet access in schools, 2004	44

Readiness Component Index (continued)	
4.05	Buyer sophistication, 2004	72
4.06	Buyer dynamism, 2004	79
4.07	Residential telephone connection charge, 2002	29
4.08	Affordability of Internet access, 2003	65
5.01	Investment in training, 2004	57
5.02	Availability of training services, 2004	56
5.03	Quality of business schools, 2004	73
5.04	Business investment in R&D, 2004	51
5.05	Business monthly telephone subscription, 2002–3	46
5.06	Business telephone connection charge, 2002–3	30
6.01	Government prioritization of ICT, 2004	58
6.02	Government procurement of ICT, 2004	56

Usage Component Index	53
Individual Usage	53
Business Usage	64
Government Usage	41

7.01	Cellular mobile subscribers, 2003	49
7.02	Telephone subscribers, 2002	53
7.03	Public payphones, 2002	50
7.04	Telephone lines, 2002	57
7.05	Television sets, 2002	52
7.06	Broadband-DSL Internet subscribers, 2002–3	53
7.07	Broadband-cable modem, 2002–3	37
7.08	Internet users per 100 inhabitants, 2002	50
8.01	Prevalence of foreign technology licensing, 2004	53
8.02	Firm-level technology absorption, 2004	73
8.03	Capacity for innovation, 2004	45
8.04	Availability of new telephone lines, 2004	76
8.05	Availability of cellular phones, 2004	68
9.01	Government success in ICT promotion, 2004	47
9.02	Government online services, 2003	38

Russia

Key Indicators

Population (mn), 2003	143
GDP per capita (PPP US$), 2003	9,195
Internet users per 100 inhabitants, 2002	4.1

Networked Readiness Index Rank

Year (number of countries)	Rank
2004 (104)	**62**
2003 (102)	63
2002 (82)	69

RANK/104

Environment Component Index		69
Market Environment		61
Political and Regulatory Environment		80
Infrastructure Environment		52
1.01	Availability of scientists and engineers, 2004	33
1.02	Venture capital availability, 2004	49
1.03	Sophistication of financial markets, 2004	72
1.04	Technological sophistication, 2004	69
1.05	State of cluster development, 2004	56
1.06	Collaboration in clusters, 2004	37
1.07	University-industry collaboration, 2004	40
1.08	Quality of scientific research institutions, 2004	19
1.09	Subsidies for firm-level R&D, 2004	44
1.10	Brain drain, 2004	53
1.11	Ease of access to loans, 2004	67
1.12	Administrative burden, 2004	77
1.13	Ease to start a new business, 2004	87
2.01	Effectiveness of lawmaking, 2004	63
2.02	Laws relating to ICT, 2004	75
2.03	Effectiveness of judiciary, 2004	84
2.04	Intellectual property protection, 2004	84
3.01	Telephone mainlines, 2002	49
3.02	Secure Internet servers, 2003	69
3.03	Internet hosts, 2003	52

Readiness Component Index		59
Individual Readiness		48
Business Readiness		66
Government Readiness		65
4.01	Quality of math and science education, 2004	23
4.02	Quality of educational system, 2004	49
4.03	Quality of public schools, 2004	45
4.04	Internet access in schools, 2004	52

Readiness Component Index (continued)		
4.05	Buyer sophistication, 2004	48
4.06	Buyer dynamism, 2004	46
4.07	Residential telephone connection charge, 2002	83
4.08	Affordability of Internet access, 2003	45
5.01	Investment in training, 2004	81
5.02	Availability of training services, 2004	41
5.03	Quality of business schools, 2004	69
5.04	Business investment in R&D, 2004	36
5.05	Business monthly telephone subscription, 2002–3	55
5.06	Business telephone connection charge, 2002–3	88
6.01	Government prioritization of ICT, 2004	68
6.02	Government procurement of ICT, 2004	57

Usage Component Index		62
Individual Usage		63
Business Usage		69
Government Usage		71
7.01	Cellular mobile subscribers, 2003	73
7.02	Telephone subscribers, 2002	60
7.03	Public payphones, 2002	73
7.04	Telephone lines, 2002	50
7.05	Television sets, 2002	13
7.06	Broadband-DSL Internet subscribers, 2002–3	65
7.07	Broadband-cable modem, 2002–3	60
7.08	Internet users per 100 inhabitants, 2002	71
8.01	Prevalence of foreign technology licensing, 2004	87
8.02	Firm-level technology absorption, 2004	56
8.03	Capacity for innovation, 2004	35
8.04	Availability of new telephone lines, 2004	74
8.05	Availability of cellular phones, 2004	69
9.01	Government success in ICT promotion, 2004	78
9.02	Government online services, 2003	47

Serbia and Montenegro

Key Indicators

Population (mn), 2003	10.5
GDP per capita (PPP US$), 2003	3,970
Internet users per 100 inhabitants, 2002	6.0

Networked Readiness Index Rank

Year (number of countries)	Rank
2004 (104)	**79**
2003 (102)	77
2002 (82)	—

RANK/104

Environment Component Index		83
Market Environment		79
Political and Regulatory Environment		92
Infrastructure Environment		54
1.01	Availability of scientists and engineers, 2004	21
1.02	Venture capital availability, 2004	70
1.03	Sophistication of financial markets, 2004	97
1.04	Technological sophistication, 2004	97
1.05	State of cluster development, 2004	94
1.06	Collaboration in clusters, 2004	73
1.07	University-industry collaboration, 2004	61
1.08	Quality of scientific research institutions, 2004	50
1.09	Subsidies for firm-level R&D, 2004	65
1.10	Brain drain, 2004	93
1.11	Ease of access to loans, 2004	70
1.12	Administrative burden, 2004	91
1.13	Ease to start a new business, 2004	42
2.01	Effectiveness of lawmaking, 2004	85
2.02	Laws relating to ICT, 2004	93
2.03	Effectiveness of judiciary, 2004	85
2.04	Intellectual property protection, 2004	96
3.01	Telephone mainlines, 2002	51
3.02	Secure Internet servers, 2003	80
3.03	Internet hosts, 2003	59

Readiness Component Index		71
Individual Readiness		67
Business Readiness		69
Government Readiness		80
4.01	Quality of math and science education, 2004	33
4.02	Quality of educational system, 2004	53
4.03	Quality of public schools, 2004	50
4.04	Internet access in schools, 2004	75

Readiness Component Index (continued)		
4.05	Buyer sophistication, 2004	84
4.06	Buyer dynamism, 2004	93
4.07	Residential telephone connection charge, 2002	75
4.08	Affordability of Internet access, 2003	58
5.01	Investment in training, 2004	92
5.02	Availability of training services, 2004	67
5.03	Quality of business schools, 2004	72
5.04	Business investment in R&D, 2004	79
5.05	Business monthly telephone subscription, 2002–3	3
5.06	Business telephone connection charge, 2002–3	78
6.01	Government prioritization of ICT, 2004	90
6.02	Government procurement of ICT, 2004	67

Usage Component Index		82
Individual Usage		56
Business Usage		89
Government Usage		79
7.01	Cellular mobile subscribers, 2003	48
7.02	Telephone subscribers, 2002	49
7.03	Public payphones, 2002	79
7.04	Telephone lines, 2002	49
7.05	Television sets, 2002	47
7.06	Broadband-DSL Internet subscribers, 2002–3	65
7.07	Broadband-cable modem, 2002–3	60
7.08	Internet users per 100 inhabitants, 2002	59
8.01	Prevalence of foreign technology licensing, 2004	91
8.02	Firm-level technology absorption, 2004	95
8.03	Capacity for innovation, 2004	70
8.04	Availability of new telephone lines, 2004	87
8.05	Availability of cellular phones, 2004	79
9.01	Government success in ICT promotion, 2004	86
9.02	Government online services, 2003	52

Singapore

Key Indicators

Population (mn), 2003	4.3
GDP per capita (PPP US$), 2003	24,480
Internet users per 100 inhabitants, 2002	50.4

RANK/104

Environment Component Index	6
Market Environment	3
Political and Regulatory Environment	3
Infrastructure Environment	14

1.01	Availability of scientists and engineers, 2004	19
1.02	Venture capital availability, 2004	15
1.03	Sophistication of financial markets, 2004	12
1.04	Technological sophistication, 2004	8
1.05	State of cluster development, 2004	6
1.06	Collaboration in clusters, 2004	11
1.07	University-industry collaboration, 2004	5
1.08	Quality of scientific research institutions, 2004	13
1.09	Subsidies for firm-level R&D, 2004	1
1.10	Brain drain, 2004	16
1.11	Ease of access to loans, 2004	15
1.12	Administrative burden, 2004	1
1.13	Ease to start a new business, 2004	4
2.01	Effectiveness of lawmaking, 2004	1
2.02	Laws relating to ICT, 2004	1
2.03	Effectiveness of judiciary, 2004	24
2.04	Intellectual property protection, 2004	13
3.01	Telephone mainlines, 2002	28
3.02	Secure Internet servers, 2003	13
3.03	Internet hosts, 2003	8

Readiness Component Index	1
Individual Readiness	2
Business Readiness	15
Government Readiness	1

4.01	Quality of math and science education, 2004	1
4.02	Quality of educational system, 2004	2
4.03	Quality of public schools, 2004	8
4.04	Internet access in schools, 2004	6

Readiness Component Index (continued)	
4.05	Buyer sophistication, 2004	15
4.06	Buyer dynamism, 2004	14
4.07	Residential telephone connection charge, 2002	1
4.08	Affordability of Internet access, 2003	3
5.01	Investment in training, 2004	14
5.02	Availability of training services, 2004	20
5.03	Quality of business schools, 2004	16
5.04	Business investment in R&D, 2004	9
5.05	Business monthly telephone subscription, 2002–3	5
5.06	Business telephone connection charge, 2002–3	2
6.01	Government prioritization of ICT, 2004	1
6.02	Government procurement of ICT, 2004	1

Usage Component Index	1
Individual Usage	12
Business Usage	9
Government Usage	1

7.01	Cellular mobile subscribers, 2003	15
7.02	Telephone subscribers, 2002	19
7.03	Public payphones, 2002	31
7.04	Telephone lines, 2002	27
7.05	Television sets, 2002	11
7.06	Broadband-DSL Internet subscribers, 2002–3	11
7.07	Broadband-cable modem, 2002–3	9
7.08	Internet users per 100 inhabitants, 2002	9
8.01	Prevalence of foreign technology licensing, 2004	1
8.02	Firm-level technology absorption, 2004	9
8.03	Capacity for innovation, 2004	18
8.04	Availability of new telephone lines, 2004	6
8.05	Availability of cellular phones, 2004	17
9.01	Government success in ICT promotion, 2004	1
9.02	Government online services, 2003	1

Slovak Republic

Key Indicators

Population (mn), 2003	5.4
GDP per capita (PPP US$), 2003	13,469
Internet users per 100 inhabitants, 2002	16.0

Networked Readiness Index Rank

Year (number of countries)	Rank
2004 (104)	**48**
2003 (102)	41
2002 (82)	40

RANK/104

Environment Component Index — 49

Market Environment		45
Political and Regulatory Environment		52
Infrastructure Environment		41
1.01	Availability of scientists and engineers, 2004	11
1.02	Venture capital availability, 2004	38
1.03	Sophistication of financial markets, 2004	68
1.04	Technological sophistication, 2004	43
1.05	State of cluster development, 2004	68
1.06	Collaboration in clusters, 2004	44
1.07	University-industry collaboration, 2004	39
1.08	Quality of scientific research institutions, 2004	63
1.09	Subsidies for firm-level R&D, 2004	63
1.10	Brain drain, 2004	63
1.11	Ease of access to loans, 2004	38
1.12	Administrative burden, 2004	64
1.13	Ease to start a new business, 2004	29
2.01	Effectiveness of lawmaking, 2004	64
2.02	Laws relating to ICT, 2004	43
2.03	Effectiveness of judiciary, 2004	62
2.04	Intellectual property protection, 2004	49
3.01	Telephone mainlines, 2002	45
3.02	Secure Internet servers, 2003	47
3.03	Internet hosts, 2003	28

Readiness Component Index — 50

Individual Readiness		35
Business Readiness		41
Government Readiness		67
4.01	Quality of math and science education, 2004	13
4.02	Quality of educational system, 2004	34
4.03	Quality of public schools, 2004	26
4.04	Internet access in schools, 2004	41

Readiness Component Index (continued)

4.05	Buyer sophistication, 2004	77
4.06	Buyer dynamism, 2004	53
4.07	Residential telephone connection charge, 2002	45
4.08	Affordability of Internet access, 2003	48
5.01	Investment in training, 2004	35
5.02	Availability of training services, 2004	38
5.03	Quality of business schools, 2004	51
5.04	Business investment in R&D, 2004	53
5.05	Business monthly telephone subscription, 2002–3	62
5.06	Business telephone connection charge, 2002–3	43
6.01	Government prioritization of ICT, 2004	66
6.02	Government procurement of ICT, 2004	65

Usage Component Index — 46

Individual Usage		41
Business Usage		29
Government Usage		63
7.01	Cellular mobile subscribers, 2003	30
7.02	Telephone subscribers, 2002	38
7.03	Public payphones, 2002	61
7.04	Telephone lines, 2002	51
7.05	Television sets, 2002	3
7.06	Broadband-DSL Internet subscribers, 2002–3	65
7.07	Broadband-cable modem, 2002–3	45
7.08	Internet users per 100 inhabitants, 2002	41
8.01	Prevalence of foreign technology licensing, 2004	31
8.02	Firm-level technology absorption, 2004	20
8.03	Capacity for innovation, 2004	41
8.04	Availability of new telephone lines, 2004	38
8.05	Availability of cellular phones, 2004	34
9.01	Government success in ICT promotion, 2004	72
9.02	Government online services, 2003	43

Slovenia

Key Indicators

Population (mn), 2003	2
GDP per capita (PPP US$), 2003	19,300
Internet users per 100 inhabitants, 2002	37.6

Year (number of countries)	Rank
2004 (104)	**32**
2003 (102)	30
2002 (82)	33

RANK/104

Environment Component Index		32
Market Environment		40
Political and Regulatory Environment		39
Infrastructure Environment		26
1.01	Availability of scientists and engineers, 2004	59
1.02	Venture capital availability, 2004	51
1.03	Sophistication of financial markets, 2004	61
1.04	Technological sophistication, 2004	48
1.05	State of cluster development, 2004	62
1.06	Collaboration in clusters, 2004	57
1.07	University-industry collaboration, 2004	30
1.08	Quality of scientific research institutions, 2004	30
1.09	Subsidies for firm-level R&D, 2004	28
1.10	Brain drain, 2004	33
1.11	Ease of access to loans, 2004	31
1.12	Administrative burden, 2004	60
1.13	Ease to start a new business, 2004	36
2.01	Effectiveness of lawmaking, 2004	59
2.02	Laws relating to ICT, 2004	27
2.03	Effectiveness of judiciary, 2004	54
2.04	Intellectual property protection, 2004	32
3.01	Telephone mainlines, 2002	20
3.02	Secure Internet servers, 2003	24
3.03	Internet hosts, 2003	27

Readiness Component Index		33
Individual Readiness		25
Business Readiness		26
Government Readiness		56
4.01	Quality of math and science education, 2004	26
4.02	Quality of educational system, 2004	25
4.03	Quality of public schools, 2004	23
4.04	Internet access in schools, 2004	21

Readiness Component Index (continued)		
4.05	Buyer sophistication, 2004	28
4.06	Buyer dynamism, 2004	32
4.07	Residential telephone connection charge, 2002	37
4.08	Affordability of Internet access, 2003	33
5.01	Investment in training, 2004	32
5.02	Availability of training services, 2004	29
5.03	Quality of business schools, 2004	35
5.04	Business investment in R&D, 2004	21
5.05	Business monthly telephone subscription, 2002–3	20
5.06	Business telephone connection charge, 2002–3	35
6.01	Government prioritization of ICT, 2004	61
6.02	Government procurement of ICT, 2004	44

Usage Component Index		33
Individual Usage		28
Business Usage		33
Government Usage		53
7.01	Cellular mobile subscribers, 2003	14
7.02	Telephone subscribers, 2002	22
7.03	Public payphones, 2002	62
7.04	Telephone lines, 2002	32
7.05	Television sets, 2002	53
7.06	Broadband-DSL Internet subscribers, 2002–3	28
7.07	Broadband-cable modem, 2002–3	23
7.08	Internet users per 100 inhabitants, 2002	19
8.01	Prevalence of foreign technology licensing, 2004	45
8.02	Firm-level technology absorption, 2004	50
8.03	Capacity for innovation, 2004	22
8.04	Availability of new telephone lines, 2004	42
8.05	Availability of cellular phones, 2004	24
9.01	Government success in ICT promotion, 2004	53
9.02	Government online services, 2003	57

189

Part 2 Country Profiles

South Africa

Key Indicators

Population (mn), 2003	45
GDP per capita (PPP US$), 2003	10,492
Internet users per 100 inhabitants, 2002	6.8

Networked Readiness Index Rank

Year (number of countries)	Rank
2004 (104)	**34**
2003 (102)	37
2002 (82)	36

RANK/104

Environment Component Index	31
Market Environment	36
Political and Regulatory Environment	19
Infrastructure Environment	63
1.01 Availability of scientists and engineers, 2004	88
1.02 Venture capital availability, 2004	33
1.03 Sophistication of financial markets, 2004	16
1.04 Technological sophistication, 2004	32
1.05 State of cluster development, 2004	38
1.06 Collaboration in clusters, 2004	30
1.07 University-industry collaboration, 2004	19
1.08 Quality of scientific research institutions, 2004	27
1.09 Subsidies for firm-level R&D, 2004	39
1.10 Brain drain, 2004	62
1.11 Ease of access to loans, 2004	41
1.12 Administrative burden, 2004	41
1.13 Ease to start a new business, 2004	38
2.01 Effectiveness of lawmaking, 2004	16
2.02 Laws relating to ICT, 2004	20
2.03 Effectiveness of judiciary, 2004	15
2.04 Intellectual property protection, 2004	22
3.01 Telephone mainlines, 2002	72
3.02 Secure Internet servers, 2003	40
3.03 Internet hosts, 2003	45

Readiness Component Index	30
Individual Readiness	57
Business Readiness	23
Government Readiness	29
4.01 Quality of math and science education, 2004	96
4.02 Quality of educational system, 2004	72
4.03 Quality of public schools, 2004	62
4.04 Internet access in schools, 2004	68

Readiness Component Index (continued)	
4.05 Buyer sophistication, 2004	27
4.06 Buyer dynamism, 2004	30
4.07 Residential telephone connection charge, 2002	44
4.08 Affordability of Internet access, 2003	64
5.01 Investment in training, 2004	24
5.02 Availability of training services, 2004	21
5.03 Quality of business schools, 2004	10
5.04 Business investment in R&D, 2004	24
5.05 Business monthly telephone subscription, 2002–3	68
5.06 Business telephone connection charge, 2002–3	41
6.01 Government prioritization of ICT, 2004	41
6.02 Government procurement of ICT, 2004	18

Usage Component Index	41
Individual Usage	62
Business Usage	30
Government Usage	37
7.01 Cellular mobile subscribers, 2003	47
7.02 Telephone subscribers, 2002	55
7.03 Public payphones, 2002	25
7.04 Telephone lines, 2002	71
7.05 Television sets, 2002	81
7.06 Broadband-DSL Internet subscribers, 2002–3	57
7.07 Broadband-cable modem, 2002–3	60
7.08 Internet users per 100 inhabitants, 2002	57
8.01 Prevalence of foreign technology licensing, 2004	2
8.02 Firm-level technology absorption, 2004	28
8.03 Capacity for innovation, 2004	42
8.04 Availability of new telephone lines, 2004	70
8.05 Availability of cellular phones, 2004	38
9.01 Government success in ICT promotion, 2004	51
9.02 Government online services, 2003	30

Spain

Key Indicators

Population (mn), 2003	41.1
GDP per capita (PPP US$), 2003	22,264
Internet users per 100 inhabitants, 2002	19.3

RANK/104

Environment Component Index — 30

Market Environment	35
Political and Regulatory Environment	34
Infrastructure Environment	24

1.01	Availability of scientists and engineers, 2004	40
1.02	Venture capital availability, 2004	26
1.03	Sophistication of financial markets, 2004	27
1.04	Technological sophistication, 2004	33
1.05	State of cluster development, 2004	39
1.06	Collaboration in clusters, 2004	43
1.07	University-industry collaboration, 2004	33
1.08	Quality of scientific research institutions, 2004	51
1.09	Subsidies for firm-level R&D, 2004	25
1.10	Brain drain, 2004	25
1.11	Ease of access to loans, 2004	38
1.12	Administrative burden, 2004	48
1.13	Ease to start a new business, 2004	71
2.01	Effectiveness of lawmaking, 2004	25
2.02	Laws relating to ICT, 2004	39
2.03	Effectiveness of judiciary, 2004	50
2.04	Intellectual property protection, 2004	31
3.01	Telephone mainlines, 2002	19
3.02	Secure Internet servers, 2003	25
3.03	Internet hosts, 2003	26

Readiness Component Index — 29

Individual Readiness	32
Business Readiness	25
Government Readiness	38

4.01	Quality of math and science education, 2004	38
4.02	Quality of educational system, 2004	30
4.03	Quality of public schools, 2004	38
4.04	Internet access in schools, 2004	38

Readiness Component Index (continued)

4.05	Buyer sophistication, 2004	33
4.06	Buyer dynamism, 2004	45
4.07	Residential telephone connection charge, 2002	30
4.08	Affordability of Internet access, 2003	24
5.01	Investment in training, 2004	30
5.02	Availability of training services, 2004	31
5.03	Quality of business schools, 2004	8
5.04	Business investment in R&D, 2004	41
5.05	Business monthly telephone subscription, 2002–3	17
5.06	Business telephone connection charge, 2002–3	31
6.01	Government prioritization of ICT, 2004	52
6.02	Government procurement of ICT, 2004	19

Usage Component Index — 34

Individual Usage	24
Business Usage	37
Government Usage	57

7.01	Cellular mobile subscribers, 2003	7
7.02	Telephone subscribers, 2002	20
7.03	Public payphones, 2002	20
7.04	Telephone lines, 2002	29
7.05	Television sets, 2002	8
7.06	Broadband-DSL Internet subscribers, 2002–3	18
7.07	Broadband-cable modem, 2002–3	21
7.08	Internet users per 100 inhabitants, 2002	38
8.01	Prevalence of foreign technology licensing, 2004	29
8.02	Firm-level technology absorption, 2004	52
8.03	Capacity for innovation, 2004	27
8.04	Availability of new telephone lines, 2004	50
8.05	Availability of cellular phones, 2004	36
9.01	Government success in ICT promotion, 2004	69
9.02	Government online services, 2003	39

Sri Lanka

Key Indicators

Population (mn), 2003	19.1
GDP per capita (PPP US$), 2003	3,776
Internet users per 100 inhabitants, 2002	1.1

Networked Readiness Index Rank

Year (number of countries)	Rank
2004 (104)	**71**
2003 (102)	66
2002 (82)	54

RANK/104

Environment Component Index — 71

Market Environment		62
Political and Regulatory Environment		67
Infrastructure Environment		83
1.01	Availability of scientists and engineers, 2004	48
1.02	Venture capital availability, 2004	60
1.03	Sophistication of financial markets, 2004	62
1.04	Technological sophistication, 2004	73
1.05	State of cluster development, 2004	67
1.06	Collaboration in clusters, 2004	68
1.07	University-industry collaboration, 2004	80
1.08	Quality of scientific research institutions, 2004	67
1.09	Subsidies for firm-level R&D, 2004	71
1.10	Brain drain, 2004	85
1.11	Ease of access to loans, 2004	59
1.12	Administrative burden, 2004	43
1.13	Ease to start a new business, 2004	27
2.01	Effectiveness of lawmaking, 2004	60
2.02	Laws relating to ICT, 2004	78
2.03	Effectiveness of judiciary, 2004	64
2.04	Intellectual property protection, 2004	63
3.01	Telephone mainlines, 2002	84
3.02	Secure Internet servers, 2003	74
3.03	Internet hosts, 2003	89

Readiness Component Index — 72

Individual Readiness		85
Business Readiness		85
Government Readiness		51
4.01	Quality of math and science education, 2004	63
4.02	Quality of educational system, 2004	88
4.03	Quality of public schools, 2004	64
4.04	Internet access in schools, 2004	80

Readiness Component Index (continued)

4.05	Buyer sophistication, 2004	57
4.06	Buyer dynamism, 2004	81
4.07	Residential telephone connection charge, 2002	98
4.08	Affordability of Internet access, 2003	73
5.01	Investment in training, 2004	65
5.02	Availability of training services, 2004	83
5.03	Quality of business schools, 2004	90
5.04	Business investment in R&D, 2004	63
5.05	Business monthly telephone subscription, 2002–3	76
5.06	Business telephone connection charge, 2002–3	97
6.01	Government prioritization of ICT, 2004	40
6.02	Government procurement of ICT, 2004	62

Usage Component Index — 73

Individual Usage		87
Business Usage		62
Government Usage		65
7.01	Cellular mobile subscribers, 2003	83
7.02	Telephone subscribers, 2002	83
7.03	Public payphones, 2002	90
7.04	Telephone lines, 2002	81
7.05	Television sets, 2002	87
7.06	Broadband-DSL Internet subscribers, 2002–3	64
7.07	Broadband-cable modem, 2002–3	58
7.08	Internet users per 100 inhabitants, 2002	90
8.01	Prevalence of foreign technology licensing, 2004	75
8.02	Firm-level technology absorption, 2004	61
8.03	Capacity for innovation, 2004	64
8.04	Availability of new telephone lines, 2004	58
8.05	Availability of cellular phones, 2004	62
9.01	Government success in ICT promotion, 2004	56
9.02	Government online services, 2003	72

Sweden

Key Indicators

Population (mn), 2003	8.9
GDP per capita (PPP US$), 2003	26,656
Internet users per 100 inhabitants, 2002	57.3

RANK/104

Environment Component Index	5
Market Environment	6
Political and Regulatory Environment	8
Infrastructure Environment	11

1.01	Availability of scientists and engineers, 2004	6
1.02	Venture capital availability, 2004	13
1.03	Sophistication of financial markets, 2004	8
1.04	Technological sophistication, 2004	4
1.05	State of cluster development, 2004	10
1.06	Collaboration in clusters, 2004	6
1.07	University-industry collaboration, 2004	3
1.08	Quality of scientific research institutions, 2004	2
1.09	Subsidies for firm-level R&D, 2004	50
1.10	Brain drain, 2004	21
1.11	Ease of access to loans, 2004	4
1.12	Administrative burden, 2004	13
1.13	Ease to start a new business, 2004	10
2.01	Effectiveness of lawmaking, 2004	8
2.02	Laws relating to ICT, 2004	12
2.03	Effectiveness of judiciary, 2004	8
2.04	Intellectual property protection, 2004	1
3.01	Telephone mainlines, 2002	3
3.02	Secure Internet servers, 2003	12
3.03	Internet hosts, 2003	9

Readiness Component Index	9
Individual Readiness	7
Business Readiness	5
Government Readiness	22

4.01	Quality of math and science education, 2004	21
4.02	Quality of educational system, 2004	8
4.03	Quality of public schools, 2004	15
4.04	Internet access in schools, 2004	5

Readiness Component Index (continued)	
4.05	Buyer sophistication, 2004	14
4.06	Buyer dynamism, 2004	3
4.07	Residential telephone connection charge, 2002	27
4.08	Affordability of Internet access, 2003	16
5.01	Investment in training, 2004	3
5.02	Availability of training services, 2004	8
5.03	Quality of business schools, 2004	12
5.04	Business investment in R&D, 2004	4
5.05	Business monthly telephone subscription, 2002–3	51
5.06	Business telephone connection charge, 2002–3	27
6.01	Government prioritization of ICT, 2004	26
6.02	Government procurement of ICT, 2004	22

Usage Component Index	3
Individual Usage	1
Business Usage	3
Government Usage	20

7.01	Cellular mobile subscribers, 2003	11
7.02	Telephone subscribers, 2002	2
7.03	Public payphones, 2002	17
7.04	Telephone lines, 2002	3
7.05	Television sets, 2002	31
7.06	Broadband-DSL Internet subscribers, 2002–3	8
7.07	Broadband-cable modem, 2002–3	1
7.08	Internet users per 100 inhabitants, 2002	2
8.01	Prevalence of foreign technology licensing, 2004	26
8.02	Firm-level technology absorption, 2004	3
8.03	Capacity for innovation, 2004	2
8.04	Availability of new telephone lines, 2004	9
8.05	Availability of cellular phones, 2004	13
9.01	Government success in ICT promotion, 2004	40
9.02	Government online services, 2003	11

Switzerland

Key Indicators

Population (mn), 2003	7.2
GDP per capita (PPP US$), 2003	30,186
Internet users per 100 inhabitants, 2002	35.1

Networked Readiness Index Rank

Year (number of countries)   **Rank**

2004 (104)	**9**
2003 (102)	7
2002 (82)	13

RANK/104

Environment Component Index		11
Market Environment		9
Political and Regulatory Environment		14
Infrastructure Environment		8
1.01	Availability of scientists and engineers, 2004	14
1.02	Venture capital availability, 2004	25
1.03	Sophistication of financial markets, 2004	4
1.04	Technological sophistication, 2004	9
1.05	State of cluster development, 2004	23
1.06	Collaboration in clusters, 2004	9
1.07	University-industry collaboration, 2004	7
1.08	Quality of scientific research institutions, 2004	6
1.09	Subsidies for firm-level R&D, 2004	43
1.10	Brain drain, 2004	4
1.11	Ease of access to loans, 2004	28
1.12	Administrative burden, 2004	12
1.13	Ease to start a new business, 2004	11
2.01	Effectiveness of lawmaking, 2004	20
2.02	Laws relating to ICT, 2004	23
2.03	Effectiveness of judiciary, 2004	7
2.04	Intellectual property protection, 2004	8
3.01	Telephone mainlines, 2002	2
3.02	Secure Internet servers, 2003	6
3.03	Internet hosts, 2003	13

Readiness Component Index		7
Individual Readiness		4
Business Readiness		2
Government Readiness		18
4.01	Quality of math and science education, 2004	8
4.02	Quality of educational system, 2004	6
4.03	Quality of public schools, 2004	5
4.04	Internet access in schools, 2004	10

Readiness Component Index (continued)		
4.05	Buyer sophistication, 2004	4
4.06	Buyer dynamism, 2004	11
4.07	Residential telephone connection charge, 2002	13
4.08	Affordability of Internet access, 2003	4
5.01	Investment in training, 2004	1
5.02	Availability of training services, 2004	6
5.03	Quality of business schools, 2004	4
5.04	Business investment in R&D, 2004	5
5.05	Business monthly telephone subscription, 2002-3	8
5.06	Business telephone connection charge, 2002-3	12
6.01	Government prioritization of ICT, 2004	47
6.02	Government procurement of ICT, 2004	11

Usage Component Index		14
Individual Usage		7
Business Usage		4
Government Usage		39
7.01	Cellular mobile subscribers, 2003	17
7.02	Telephone subscribers, 2002	5
7.03	Public payphones, 2002	10
7.04	Telephone lines, 2002	2
7.05	Television sets, 2002	5
7.06	Broadband-DSL Internet subscribers, 2002-3	15
7.07	Broadband-cable modem, 2002-3	6
7.08	Internet users per 100 inhabitants, 2002	22
8.01	Prevalence of foreign technology licensing, 2004	11
8.02	Firm-level technology absorption, 2004	8
8.03	Capacity for innovation, 2004	6
8.04	Availability of new telephone lines, 2004	4
8.05	Availability of cellular phones, 2004	12
9.01	Government success in ICT promotion, 2004	43
9.02	Government online services, 2003	36

Taiwan

Key Indicators

Population (mn), 2003	22.6
GDP per capita (PPP US$), 2003	24,560
Internet users per 100 inhabitants, 2002	38.1

Networked Readiness Index Rank

Year (number of countries)	Rank
2004 (104)	**15**
2003 (102)	17
2002 (82)	9

RANK/104

Environment Component Index — 23

Market Environment		7
Political and Regulatory Environment		27
Infrastructure Environment		37
1.01	Availability of scientists and engineers, 2004	12
1.02	Venture capital availability, 2004	11
1.03	Sophistication of financial markets, 2004	34
1.04	Technological sophistication, 2004	11
1.05	State of cluster development, 2004	2
1.06	Collaboration in clusters, 2004	4
1.07	University-industry collaboration, 2004	6
1.08	Quality of scientific research institutions, 2004	21
1.09	Subsidies for firm-level R&D, 2004	3
1.10	Brain drain, 2004	15
1.11	Ease of access to loans, 2004	13
1.12	Administrative burden, 2004	6
1.13	Ease to start a new business, 2004	13
2.01	Effectiveness of lawmaking, 2004	41
2.02	Laws relating to ICT, 2004	21
2.03	Effectiveness of judiciary, 2004	39
2.04	Intellectual property protection, 2004	24
3.01	Telephone mainlines, 2002	37
3.02	Secure Internet servers, 2003	29
3.03	Internet hosts, 2003	40

Readiness Component Index — 4

Individual Readiness		11
Business Readiness		18
Government Readiness		3
4.01	Quality of math and science education, 2004	7
4.02	Quality of educational system, 2004	11
4.03	Quality of public schools, 2004	20
4.04	Internet access in schools, 2004	8

Readiness Component Index (continued)

4.05	Buyer sophistication, 2004	21
4.06	Buyer dynamism, 2004	9
4.07	Residential telephone connection charge, 2002	32
4.08	Affordability of Internet access, 2003	4
5.01	Investment in training, 2004	15
5.02	Availability of training services, 2004	16
5.03	Quality of business schools, 2004	32
5.04	Business investment in R&D, 2004	12
5.05	Business monthly telephone subscription, 2002–3	15
5.06	Business telephone connection charge, 2002–3	33
6.01	Government prioritization of ICT, 2004	6
6.02	Government procurement of ICT, 2004	3

Usage Component Index — 11

Individual Usage		27
Business Usage		11
Government Usage		5
7.01	Cellular mobile subscribers, 2003	104
7.02	Telephone subscribers, 2002	30
7.03	Public payphones, 2002	12
7.04	Telephone lines, 2002	13
7.05	Television sets, 2002	40
7.06	Broadband-DSL Internet subscribers, 2002–3	13
7.07	Broadband-cable modem, 2002–3	14
7.08	Internet users per 100 inhabitants, 2002	18
8.01	Prevalence of foreign technology licensing, 2004	4
8.02	Firm-level technology absorption, 2004	5
8.03	Capacity for innovation, 2004	14
8.04	Availability of new telephone lines, 2004	23
8.05	Availability of cellular phones, 2004	52
9.01	Government success in ICT promotion, 2004	4
9.02	Government online services, 2003	11

Tanzania

Key Indicators

Population (mn), 2003	37
GDP per capita (PPP US$), 2003	611
Internet users per 100 inhabitants, 2002	0.2

Networked Readiness Index Rank

Year (number of countries)	Rank
2004 (104)	**83**
2003 (102)	71
2002 (82)	–

RANK/104

Environment Component Index		60
Market Environment		53
Political and Regulatory Environment		56
Infrastructure Environment		100
1.01	Availability of scientists and engineers, 2004	83
1.02	Venture capital availability, 2004	42
1.03	Sophistication of financial markets, 2004	92
1.04	Technological sophistication, 2004	80
1.05	State of cluster development, 2004	40
1.06	Collaboration in clusters, 2004	54
1.07	University-industry collaboration, 2004	50
1.08	Quality of scientific research institutions, 2004	45
1.09	Subsidies for firm-level R&D, 2004	46
1.10	Brain drain, 2004	48
1.11	Ease of access to loans, 2004	54
1.12	Administrative burden, 2004	26
1.13	Ease to start a new business, 2004	56
2.01	Effectiveness of lawmaking, 2004	31
2.02	Laws relating to ICT, 2004	69
2.03	Effectiveness of judiciary, 2004	57
2.04	Intellectual property protection, 2004	74
3.01	Telephone mainlines, 2002	100
3.02	Secure Internet servers, 2003	100
3.03	Internet hosts, 2003	85

Readiness Component Index		90
Individual Readiness		99
Business Readiness		96
Government Readiness		52
4.01	Quality of math and science education, 2004	75
4.02	Quality of educational system, 2004	82
4.03	Quality of public schools, 2004	84
4.04	Internet access in schools, 2004	88

Readiness Component Index (continued)		
4.05	Buyer sophistication, 2004	59
4.06	Buyer dynamism, 2004	92
4.07	Residential telephone connection charge, 2002	97
4.08	Affordability of Internet access, 2003	104
5.01	Investment in training, 2004	80
5.02	Availability of training services, 2004	72
5.03	Quality of business schools, 2004	95
5.04	Business investment in R&D, 2004	69
5.05	Business monthly telephone subscription, 2002–3	96
5.06	Business telephone connection charge, 2002–3	96
6.01	Government prioritization of ICT, 2004	49
6.02	Government procurement of ICT, 2004	51

Usage Component Index		80
Individual Usage		98
Business Usage		78
Government Usage		58
7.01	Cellular mobile subscribers, 2003	92
7.02	Telephone subscribers, 2002	93
7.03	Public payphones, 2002	100
7.04	Telephone lines, 2002	101
7.05	Television sets, 2002	96
7.06	Broadband-DSL Internet subscribers, 2002–3	65
7.07	Broadband-cable modem, 2002–3	60
7.08	Internet users per 100 inhabitants, 2002	101
8.01	Prevalence of foreign technology licensing, 2004	65
8.02	Firm-level technology absorption, 2004	69
8.03	Capacity for innovation, 2004	95
8.04	Availability of new telephone lines, 2004	79
8.05	Availability of cellular phones, 2004	88
9.01	Government success in ICT promotion, 2004	42
9.02	Government online services, 2003	79

Thailand

Key Indicators

Population (mn), 2003	62.8
GDP per capita (PPP US$), 2003	7,580
Internet users per 100 inhabitants, 2002	7.8

Networked Readiness Index Rank

Year (number of countries)	Rank
2004 (104)	**36**
2003 (102)	38
2002 (82)	41

RANK/104

Environment Component Index		40
Market Environment		31
Political and Regulatory Environment		38
Infrastructure Environment		71
1.01	Availability of scientists and engineers, 2004	69
1.02	Venture capital availability, 2004	46
1.03	Sophistication of financial markets, 2004	39
1.04	Technological sophistication, 2004	39
1.05	State of cluster development, 2004	22
1.06	Collaboration in clusters, 2004	33
1.07	University-industry collaboration, 2004	31
1.08	Quality of scientific research institutions, 2004	53
1.09	Subsidies for firm-level R&D, 2004	32
1.10	Brain drain, 2004	21
1.11	Ease of access to loans, 2004	44
1.12	Administrative burden, 2004	22
1.13	Ease to start a new business, 2004	25
2.01	Effectiveness of lawmaking, 2004	36
2.02	Laws relating to ICT, 2004	45
2.03	Effectiveness of judiciary, 2004	44
2.04	Intellectual property protection, 2004	39
3.01	Telephone mainlines, 2002	73
3.02	Secure Internet servers, 2003	61
3.03	Internet hosts, 2003	65

Readiness Component Index		32
Individual Readiness		46
Business Readiness		37
Government Readiness		15
4.01	Quality of math and science education, 2004	53
4.02	Quality of educational system, 2004	66
4.03	Quality of public schools, 2004	56
4.04	Internet access in schools, 2004	42

Readiness Component Index (continued)		
4.05	Buyer sophistication, 2004	44
4.06	Buyer dynamism, 2004	35
4.07	Residential telephone connection charge, 2002	72
4.08	Affordability of Internet access, 2003	39
5.01	Investment in training, 2004	33
5.02	Availability of training services, 2004	47
5.03	Quality of business schools, 2004	38
5.04	Business investment in R&D, 2004	43
5.05	Business monthly telephone subscription, 2002–3	33
5.06	Business telephone connection charge, 2002–3	65
6.01	Government prioritization of ICT, 2004	16
6.02	Government procurement of ICT, 2004	25

Usage Component Index		40
Individual Usage		58
Business Usage		35
Government Usage		24
7.01	Cellular mobile subscribers, 2003	58
7.02	Telephone subscribers, 2002	59
7.03	Public payphones, 2002	29
7.04	Telephone lines, 2002	72
7.05	Television sets, 2002	46
7.06	Broadband-DSL Internet subscribers, 2002–3	51
7.07	Broadband-cable modem, 2002–3	35
7.08	Internet users per 100 inhabitants, 2002	55
8.01	Prevalence of foreign technology licensing, 2004	12
8.02	Firm-level technology absorption, 2004	26
8.03	Capacity for innovation, 2004	52
8.04	Availability of new telephone lines, 2004	47
8.05	Availability of cellular phones, 2004	57
9.01	Government success in ICT promotion, 2004	12
9.02	Government online services, 2003	43

Trinidad and Tobago

Key Indicators

Population (mn), 2003	1.3
GDP per capita (PPP US$), 2003	9,975
Internet users per 100 inhabitants, 2002	10.6

Networked Readiness Index Rank

Year (number of countries)	Rank
2004 (104)	**59**
2003 (102)	52
2002 (82)	58

RANK/104

Environment Component Index		54
Market Environment		54
Political and Regulatory Environment		60
Infrastructure Environment		50
1.01	Availability of scientists and engineers, 2004	73
1.02	Venture capital availability, 2004	48
1.03	Sophistication of financial markets, 2004	44
1.04	Technological sophistication, 2004	62
1.05	State of cluster development, 2004	61
1.06	Collaboration in clusters, 2004	80
1.07	University-industry collaboration, 2004	54
1.08	Quality of scientific research institutions, 2004	62
1.09	Subsidies for firm-level R&D, 2004	60
1.10	Brain drain, 2004	55
1.11	Ease of access to loans, 2004	47
1.12	Administrative burden, 2004	42
1.13	Ease to start a new business, 2004	49
2.01	Effectiveness of lawmaking, 2004	67
2.02	Laws relating to ICT, 2004	79
2.03	Effectiveness of judiciary, 2004	35
2.04	Intellectual property protection, 2004	68
3.01	Telephone mainlines, 2002	48
3.02	Secure Internet servers, 2003	45
3.03	Internet hosts, 2003	47

Readiness Component Index		52
Individual Readiness		56
Business Readiness		52
Government Readiness		62
4.01	Quality of math and science education, 2004	61
4.02	Quality of educational system, 2004	64
4.03	Quality of public schools, 2004	61
4.04	Internet access in schools, 2004	78

Readiness Component Index (continued)		
4.05	Buyer sophistication, 2004	50
4.06	Buyer dynamism, 2004	61
4.07	Residential telephone connection charge, 2002	7
4.08	Affordability of Internet access, 2003	31
5.01	Investment in training, 2004	44
5.02	Availability of training services, 2004	79
5.03	Quality of business schools, 2004	49
5.04	Business investment in R&D, 2004	61
5.05	Business monthly telephone subscription, 2002–3	66
5.06	Business telephone connection charge, 2002–3	19
6.01	Government prioritization of ICT, 2004	72
6.02	Government procurement of ICT, 2004	40

Usage Component Index		70
Individual Usage		54
Business Usage		76
Government Usage		75
7.01	Cellular mobile subscribers, 2003	53
7.02	Telephone subscribers, 2002	48
7.03	Public payphones, 2002	65
7.04	Telephone lines, 2002	47
7.05	Television sets, 2002	62
7.06	Broadband-DSL Internet subscribers, 2002–3	55
7.07	Broadband-cable modem, 2002–3	60
7.08	Internet users per 100 inhabitants, 2002	48
8.01	Prevalence of foreign technology licensing, 2004	51
8.02	Firm-level technology absorption, 2004	54
8.03	Capacity for innovation, 2004	69
8.04	Availability of new telephone lines, 2004	88
8.05	Availability of cellular phones, 2004	93
9.01	Government success in ICT promotion, 2004	70
9.02	Government online services, 2003	72

Tunisia

Key Indicators

Population (mn), 2003	9.8
GDP per capita (PPP US$), 2003	7,083
Internet users per 100 inhabitants, 2002	5.2

Year (number of countries)	Rank
2004 (104)	**31**
2003 (102)	40
2002 (82)	34

RANK/104

Environment Component Index		33
Market Environment		29
Political and Regulatory Environment		28
Infrastructure Environment		69
1.01	Availability of scientists and engineers, 2004	23
1.02	Venture capital availability, 2004	29
1.03	Sophistication of financial markets, 2004	60
1.04	Technological sophistication, 2004	34
1.05	State of cluster development, 2004	41
1.06	Collaboration in clusters, 2004	45
1.07	University-industry collaboration, 2004	32
1.08	Quality of scientific research institutions, 2004	44
1.09	Subsidies for firm-level R&D, 2004	13
1.10	Brain drain, 2004	39
1.11	Ease of access to loans, 2004	51
1.12	Administrative burden, 2004	20
1.13	Ease to start a new business, 2004	12
2.01	Effectiveness of lawmaking, 2004	26
2.02	Laws relating to ICT, 2004	32
2.03	Effectiveness of judiciary, 2004	38
2.04	Intellectual property protection, 2004	26
3.01	Telephone mainlines, 2002	67
3.02	Secure Internet servers, 2003	72
3.03	Internet hosts, 2003	94

Readiness Component Index		23
Individual Readiness		27
Business Readiness		30
Government Readiness		9
4.01	Quality of math and science education, 2004	9
4.02	Quality of educational system, 2004	15
4.03	Quality of public schools, 2004	24
4.04	Internet access in schools, 2004	37

Readiness Component Index (continued)		
4.05	Buyer sophistication, 2004	25
4.06	Buyer dynamism, 2004	33
4.07	Residential telephone connection charge, 2002	62
4.08	Affordability of Internet access, 2003	54
5.01	Investment in training, 2004	42
5.02	Availability of training services, 2004	24
5.03	Quality of business schools, 2004	21
5.04	Business investment in R&D, 2004	37
5.05	Business monthly telephone subscription, 2002–3	28
5.06	Business telephone connection charge, 2002–3	59
6.01	Government prioritization of ICT, 2004	12
6.02	Government procurement of ICT, 2004	14

Usage Component Index		45
Individual Usage		67
Business Usage		40
Government Usage		28
7.01	Cellular mobile subscribers, 2003	64
7.02	Telephone subscribers, 2002	77
7.03	Public payphones, 2002	36
7.04	Telephone lines, 2002	66
7.05	Television sets, 2002	49
7.06	Broadband-DSL Internet subscribers, 2002–3	62
7.07	Broadband-cable modem, 2002–3	60
7.08	Internet users per 100 inhabitants, 2002	61
8.01	Prevalence of foreign technology licensing, 2004	30
8.02	Firm-level technology absorption, 2004	24
8.03	Capacity for innovation, 2004	48
8.04	Availability of new telephone lines, 2004	41
8.05	Availability of cellular phones, 2004	63
9.01	Government success in ICT promotion, 2004	3
9.02	Government online services, 2003	86

Turkey

Key Indicators

Population (mn), 2003	71.3
GDP per capita (PPP US$), 2003	6,749
Internet users per 100 inhabitants, 2002	7.3

Networked Readiness Index Rank

Year (number of countries)	Rank
2004 (104)	**52**
2003 (102)	56
2002 (82)	50

RANK/104

Environment Component Index — 57

Market Environment		63
Political and Regulatory Environment		62
Infrastructure Environment		44
1.01	Availability of scientists and engineers, 2004	42
1.02	Venture capital availability, 2004	84
1.03	Sophistication of financial markets, 2004	55
1.04	Technological sophistication, 2004	54
1.05	State of cluster development, 2004	45
1.06	Collaboration in clusters, 2004	48
1.07	University-industry collaboration, 2004	62
1.08	Quality of scientific research institutions, 2004	70
1.09	Subsidies for firm-level R&D, 2004	37
1.10	Brain drain, 2004	59
1.11	Ease of access to loans, 2004	88
1.12	Administrative burden, 2004	83
1.13	Ease to start a new business, 2004	43
2.01	Effectiveness of lawmaking, 2004	34
2.02	Laws relating to ICT, 2004	77
2.03	Effectiveness of judiciary, 2004	60
2.04	Intellectual property protection, 2004	80
3.01	Telephone mainlines, 2002	40
3.02	Secure Internet servers, 2003	52
3.03	Internet hosts, 2003	50

Readiness Component Index — 60

Individual Readiness		58
Business Readiness		43
Government Readiness		76
4.01	Quality of math and science education, 2004	56
4.02	Quality of educational system, 2004	81
4.03	Quality of public schools, 2004	70
4.04	Internet access in schools, 2004	67

Readiness Component Index (continued)

4.05	Buyer sophistication, 2004	64
4.06	Buyer dynamism, 2004	37
4.07	Residential telephone connection charge, 2002	4
4.08	Affordability of Internet access, 2003	53
5.01	Investment in training, 2004	54
5.02	Availability of training services, 2004	50
5.03	Quality of business schools, 2004	53
5.04	Business investment in R&D, 2004	76
5.05	Business monthly telephone subscription, 2002–3	35
5.06	Business telephone connection charge, 2002–3	4
6.01	Government prioritization of ICT, 2004	79
6.02	Government procurement of ICT, 2004	73

Usage Component Index — 49

Individual Usage		49
Business Usage		42
Government Usage		51
7.01	Cellular mobile subscribers, 2003	44
7.02	Telephone subscribers, 2002	45
7.03	Public payphones, 2002	77
7.04	Telephone lines, 2002	42
7.05	Television sets, 2002	1
7.06	Broadband-DSL Internet subscribers, 2002–3	59
7.07	Broadband-cable modem, 2002–3	46
7.08	Internet users per 100 inhabitants, 2002	56
8.01	Prevalence of foreign technology licensing, 2004	33
8.02	Firm-level technology absorption, 2004	29
8.03	Capacity for innovation, 2004	57
8.04	Availability of new telephone lines, 2004	49
8.05	Availability of cellular phones, 2004	42
9.01	Government success in ICT promotion, 2004	81
9.02	Government online services, 2003	20

Uganda

Key Indicators

Population (mn), 2003	25.8
GDP per capita (PPP US$), 2003	1,471
Internet users per 100 inhabitants, 2002	0.4

RANK/104

Environment Component Index — 59

Market Environment		51
Political and Regulatory Environment		54
Infrastructure Environment		103
1.01	Availability of scientists and engineers, 2004	72
1.02	Venture capital availability, 2004	53
1.03	Sophistication of financial markets, 2004	93
1.04	Technological sophistication, 2004	79
1.05	State of cluster development, 2004	46
1.06	Collaboration in clusters, 2004	55
1.07	University-industry collaboration, 2004	35
1.08	Quality of scientific research institutions, 2004	33
1.09	Subsidies for firm-level R&D, 2004	41
1.10	Brain drain, 2004	81
1.11	Ease of access to loans, 2004	57
1.12	Administrative burden, 2004	27
1.13	Ease to start a new business, 2004	34
2.01	Effectiveness of lawmaking, 2004	38
2.02	Laws relating to ICT, 2004	55
2.03	Effectiveness of judiciary, 2004	53
2.04	Intellectual property protection, 2004	85
3.01	Telephone mainlines, 2002	103
3.02	Secure Internet servers, 2003	92
3.03	Internet hosts, 2003	88

Readiness Component Index — 91

Individual Readiness		100
Business Readiness		97
Government Readiness		37
4.01	Quality of math and science education, 2004	70
4.02	Quality of educational system, 2004	61
4.03	Quality of public schools, 2004	78
4.04	Internet access in schools, 2004	85

Readiness Component Index (continued)

4.05	Buyer sophistication, 2004	79
4.06	Buyer dynamism, 2004	75
4.07	Residential telephone connection charge, 2002	100
4.08	Affordability of Internet access, 2003	102
5.01	Investment in training, 2004	77
5.02	Availability of training services, 2004	46
5.03	Quality of business schools, 2004	68
5.04	Business investment in R&D, 2004	38
5.05	Business monthly telephone subscription, 2002–3	100
5.06	Business telephone connection charge, 2002–3	100
6.01	Government prioritization of ICT, 2004	32
6.02	Government procurement of ICT, 2004	32

Usage Component Index — 67

Individual Usage		100
Business Usage		67
Government Usage		47
7.01	Cellular mobile subscribers, 2003	90
7.02	Telephone subscribers, 2002	97
7.03	Public payphones, 2002	95
7.04	Telephone lines, 2002	103
7.05	Television sets, 2002	101
7.06	Broadband-DSL Internet subscribers, 2002–3	65
7.07	Broadband-cable modem, 2002–3	60
7.08	Internet users per 100 inhabitants, 2002	94
8.01	Prevalence of foreign technology licensing, 2004	55
8.02	Firm-level technology absorption, 2004	66
8.03	Capacity for innovation, 2004	62
8.04	Availability of new telephone lines, 2004	72
8.05	Availability of cellular phones, 2004	77
9.01	Government success in ICT promotion, 2004	30
9.02	Government online services, 2003	62

Ukraine

Key Indicators

Population (mn), 2003	48.5
GDP per capita (PPP US$), 2003	5,472
Internet users per 100 inhabitants, 2002	1.8

Networked Readiness Index Rank

Year (number of countries)	Rank
2004 (104)	**82**
2003 (102)	78
2002 (82)	70

RANK/104

Environment Component Index — 84

Market Environment		82
Political and Regulatory Environment		86
Infrastructure Environment		55
1.01	Availability of scientists and engineers, 2004	49
1.02	Venture capital availability, 2004	58
1.03	Sophistication of financial markets, 2004	81
1.04	Technological sophistication, 2004	85
1.05	State of cluster development, 2004	55
1.06	Collaboration in clusters, 2004	39
1.07	University-industry collaboration, 2004	60
1.08	Quality of scientific research institutions, 2004	42
1.09	Subsidies for firm-level R&D, 2004	58
1.10	Brain drain, 2004	82
1.11	Ease of access to loans, 2004	94
1.12	Administrative burden, 2004	102
1.13	Ease to start a new business, 2004	97
2.01	Effectiveness of lawmaking, 2004	83
2.02	Laws relating to ICT, 2004	81
2.03	Effectiveness of judiciary, 2004	95
2.04	Intellectual property protection, 2004	83
3.01	Telephone mainlines, 2002	55
3.02	Secure Internet servers, 2003	79
3.03	Internet hosts, 2003	60

Readiness Component Index — 77

Individual Readiness		59
Business Readiness		74
Government Readiness		88
4.01	Quality of math and science education, 2004	34
4.02	Quality of educational system, 2004	47
4.03	Quality of public schools, 2004	52
4.04	Internet access in schools, 2004	84

Readiness Component Index (continued)

4.05	Buyer sophistication, 2004	61
4.06	Buyer dynamism, 2004	67
4.07	Residential telephone connection charge, 2002	70
4.08	Affordability of Internet access, 2003	77
5.01	Investment in training, 2004	85
5.02	Availability of training services, 2004	59
5.03	Quality of business schools, 2004	66
5.04	Business investment in R&D, 2004	48
5.05	Business monthly telephone subscription, 2002–3	59
5.06	Business telephone connection charge, 2002–3	91
6.01	Government prioritization of ICT, 2004	82
6.02	Government procurement of ICT, 2004	94

Usage Component Index — 81

Individual Usage		69
Business Usage		83
Government Usage		82
7.01	Cellular mobile subscribers, 2003	80
7.02	Telephone subscribers, 2002	68
7.03	Public payphones, 2002	70
7.04	Telephone lines, 2002	55
7.05	Television sets, 2002	17
7.06	Broadband-DSL Internet subscribers, 2002–3	65
7.07	Broadband-cable modem, 2002–3	60
7.08	Internet users per 100 inhabitants, 2002	83
8.01	Prevalence of foreign technology licensing, 2004	88
8.02	Firm-level technology absorption, 2004	77
8.03	Capacity for innovation, 2004	28
8.04	Availability of new telephone lines, 2004	84
8.05	Availability of cellular phones, 2004	94
9.01	Government success in ICT promotion, 2004	76
9.02	Government online services, 2003	79

United Arab Emirates

Key Indicators

Population (mn), 2003	3
GDP per capita (PPP US$), 2003	17,520
Internet users per 100 inhabitants, 2002	27.1

RANK/104

Environment Component Index	26
Market Environment	15
Political and Regulatory Environment	26
Infrastructure Environment	35

1.01	Availability of scientists and engineers, 2004	70
1.02	Venture capital availability, 2004	20
1.03	Sophistication of financial markets, 2004	41
1.04	Technological sophistication, 2004	12
1.05	State of cluster development, 2004	8
1.06	Collaboration in clusters, 2004	28
1.07	University-industry collaboration, 2004	65
1.08	Quality of scientific research institutions, 2004	67
1.09	Subsidies for firm-level R&D, 2004	75
1.10	Brain drain, 2004	3
1.11	Ease of access to loans, 2004	18
1.12	Administrative burden, 2004	3
1.13	Ease to start a new business, 2004	9
2.01	Effectiveness of lawmaking, 2004	33
2.02	Laws relating to ICT, 2004	25
2.03	Effectiveness of judiciary, 2004	30
2.04	Intellectual property protection, 2004	27
3.01	Telephone mainlines, 2002	36
3.02	Secure Internet servers, 2003	30
3.03	Internet hosts, 2003	33

Readiness Component Index	18
Individual Readiness	28
Business Readiness	42
Government Readiness	4

4.01	Quality of math and science education, 2004	40
4.02	Quality of educational system, 2004	23
4.03	Quality of public schools, 2004	53
4.04	Internet access in schools, 2004	24

Readiness Component Index (continued)	
4.05	Buyer sophistication, 2004	32
4.06	Buyer dynamism, 2004	20
4.07	Residential telephone connection charge, 2002	11
4.08	Affordability of Internet access, 2003	9
5.01	Investment in training, 2004	43
5.02	Availability of training services, 2004	60
5.03	Quality of business schools, 2004	86
5.04	Business investment in R&D, 2004	45
5.05	Business monthly telephone subscription, 2002–3	1
5.06	Business telephone connection charge, 2002–3	11
6.01	Government prioritization of ICT, 2004	3
6.02	Government procurement of ICT, 2004	6

Usage Component Index	20
Individual Usage	29
Business Usage	24
Government Usage	7

7.01	Cellular mobile subscribers, 2003	25
7.02	Telephone subscribers, 2002	36
7.03	Public payphones, 2002	2
7.04	Telephone lines, 2002	40
7.05	Television sets, 2002	64
7.06	Broadband-DSL Internet subscribers, 2002–3	33
7.07	Broadband-cable modem, 2002–3	60
7.08	Internet users per 100 inhabitants, 2002	31
8.01	Prevalence of foreign technology licensing, 2004	9
8.02	Firm-level technology absorption, 2004	13
8.03	Capacity for innovation, 2004	77
8.04	Availability of new telephone lines, 2004	20
8.05	Availability of cellular phones, 2004	30
9.01	Government success in ICT promotion, 2004	2
9.02	Government online services, 2003	33

Part 2 Country Profiles

203

United Kingdom

Key Indicators

Population (mn), 2003	59.3
GDP per capita (PPP US$), 2003	27,106
Internet users per 100 inhabitants, 2002	42.3

Networked Readiness Index Rank

Year (number of countries)   Rank

2004 (104)	**12**
2003 (102)	15
2002 (82)	7

RANK/104

Environment Component Index		7
Market Environment		5
Political and Regulatory Environment		2
Infrastructure Environment		13
1.01	Availability of scientists and engineers, 2004	34
1.02	Venture capital availability, 2004	3
1.03	Sophistication of financial markets, 2004	1
1.04	Technological sophistication, 2004	17
1.05	State of cluster development, 2004	14
1.06	Collaboration in clusters, 2004	12
1.07	University-industry collaboration, 2004	8
1.08	Quality of scientific research institutions, 2004	5
1.09	Subsidies for firm-level R&D, 2004	11
1.10	Brain drain, 2004	9
1.11	Ease of access to loans, 2004	1
1.12	Administrative burden, 2004	30
1.13	Ease to start a new business, 2004	6
2.01	Effectiveness of lawmaking, 2004	2
2.02	Laws relating to ICT, 2004	5
2.03	Effectiveness of judiciary, 2004	9
2.04	Intellectual property protection, 2004	6
3.01	Telephone mainlines, 2002	12
3.02	Secure Internet servers, 2003	8
3.03	Internet hosts, 2003	17

Readiness Component Index		13
Individual Readiness		21
Business Readiness		7
Government Readiness		24
4.01	Quality of math and science education, 2004	51
4.02	Quality of educational system, 2004	22
4.03	Quality of public schools, 2004	31
4.04	Internet access in schools, 2004	20

Readiness Component Index (continued)		
4.05	Buyer sophistication, 2004	3
4.06	Buyer dynamism, 2004	12
4.07	Residential telephone connection charge, 2002	24
4.08	Affordability of Internet access, 2003	16
5.01	Investment in training, 2004	16
5.02	Availability of training services, 2004	4
5.03	Quality of business schools, 2004	3
5.04	Business investment in R&D, 2004	11
5.05	Business monthly telephone subscription, 2002–3	27
5.06	Business telephone connection charge, 2002–3	36
6.01	Government prioritization of ICT, 2004	28
6.02	Government procurement of ICT, 2004	24

Usage Component Index		19
Individual Usage		30
Business Usage		13
Government Usage		17
7.01	Cellular mobile subscribers, 2003	18
7.02	Telephone subscribers, 2002	8
7.03	Public payphones, 2002	56
7.04	Telephone lines, 2002	12
7.05	Television sets, 2002	26
7.06	Broadband-DSL Internet subscribers, 2002–3	25
7.07	Broadband-cable modem, 2002–3	18
7.08	Internet users per 100 inhabitants, 2002	16
8.01	Prevalence of foreign technology licensing, 2004	15
8.02	Firm-level technology absorption, 2004	21
8.03	Capacity for innovation, 2004	10
8.04	Availability of new telephone lines, 2004	19
8.05	Availability of cellular phones, 2004	21
9.01	Government success in ICT promotion, 2004	38
9.02	Government online services, 2003	8

United States

Key Indicators

Population (mn), 2003	294
GDP per capita (PPP US$), 2003	37,352
Internet users per 100 inhabitants, 2002	55.1

RANK/104

Environment Component Index		4
Market Environment		2
Political and Regulatory Environment		10
Infrastructure Environment		6
1.01	Availability of scientists and engineers, 2004	8
1.02	Venture capital availability, 2004	1
1.03	Sophistication of financial markets, 2004	2
1.04	Technological sophistication, 2004	2
1.05	State of cluster development, 2004	4
1.06	Collaboration in clusters, 2004	3
1.07	University-industry collaboration, 2004	2
1.08	Quality of scientific research institutions, 2004	1
1.09	Subsidies for firm-level R&D, 2004	16
1.10	Brain drain, 2004	1
1.11	Ease of access to loans, 2004	11
1.12	Administrative burden, 2004	36
1.13	Ease to start a new business, 2004	7
2.01	Effectiveness of lawmaking, 2004	8
2.02	Laws relating to ICT, 2004	11
2.03	Effectiveness of judiciary, 2004	26
2.04	Intellectual property protection, 2004	3
3.01	Telephone mainlines, 2002	9
3.02	Secure Internet servers, 2003	2
3.03	Internet hosts, 2003	48

Readiness Component Index		5
Individual Readiness		17
Business Readiness		1
Government Readiness		7
4.01	Quality of math and science education, 2004	46
4.02	Quality of educational system, 2004	27
4.03	Quality of public schools, 2004	27
4.04	Internet access in schools, 2004	14

Readiness Component Index (continued)		
4.05	Buyer sophistication, 2004	1
4.06	Buyer dynamism, 2004	4
4.07	Residential telephone connection charge, 2002	3
4.08	Affordability of Internet access, 2003	2
5.01	Investment in training, 2004	8
5.02	Availability of training services, 2004	1
5.03	Quality of business schools, 2004	1
5.04	Business investment in R&D, 2004	2
5.05	Business monthly telephone subscription, 2002–3	34
5.06	Business telephone connection charge, 2002–3	8
6.01	Government prioritization of ICT, 2004	20
6.02	Government procurement of ICT, 2004	8

Usage Component Index		5
Individual Usage		11
Business Usage		10
Government Usage		3
7.01	Cellular mobile subscribers, 2003	37
7.02	Telephone subscribers, 2002	29
7.03	Public payphones, 2002	9
7.04	Telephone lines, 2002	10
7.05	Television sets, 2002	14
7.06	Broadband-DSL Internet subscribers, 2002–3	21
7.07	Broadband-cable modem, 2002–3	5
7.08	Internet users per 100 inhabitants, 2002	4
8.01	Prevalence of foreign technology licensing, 2004	28
8.02	Firm-level technology absorption, 2004	2
8.03	Capacity for innovation, 2004	7
8.04	Availability of new telephone lines, 2004	27
8.05	Availability of cellular phones, 2004	48
9.01	Government success in ICT promotion, 2004	34
9.02	Government online services, 2003	2

Uruguay

Key Indicators

Population (mn), 2003	3.4
GDP per capita (PPP US$), 2003	8,280
Internet users per 100 inhabitants, 2002	11.9

Networked Readiness Index Rank

Year (number of countries)	Rank
2004 (104)	**64**
2003 (102)	54
2002 (82)	55

RANK/104

Environment Component Index		62
Market Environment		85
Political and Regulatory Environment		57
Infrastructure Environment		38
1.01	Availability of scientists and engineers, 2004	51
1.02	Venture capital availability, 2004	100
1.03	Sophistication of financial markets, 2004	73
1.04	Technological sophistication, 2004	53
1.05	State of cluster development, 2004	101
1.06	Collaboration in clusters, 2004	97
1.07	University-industry collaboration, 2004	72
1.08	Quality of scientific research institutions, 2004	73
1.09	Subsidies for firm-level R&D, 2004	83
1.10	Brain drain, 2004	66
1.11	Ease of access to loans, 2004	97
1.12	Administrative burden, 2004	56
1.13	Ease to start a new business, 2004	90
2.01	Effectiveness of lawmaking, 2004	74
2.02	Laws relating to ICT, 2004	83
2.03	Effectiveness of judiciary, 2004	29
2.04	Intellectual property protection, 2004	53
3.01	Telephone mainlines, 2002	41
3.02	Secure Internet servers, 2003	43
3.03	Internet hosts, 2003	24

Readiness Component Index		67
Individual Readiness		64
Business Readiness		65
Government Readiness		71
4.01	Quality of math and science education, 2004	69
4.02	Quality of educational system, 2004	54
4.03	Quality of public schools, 2004	55
4.04	Internet access in schools, 2004	48

Readiness Component Index (continued)		
4.05	Buyer sophistication, 2004	76
4.06	Buyer dynamism, 2004	95
4.07	Residential telephone connection charge, 2002	57
4.08	Affordability of Internet access, 2003	50
5.01	Investment in training, 2004	84
5.02	Availability of training services, 2004	70
5.03	Quality of business schools, 2004	47
5.04	Business investment in R&D, 2004	86
5.05	Business monthly telephone subscription, 2002–3	74
5.06	Business telephone connection charge, 2002–3	62
6.01	Government prioritization of ICT, 2004	70
6.02	Government procurement of ICT, 2004	69

Usage Component Index		64
Individual Usage		52
Business Usage		72
Government Usage		78
7.01	Cellular mobile subscribers, 2003	63
7.02	Telephone subscribers, 2002	50
7.03	Public payphones, 2002	43
7.04	Telephone lines, 2002	41
7.05	Television sets, 2002	65
7.06	Broadband-DSL Internet subscribers, 2002–3	31
7.07	Broadband-cable modem, 2002–3	41
7.08	Internet users per 100 inhabitants, 2002	46
8.01	Prevalence of foreign technology licensing, 2004	86
8.02	Firm-level technology absorption, 2004	87
8.03	Capacity for innovation, 2004	76
8.04	Availability of new telephone lines, 2004	29
8.05	Availability of cellular phones, 2004	59
9.01	Government success in ICT promotion, 2004	71
9.02	Government online services, 2003	78

Venezuela

Key Indicators

Population (mn), 2003	25.7
GDP per capita (PPP US$), 2003	4,909
Internet users per 100 inhabitants, 2002	5.1

Networked Readiness Index Rank

Year (number of countries)	Rank
2004 (104)	**84**
2003 (102)	72
2002 (82)	66

RANK/104

Environment Component Index — 96

Market Environment		95
Political and Regulatory Environment		100
Infrastructure Environment		67
1.01	Availability of scientists and engineers, 2004	65
1.02	Venture capital availability, 2004	88
1.03	Sophistication of financial markets, 2004	71
1.04	Technological sophistication, 2004	56
1.05	State of cluster development, 2004	100
1.06	Collaboration in clusters, 2004	103
1.07	University-industry collaboration, 2004	67
1.08	Quality of scientific research institutions, 2004	78
1.09	Subsidies for firm-level R&D, 2004	91
1.10	Brain drain, 2004	91
1.11	Ease of access to loans, 2004	85
1.12	Administrative burden, 2004	104
1.13	Ease to start a new business, 2004	101
2.01	Effectiveness of lawmaking, 2004	104
2.02	Laws relating to ICT, 2004	72
2.03	Effectiveness of judiciary, 2004	103
2.04	Intellectual property protection, 2004	92
3.01	Telephone mainlines, 2002	68
3.02	Secure Internet servers, 2003	55
3.03	Internet hosts, 2003	68

Readiness Component Index — 78

Individual Readiness		74
Business Readiness		67
Government Readiness		90
4.01	Quality of math and science education, 2004	82
4.02	Quality of educational system, 2004	93
4.03	Quality of public schools, 2004	96
4.04	Internet access in schools, 2004	71

Readiness Component Index (continued)

4.05	Buyer sophistication, 2004	82
4.06	Buyer dynamism, 2004	63
4.07	Residential telephone connection charge, 2002	54
4.08	Affordability of Internet access, 2003	46
5.01	Investment in training, 2004	83
5.02	Availability of training services, 2004	94
5.03	Quality of business schools, 2004	48
5.04	Business investment in R&D, 2004	68
5.05	Business monthly telephone subscription, 2002–3	77
5.06	Business telephone connection charge, 2002–3	54
6.01	Government prioritization of ICT, 2004	91
6.02	Government procurement of ICT, 2004	87

Usage Component Index — 68

Individual Usage		59
Business Usage		71
Government Usage		80
7.01	Cellular mobile subscribers, 2003	59
7.02	Telephone subscribers, 2002	58
7.03	Public payphones, 2002	22
7.04	Telephone lines, 2002	70
7.05	Television sets, 2002	67
7.06	Broadband-DSL Internet subscribers, 2002–3	42
7.07	Broadband-cable modem, 2002–3	32
7.08	Internet users per 100 inhabitants, 2002	62
8.01	Prevalence of foreign technology licensing, 2004	69
8.02	Firm-level technology absorption, 2004	81
8.03	Capacity for innovation, 2004	86
8.04	Availability of new telephone lines, 2004	64
8.05	Availability of cellular phones, 2004	47
9.01	Government success in ICT promotion, 2004	92
9.02	Government online services, 2003	50

Vietnam

Key Indicators

Population (mn), 2003	81.4
GDP per capita (PPP US$), 2003	2,490
Internet users per 100 inhabitants, 2002	1.8

Networked Readiness Index Rank

Year (number of countries)	Rank
2004 (104)	**68**
2003 (102)	68
2002 (82)	71

RANK/104

Environment Component Index	79
Market Environment	74
Political and Regulatory Environment	74
Infrastructure Environment	84

1.01	Availability of scientists and engineers, 2004	41
1.02	Venture capital availability, 2004	66
1.03	Sophistication of financial markets, 2004	90
1.04	Technological sophistication, 2004	81
1.05	State of cluster development, 2004	87
1.06	Collaboration in clusters, 2004	77
1.07	University-industry collaboration, 2004	82
1.08	Quality of scientific research institutions, 2004	93
1.09	Subsidies for firm-level R&D, 2004	76
1.10	Brain drain, 2004	61
1.11	Ease of access to loans, 2004	93
1.12	Administrative burden, 2004	59
1.13	Ease to start a new business, 2004	40
2.01	Effectiveness of lawmaking, 2004	43
2.02	Laws relating to ICT, 2004	94
2.03	Effectiveness of judiciary, 2004	59
2.04	Intellectual property protection, 2004	93
3.01	Telephone mainlines, 2002	81
3.02	Secure Internet servers, 2003	96
3.03	Internet hosts, 2003	99

Readiness Component Index	65
Individual Readiness	81
Business Readiness	87
Government Readiness	40

4.01	Quality of math and science education, 2004	60
4.02	Quality of educational system, 2004	89
4.03	Quality of public schools, 2004	63
4.04	Internet access in schools, 2004	55

Readiness Component Index (continued)	
4.05	Buyer sophistication, 2004	65
4.06	Buyer dynamism, 2004	56
4.07	Residential telephone connection charge, 2002	94
4.08	Affordability of Internet access, 2003	86
5.01	Investment in training, 2004	78
5.02	Availability of training services, 2004	84
5.03	Quality of business schools, 2004	100
5.04	Business investment in R&D, 2004	71
5.05	Business monthly telephone subscription, 2002–3	70
5.06	Business telephone connection charge, 2002–3	93
6.01	Government prioritization of ICT, 2004	33
6.02	Government procurement of ICT, 2004	36

Usage Component Index	63
Individual Usage	82
Business Usage	63
Government Usage	54

7.01	Cellular mobile subscribers, 2003	88
7.02	Telephone subscribers, 2002	86
7.03	Public payphones, 2002	96
7.04	Telephone lines, 2002	79
7.05	Television sets, 2002	69
7.06	Broadband-DSL Internet subscribers, 2002–3	65
7.07	Broadband-cable modem, 2002–3	60
7.08	Internet users per 100 inhabitants, 2002	82
8.01	Prevalence of foreign technology licensing, 2004	99
8.02	Firm-level technology absorption, 2004	38
8.03	Capacity for innovation, 2004	47
8.04	Availability of new telephone lines, 2004	59
8.05	Availability of cellular phones, 2004	57
9.01	Government success in ICT promotion, 2004	28
9.02	Government online services, 2003	86

Zambia

Key Indicators

Population (mn), 2003	10.8
GDP per capita (PPP US$), 2003	883
Internet users per 100 inhabitants, 2002	0.5

Networked Readiness Index Rank

Year (number of countries)   Rank

2004 (104)   **81**

2003 (102)   85

2002 (82)   —

RANK/104

Environment Component Index		68
Market Environment		67
Political and Regulatory Environment		63
Infrastructure Environment		70
1.01	Availability of scientists and engineers, 2004	64
1.02	Venture capital availability, 2004	64
1.03	Sophistication of financial markets, 2004	85
1.04	Technological sophistication, 2004	92
1.05	State of cluster development, 2004	35
1.06	Collaboration in clusters, 2004	35
1.07	University-industry collaboration, 2004	87
1.08	Quality of scientific research institutions, 2004	64
1.09	Subsidies for firm-level R&D, 2004	72
1.10	Brain drain, 2004	103
1.11	Ease of access to loans, 2004	78
1.12	Administrative burden, 2004	35
1.13	Ease to start a new business, 2004	41
2.01	Effectiveness of lawmaking, 2004	51
2.02	Laws relating to ICT, 2004	74
2.03	Effectiveness of judiciary, 2004	56
2.04	Intellectual property protection, 2004	75
3.01	Telephone mainlines, 2002	66
3.02	Secure Internet servers, 2003	100
3.03	Internet hosts, 2003	83

Readiness Component Index		83
Individual Readiness		80
Business Readiness		79
Government Readiness		85
4.01	Quality of math and science education, 2004	71
4.02	Quality of educational system, 2004	70
4.03	Quality of public schools, 2004	82
4.04	Internet access in schools, 2004	86

Readiness Component Index (continued)		
4.05	Buyer sophistication, 2004	58
4.06	Buyer dynamism, 2004	60
4.07	Residential telephone connection charge, 2002	67
4.08	Affordability of Internet access, 2003	90
5.01	Investment in training, 2004	73
5.02	Availability of training services, 2004	51
5.03	Quality of business schools, 2004	77
5.04	Business investment in R&D, 2004	80
5.05	Business monthly telephone subscription, 2002–3	80
5.06	Business telephone connection charge, 2002–3	79
6.01	Government prioritization of ICT, 2004	84
6.02	Government procurement of ICT, 2004	83

Usage Component Index		89
Individual Usage		93
Business Usage		75
Government Usage		83
7.01	Cellular mobile subscribers, 2003	96
7.02	Telephone subscribers, 2002	94
7.03	Public payphones, 2002	98
7.04	Telephone lines, 2002	94
7.05	Television sets, 2002	90
7.06	Broadband-DSL Internet subscribers, 2002–3	65
7.07	Broadband-cable modem, 2002–3	60
7.08	Internet users per 100 inhabitants, 2002	93
8.01	Prevalence of foreign technology licensing, 2004	73
8.02	Firm-level technology absorption, 2004	64
8.03	Capacity for innovation, 2004	71
8.04	Availability of new telephone lines, 2004	82
8.05	Availability of cellular phones, 2004	80
9.01	Government success in ICT promotion, 2004	79
9.02	Government online services, 2003	79

Zimbabwe

Key Indicators

Population (mn), 2003	12.9
GDP per capita (PPP US$), 2003	1,892
Internet users per 100 inhabitants, 2002	4.3

Networked Readiness Index Rank

Year (number of countries)	Rank
2004 (104)	**94**
2003 (102)	95
2002 (82)	80

RANK/104

Environment Component Index		89
Market Environment		88
Political and Regulatory Environment		87
Infrastructure Environment		68
1.01	Availability of scientists and engineers, 2004	84
1.02	Venture capital availability, 2004	72
1.03	Sophistication of financial markets, 2004	57
1.04	Technological sophistication, 2004	88
1.05	State of cluster development, 2004	74
1.06	Collaboration in clusters, 2004	81
1.07	University-industry collaboration, 2004	78
1.08	Quality of scientific research institutions, 2004	60
1.09	Subsidies for firm-level R&D, 2004	79
1.10	Brain drain, 2004	104
1.11	Ease of access to loans, 2004	75
1.12	Administrative burden, 2004	80
1.13	Ease to start a new business, 2004	76
2.01	Effectiveness of lawmaking, 2004	94
2.02	Laws relating to ICT, 2004	84
2.03	Effectiveness of judiciary, 2004	101
2.04	Intellectual property protection, 2004	70
3.01	Telephone mainlines, 2002	65
3.02	Secure Internet servers, 2003	81
3.03	Internet hosts, 2003	78

Readiness Component Index		82
Individual Readiness		73
Business Readiness		56
Government Readiness		96
4.01	Quality of math and science education, 2004	58
4.02	Quality of educational system, 2004	63
4.03	Quality of public schools, 2004	69
4.04	Internet access in schools, 2004	93

Readiness Component Index (continued)		
4.05	Buyer sophistication, 2004	68
4.06	Buyer dynamism, 2004	89
4.07	Residential telephone connection charge, 2002	38
4.08	Affordability of Internet access, 2003	87
5.01	Investment in training, 2004	38
5.02	Availability of training services, 2004	87
5.03	Quality of business schools, 2004	83
5.04	Business investment in R&D, 2004	62
5.05	Business monthly telephone subscription, 2002–3	30
5.06	Business telephone connection charge, 2002–3	50
6.01	Government prioritization of ICT, 2004	98
6.02	Government procurement of ICT, 2004	92

Usage Component Index		102
Individual Usage		89
Business Usage		101
Government Usage		103
7.01	Cellular mobile subscribers, 2003	89
7.02	Telephone subscribers, 2002	88
7.03	Public payphones, 2002	94
7.04	Telephone lines, 2002	90
7.05	Television sets, 2002	89
7.06	Broadband-DSL Internet subscribers, 2002–3	65
7.07	Broadband-cable modem, 2002–3	53
7.08	Internet users per 100 inhabitants, 2002	68
8.01	Prevalence of foreign technology licensing, 2004	76
8.02	Firm-level technology absorption, 2004	90
8.03	Capacity for innovation, 2004	99
8.04	Availability of new telephone lines, 2004	102
8.05	Availability of cellular phones, 2004	101
9.01	Government success in ICT promotion, 2004	95
9.02	Government online services, 2003	102

Part 3

Data Presentation

How to Read the Data Tables

The data ranking section provides a list of all the variables with detailed data for all 104 countries included in the study.

The data are divided into nine sections, according to the Networked Readiness Framework.

i.   **Environment–Market**

ii. **Environment–Policy and Regulation**

iii. **Environment–Infrastructure**

iv. **Readiness–Individual**

v.   **Readiness–Business**

vi. **Readiness–Government**

vii. **Usage–Individual**

viii. **Usage–Business**

ix. **Usage–Government**

Two types of variables are used in our analysis: hard variables and soft variables. For each variable, the short name and a description are listed in the beginning of each table.

Soft variables: For each soft variable, the original question is included in the description of the variable. The values for these variables range from 1 to 7; a response of 1 corresponds to a lower relative performance, and a response of 7 corresponds to the highest level of relative performance. The values are responses to questionnaires and represent the average score of different respondents in a country. Variable 9.01, for example, corresponds to a question about the level of success of governments in their efforts to promote ICT; here, a high score means that the governments have been relatively successful in their efforts, whereas a low score implies the opposite.

Hard variables: Some hard variables had to be "transformed" to ensure that they were comparable across countries; the hard data presented in the tables are transformed data.

Missing data: The missing data were estimated to complete the data set. This was done primarily because the missing

values would have led to a bias in calculating the Index, and would limit our cross-country comparisons. In the tables, estimated data are indicated with an asterisk (*) superscript.

Ranking: The countries have been ranked with the use of the complete data set for each variable. The country responses shown in the tables are rounded off to two decimal places. Two countries with the same listed variable value can have different rankings because exact figures, not rounded numbers, were used to rank the countries. In the case of variable 1.02, for example, Australia's average score

was 4.692, and New Zealand's was 4.690. These countries are therefore ranked 8th and 9th, respectively, even though they are both listed with the same rounded score of 4.69.

In the event that two countries have exactly the same value, they have the same rank; the subsequent rank is two less to ensure that the last rank is 104. In the case of variable 1.04, for example, Ireland's average score was 5.00 and the Netherland's was also 5.00. Both these countries are therefore ranked 20th and the next country, New Zealand, with a score of 4.98, is ranked 22nd.

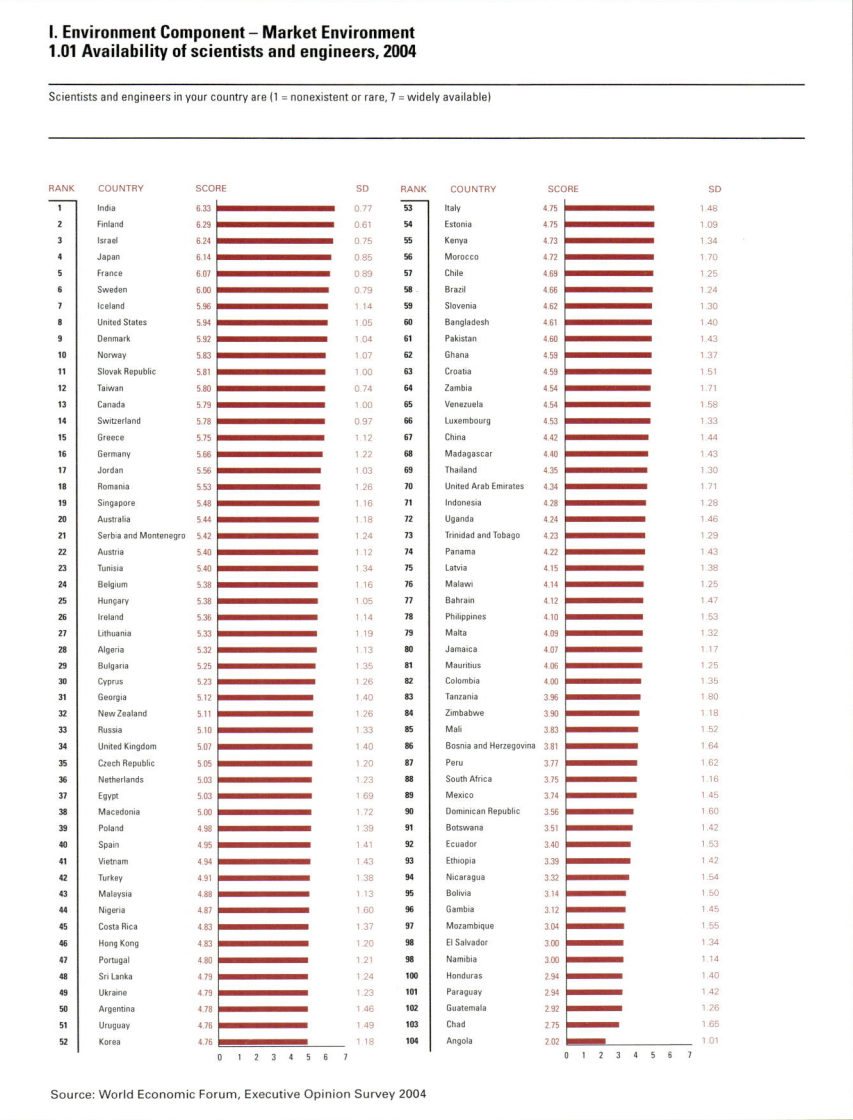

I. Environment Component – Market Environment
1.01 Availability of scientists and engineers, 2004

Scientists and engineers in your country are (1 = nonexistent or rare, 7 = widely available)

RANK	COUNTRY	SCORE	SD	RANK	COUNTRY	SCORE	SD
1	India	6.33	0.77	53	Italy	4.75	1.48
2	Finland	6.29	0.61	54	Estonia	4.75	1.09
3	Israel	6.24	0.75	55	Kenya	4.73	1.34
4	Japan	6.14	0.85	56	Morocco	4.72	1.70
5	France	6.07	0.89	57	Chile	4.69	1.25
6	Sweden	6.00	0.79	58	Brazil	4.66	1.24
7	Iceland	5.96	1.14	59	Slovenia	4.62	1.30
8	United States	5.94	1.05	60	Bangladesh	4.61	1.40
9	Denmark	5.92	1.04	61	Pakistan	4.60	1.43
10	Norway	5.83	1.07	62	Ghana	4.59	1.37
11	Slovak Republic	5.81	1.00	63	Croatia	4.59	1.51
12	Taiwan	5.80	0.74	64	Zambia	4.54	1.71
13	Canada	5.79	1.00	65	Venezuela	4.54	1.58
14	Switzerland	5.78	0.97	66	Luxembourg	4.53	1.33
15	Greece	5.75	1.12	67	China	4.42	1.44
16	Germany	5.66	1.22	68	Madagascar	4.40	1.43
17	Jordan	5.56	1.03	69	Thailand	4.35	1.30
18	Romania	5.53	1.26	70	United Arab Emirates	4.34	1.71
19	Singapore	5.48	1.16	71	Indonesia	4.28	1.28
20	Australia	5.44	1.18	72	Uganda	4.24	1.46
21	Serbia and Montenegro	5.42	1.24	73	Trinidad and Tobago	4.23	1.29
22	Austria	5.40	1.12	74	Panama	4.22	1.43
23	Tunisia	5.40	1.34	75	Latvia	4.15	1.38
24	Belgium	5.38	1.16	76	Malawi	4.14	1.25
25	Hungary	5.38	1.05	77	Bahrain	4.12	1.47
26	Ireland	5.36	1.14	78	Philippines	4.10	1.53
27	Lithuania	5.33	1.19	79	Malta	4.09	1.32
28	Algeria	5.32	1.13	80	Jamaica	4.07	1.17
29	Bulgaria	5.25	1.35	81	Mauritius	4.06	1.25
30	Cyprus	5.23	1.26	82	Colombia	4.00	1.35
31	Georgia	5.12	1.40	83	Tanzania	3.96	1.80
32	New Zealand	5.11	1.26	84	Zimbabwe	3.90	1.18
33	Russia	5.10	1.33	85	Mali	3.83	1.52
34	United Kingdom	5.07	1.40	86	Bosnia and Herzegovina	3.81	1.64
35	Czech Republic	5.05	1.20	87	Peru	3.77	1.62
36	Netherlands	5.03	1.23	88	South Africa	3.75	1.16
37	Egypt	5.03	1.69	89	Mexico	3.74	1.45
38	Macedonia	5.00	1.72	90	Dominican Republic	3.56	1.60
39	Poland	4.98	1.39	91	Botswana	3.51	1.42
40	Spain	4.95	1.41	92	Ecuador	3.40	1.53
41	Vietnam	4.94	1.43	93	Ethiopia	3.39	1.42
42	Turkey	4.91	1.38	94	Nicaragua	3.32	1.54
43	Malaysia	4.88	1.13	95	Bolivia	3.14	1.50
44	Nigeria	4.87	1.60	96	Gambia	3.12	1.45
45	Costa Rica	4.83	1.37	97	Mozambique	3.04	1.55
46	Hong Kong	4.83	1.20	98	El Salvador	3.00	1.34
47	Portugal	4.80	1.21	99	Namibia	3.00	1.14
48	Sri Lanka	4.79	1.24	100	Honduras	2.94	1.40
49	Ukraine	4.79	1.23	101	Paraguay	2.94	1.42
50	Argentina	4.78	1.46	102	Guatemala	2.92	1.26
51	Uruguay	4.76	1.49	103	Chad	2.75	1.65
52	Korea	4.76	1.18	104	Angola	2.02	1.01

Source: World Economic Forum, Executive Opinion Survey 2004

List of Data Tables

Market Environment

I. Environment Component – Market Environment
1.01 Availability of scientists and engineers, 2004

Scientists and engineers in your country are (1 = nonexistent or rare, 7 = widely available)

RANK	COUNTRY	SCORE	SD	RANK	COUNTRY	SCORE	SD
1	India	6.33	0.77	53	Italy	4.75	1.48
2	Finland	6.29	0.61	54	Estonia	4.75	1.09
3	Israel	6.24	0.75	55	Kenya	4.73	1.34
4	Japan	6.14	0.85	56	Morocco	4.72	1.70
5	France	6.07	0.89	57	Chile	4.69	1.25
6	Sweden	6.00	0.79	58	Brazil	4.66	1.24
7	Iceland	5.96	1.14	59	Slovenia	4.62	1.30
8	United States	5.94	1.05	60	Bangladesh	4.61	1.40
9	Denmark	5.92	1.04	61	Pakistan	4.60	1.43
10	Norway	5.83	1.07	62	Ghana	4.59	1.37
11	Slovak Republic	5.81	1.00	63	Croatia	4.59	1.51
12	Taiwan	5.80	0.74	64	Zambia	4.54	1.71
13	Canada	5.79	1.00	65	Venezuela	4.54	1.58
14	Switzerland	5.78	0.97	66	Luxembourg	4.53	1.33
15	Greece	5.75	1.12	67	China	4.42	1.44
16	Germany	5.66	1.22	68	Madagascar	4.40	1.43
17	Jordan	5.56	1.03	69	Thailand	4.35	1.30
18	Romania	5.53	1.26	70	United Arab Emirates	4.34	1.71
19	Singapore	5.48	1.16	71	Indonesia	4.28	1.28
20	Australia	5.44	1.18	72	Uganda	4.24	1.46
21	Serbia and Montenegro	5.42	1.24	73	Trinidad and Tobago	4.23	1.29
22	Austria	5.40	1.12	74	Panama	4.22	1.43
23	Tunisia	5.40	1.34	75	Latvia	4.15	1.38
24	Belgium	5.38	1.16	76	Malawi	4.14	1.25
25	Hungary	5.38	1.05	77	Bahrain	4.12	1.47
26	Ireland	5.36	1.14	78	Philippines	4.10	1.53
27	Lithuania	5.33	1.19	79	Malta	4.09	1.32
28	Algeria	5.32	1.13	80	Jamaica	4.07	1.17
29	Bulgaria	5.25	1.35	81	Mauritius	4.06	1.25
30	Cyprus	5.23	1.26	82	Colombia	4.00	1.35
31	Georgia	5.12	1.40	83	Tanzania	3.96	1.80
32	New Zealand	5.11	1.26	84	Zimbabwe	3.90	1.18
33	Russia	5.10	1.33	85	Mali	3.83	1.52
34	United Kingdom	5.07	1.40	86	Bosnia and Herzegovina	3.81	1.64
35	Czech Republic	5.05	1.20	87	Peru	3.77	1.62
36	Netherlands	5.03	1.23	88	South Africa	3.75	1.16
37	Egypt	5.03	1.69	89	Mexico	3.74	1.45
38	Macedonia	5.00	1.72	90	Dominican Republic	3.56	1.60
39	Poland	4.98	1.39	91	Botswana	3.51	1.42
40	Spain	4.95	1.41	92	Ecuador	3.40	1.53
41	Vietnam	4.94	1.43	93	Ethiopia	3.39	1.42
42	Turkey	4.91	1.38	94	Nicaragua	3.32	1.54
43	Malaysia	4.88	1.13	95	Bolivia	3.14	1.50
44	Nigeria	4.87	1.60	96	Gambia	3.12	1.45
45	Costa Rica	4.83	1.37	97	Mozambique	3.04	1.55
46	Hong Kong	4.83	1.20	98	El Salvador	3.00	1.34
47	Portugal	4.80	1.21	98	Namibia	3.00	1.14
48	Sri Lanka	4.79	1.24	100	Honduras	2.94	1.40
49	Ukraine	4.79	1.23	101	Paraguay	2.94	1.42
50	Argentina	4.78	1.46	102	Guatemala	2.92	1.26
51	Uruguay	4.76	1.49	103	Chad	2.75	1.65
52	Korea	4.76	1.18	104	Angola	2.02	1.01

Source: World Economic Forum, Executive Opinion Survey 2004

I. Environment Component – Market Environment
1.02 Venture capital availability, 2004

Entrepreneurs with innovative but risky projects can generally find venture capital in your country (1 = not true, 7 = true)

RANK	COUNTRY	SCORE	SD	RANK	COUNTRY	SCORE	SD
1	United States	5.76	1.24	53	Uganda	3.09	1.77
2	Israel	5.44	1.26	54	Kenya	3.09	1.67
3	United Kingdom	5.38	1.19	55	Latvia	3.07	1.46
4	Hong Kong	5.36	1.40	56	Brazil	3.06	1.50
5	Finland	5.05	1.35	57	Bulgaria	3.05	1.67
6	Netherlands	4.89	1.33	58	Ukraine	3.02	1.48
7	Ireland	4.70	1.42	59	China	3.02	1.27
8	Australia	4.69	1.37	60	Costa Rica	2.99	1.43
9	New Zealand	4.69	1.51	60	Sri Lanka	2.99	1.46
10	Norway	4.68	0.89	62	Gambia	2.99	1.77
11	Taiwan	4.63	1.35	63	Namibia	2.97	1.81
12	Denmark	4.63	1.24	64	Zambia	2.93	1.86
13	Sweden	4.58	1.30	65	Italy	2.93	1.48
14	Luxembourg	4.53	1.83	66	Vietnam	2.91	1.69
15	Singapore	4.49	1.56	67	Nigeria	2.83	1.70
16	Canada	4.47	1.63	68	Egypt	2.83	2.03
17	France	4.35	1.55	69	Jamaica	2.80	1.26
18	Lithuania	4.33	1.31	70	Serbia and Montenegro	2.80	1.44
19	Malaysia	4.04	1.57	71	Jordan	2.79	1.53
20	Indonesia	4.03	1.14	72	Zimbabwe	2.77	1.57
20	United Arab Emirates	4.03	1.58	73	Malta	2.76	1.54
22	Portugal	4.00	1.41	74	Poland	2.71	1.36
23	Japan	3.95	1.26	75	Colombia	2.70	1.15
24	Belgium	3.95	1.58	76	Pakistan	2.69	1.08
25	Switzerland	3.93	1.60	77	Philippines	2.69	1.42
26	Spain	3.91	1.63	78	Madagascar	2.65	1.74
27	India	3.84	1.49	79	Ghana	2.65	1.56
28	Estonia	3.81	1.46	80	Peru	2.63	1.32
29	Germany	3.77	1.46	81	Nicaragua	2.62	1.67
29	Tunisia	3.77	1.69	82	Guatemala	2.61	1.46
31	Austria	3.74	1.72	83	Romania	2.60	1.44
32	Bahrain	3.73	1.70	84	Turkey	2.54	1.55
33	South Africa	3.72	1.68	85	Mexico	2.51	1.38
34	Iceland	3.68	1.41	86	Georgia	2.44	1.49
35	Cyprus	3.67	1.57	87	Bosnia and Herzegovina	2.40	1.32
36	Macedonia	3.65	1.92	88	Venezuela	2.35	1.28
37	Panama	3.63	1.69	89	Mozambique	2.32	1.67
38	Slovak Republic	3.60	1.50	90	Honduras	2.32	1.43
39	Chile	3.53	1.56	91	Croatia	2.31	1.52
40	Morocco	3.43	1.68	92	Ecuador	2.30	1.40
41	Hungary	3.43	1.43	93	Dominican Republic	2.30	1.32
42	Tanzania	3.40	1.97	94	Mali	2.29	1.45
43	Botswana	3.39	1.52	95	Argentina	2.22	1.23
44	Greece	3.39	1.24	96	Malawi	2.19	1.24
45	Czech Republic	3.35	1.49	97	Bolivia	2.16	1.38
46	Thailand	3.31	1.31	98	Bangladesh	2.14	1.34
47	Mauritius	3.30	1.49	99	Paraguay	2.12	1.29
48	Trinidad and Tobago	3.26	1.48	100	Uruguay	2.07	1.00
49	Russia	3.22	1.53	101	Chad	2.01	1.51
50	El Salvador	3.19	1.76	102	Algeria	2.01	1.34
51	Slovenia	3.15	1.32	103	Ethiopia	1.70	1.13
52	Korea	3.13	1.39	104	Angola	1.37	0.95

Source: World Economic Forum, Executive Opinion Survey 2004

I. Environment Component – Market Environment
1.03 Sophistication of financial markets, 2004

The level of sophistication of financial markets in your country is (1 = lower than international norms, 7 = higher than international norms)

RANK	COUNTRY	SCORE	SD	RANK	COUNTRY	SCORE	SD
1	United Kingdom	6.77	0.79	53	Peru	3.87	1.23
2	United States	6.66	0.85	54	Malta	3.83	1.39
3	Luxembourg	6.53	0.57	55	Turkey	3.81	1.42
4	Switzerland	6.46	0.80	56	Kenya	3.79	1.27
5	Hong Kong	6.38	0.88	57	Zimbabwe	3.72	1.41
6	Canada	6.20	0.86	58	Mauritius	3.67	1.31
7	Australia	6.08	1.13	59	Poland	3.66	1.11
8	Sweden	6.05	0.91	60	Tunisia	3.63	1.50
9	Finland	6.03	0.82	61	Slovenia	3.62	1.16
10	Netherlands	5.95	0.83	62	Sri Lanka	3.55	1.29
11	New Zealand	5.82	0.98	63	Czech Republic	3.45	1.12
12	Singapore	5.80	1.01	64	Pakistan	3.44	1.67
13	Denmark	5.79	0.78	65	Guatemala	3.44	1.34
14	Germany	5.72	0.89	66	Ghana	3.43	1.39
15	Belgium	5.71	0.87	67	Philippines	3.43	1.16
16	South Africa	5.68	0.92	68	Slovak Republic	3.42	1.32
17	France	5.62	1.05	69	Dominican Republic	3.41	1.45
18	Israel	5.50	0.63	70	Romania	3.41	1.33
19	Japan	5.49	1.20	71	Venezuela	3.39	1.33
20	Norway	5.48	0.95	72	Russia	3.39	1.26
21	Ireland	5.45	1.11	73	Uruguay	3.39	1.23
22	Chile	5.36	0.97	74	Botswana	3.32	1.27
23	Bahrain	5.35	1.09	75	Nigeria	3.27	1.49
24	Estonia	5.26	1.00	76	Argentina	3.23	1.41
25	Panama	5.22	1.32	77	China	3.12	1.23
26	Brazil	5.19	1.11	78	Madagascar	3.11	1.78
27	Spain	5.08	1.18	79	Ecuador	3.10	1.29
28	Portugal	4.97	1.04	80	Honduras	3.09	1.47
29	Malaysia	4.97	1.17	81	Ukraine	3.08	1.35
30	Iceland	4.92	1.50	82	Mali	3.03	1.71
31	El Salvador	4.68	1.34	83	Egypt	3.00	1.46
32	India	4.68	1.25	84	Croatia	2.94	1.30
33	Austria	4.65	1.60	85	Zambia	2.88	1.31
34	Taiwan	4.63	1.13	86	Nicaragua	2.85	1.42
35	Mexico	4.55	1.23	87	Bolivia	2.74	1.26
36	Hungary	4.51	0.97	88	Paraguay	2.71	1.23
37	Jamaica	4.48	1.23	89	Gambia	2.67	1.63
38	Greece	4.41	1.15	90	Vietnam	2.64	1.12
39	Thailand	4.40	1.01	91	Bangladesh	2.55	1.21
40	Indonesia	4.31	1.09	92	Tanzania	2.53	1.32
41	United Arab Emirates	4.28	1.48	93	Uganda	2.50	1.26
42	Cyprus	4.25	1.29	94	Macedonia	2.48	1.11
43	Italy	4.21	1.52	95	Mozambique	2.43	1.36
44	Trinidad and Tobago	4.21	1.44	96	Georgia	2.41	1.19
45	Namibia	4.20	1.45	97	Serbia and Montenegro	2.39	1.10
46	Colombia	4.18	1.07	98	Bosnia and Herzegovina	2.37	1.26
47	Costa Rica	4.11	1.23	99	Malawi	2.28	1.03
48	Jordan	4.10	1.19	100	Bulgaria	2.23	1.09
49	Korea	4.10	1.36	101	Algeria	2.02	1.21
50	Lithuania	3.99	1.17	102	Ethiopia	2.01	1.13
51	Latvia	3.94	1.20	103	Chad	1.94	1.18
52	Morocco	3.94	1.57	104	Angola	1.72	0.96

Source: World Economic Forum, Executive Opinion Survey 2004

I. Environment Component – Market Environment
1.04 Technological sophistication, 2004

Your country's level of technology readiness is (1 = generally lags behind most other countries, 7 = is among the world leaders)

RANK	COUNTRY	SCORE		SD	RANK	COUNTRY	SCORE		SD
1	Israel	6.53		0.72	53	Uruguay	3.85		1.24
2	United States	6.47		0.90	54	Turkey	3.79		1.35
3	Finland	6.46		0.82	55	Mauritius	3.79		1.37
4	Sweden	6.45		0.51	56	Venezuela	3.78		1.32
5	Denmark	6.36		0.57	57	Indonesia	3.77		1.20
6	Iceland	6.24		0.72	58	Botswana	3.68		1.36
7	Japan	6.21		1.13	59	Cyprus	3.68		1.29
8	Singapore	5.90		0.97	60	China	3.68		1.24
9	Switzerland	5.75		1.14	61	Poland	3.67		1.20
10	Hong Kong	5.58		0.84	62	Trinidad and Tobago	3.67		1.35
11	Taiwan	5.52		0.98	63	Latvia	3.57		1.27
12	United Arab Emirates	5.51		1.06	64	Lithuania	3.57		1.24
13	Australia	5.42		1.12	65	Guatemala	3.54		1.41
14	Canada	5.42		1.08	66	Colombia	3.50		0.91
15	Norway	5.35		1.43	67	Peru	3.33		1.34
16	Germany	5.28		1.38	68	Kenya	3.30		1.26
17	United Kingdom	5.28		1.12	69	Russia	3.30		1.47
18	Belgium	5.26		1.13	70	Greece	3.25		1.12
19	France	5.01		1.47	71	Morocco	3.21		1.42
20	Ireland	5.00		1.36	72	Ghana	3.20		1.23
20	Netherlands	5.00		1.24	73	Sri Lanka	3.20		1.26
22	New Zealand	4.98		1.18	74	Romania	2.98		1.35
23	Chile	4.97		1.09	75	Ecuador	2.93		1.26
24	Korea	4.89		0.85	76	Paraguay	2.92		1.23
25	Luxembourg	4.87		1.25	77	Philippines	2.90		1.18
26	India	4.80		1.21	78	Algeria	2.89		1.27
27	Panama	4.73		1.38	79	Uganda	2.88		1.55
28	Malaysia	4.73		1.07	80	Tanzania	2.77		1.71
29	Austria	4.72		1.24	81	Vietnam	2.72		1.18
30	Estonia	4.69		1.31	82	Gambia	2.72		1.54
31	Bahrain	4.50		1.36	83	Nigeria	2.71		1.35
32	South Africa	4.49		1.46	84	Pakistan	2.71		0.90
33	Spain	4.47		1.50	85	Ukraine	2.67		1.32
34	Tunisia	4.44		1.16	86	Mali	2.63		1.11
35	Malta	4.42		1.30	87	Bulgaria	2.60		1.14
36	Brazil	4.33		1.17	88	Zimbabwe	2.60		1.19
37	Costa Rica	4.33		1.23	89	Croatia	2.58		1.30
38	Jordan	4.24		1.26	90	Madagascar	2.57		1.08
39	Thailand	4.24		1.19	91	Honduras	2.42		1.22
40	Namibia	4.13		1.50	92	Zambia	2.42		1.27
41	Czech Republic	4.12		1.29	93	Bangladesh	2.35		1.11
42	Dominican Republic	4.11		1.39	94	Bolivia	2.33		1.20
43	Slovak Republic	4.08		1.31	95	Malawi	2.31		1.01
44	Jamaica	4.07		1.41	96	Nicaragua	2.30		1.20
45	Italy	4.05		1.49	97	Serbia and Montenegro	2.26		1.31
46	Mexico	4.03		1.27	98	Georgia	2.17		0.97
47	Hungary	3.97		1.07	99	Mozambique	2.03		1.14
48	Slovenia	3.97		1.19	100	Macedonia	2.00		1.27
49	El Salvador	3.96		1.20	101	Angola	1.93		0.98
50	Argentina	3.94		1.28	102	Bosnia and Herzegovina	1.87		0.96
51	Portugal	3.93		1.21	103	Ethiopia	1.80		0.87
52	Egypt	3.91		1.60	104	Chad	1.67		0.96

0 1 2 3 4 5 6 7

Source: World Economic Forum, Executive Opinion Survey 2004

I. Environment Component – Market Environment
1.05 State of cluster development, 2004

How common are clusters in your country? (1 = limited and shallow, 7 = clusters are present in many fields and include many specialized suppliers and service providers)

RANK	COUNTRY	SCORE		SD	RANK	COUNTRY	SCORE		SD
1	Japan	5.46		1.10	53	Jamaica	3.28		1.46
2	Taiwan	5.39		1.08	54	Madagascar	3.26		1.79
3	Finland	5.33		1.37	55	Ukraine	3.22		1.40
4	United States	5.19		1.05	56	Russia	3.20		1.48
5	Italy	5.16		1.88	57	Gambia	3.18		1.71
6	Singapore	5.15		1.29	58	Botswana	3.16		1.61
7	India	4.83		1.31	59	Czech Republic	3.14		1.46
8	United Arab Emirates	4.78		1.66	60	Chile	3.13		1.53
9	Denmark	4.75		1.03	61	Trinidad and Tobago	3.11		1.48
10	Sweden	4.68		0.89	62	Slovenia	3.08		1.30
11	Hong Kong	4.68		1.73	63	Mauritius	3.06		1.52
12	Egypt	4.66		1.74	64	Costa Rica	3.06		1.32
13	Ireland	4.64		1.33	65	Jordan	3.03		1.49
14	United Kingdom	4.63		1.32	66	Poland	3.03		1.14
15	Canada	4.54		1.44	67	Sri Lanka	2.99		1.57
16	Israel	4.47		1.60	68	Slovak Republic	2.98		1.30
17	Germany	4.45		1.69	69	Colombia	2.98		1.19
18	Malaysia	4.42		1.23	70	Malawi	2.94		1.68
19	Pakistan	4.41		1.38	71	Namibia	2.93		1.84
20	Indonesia	4.41		1.26	72	Latvia	2.83		1.50
21	Korea	4.38		1.27	73	Greece	2.79		1.21
22	Thailand	4.36		1.31	74	Zimbabwe	2.76		1.68
23	Switzerland	4.31		1.73	75	Argentina	2.76		1.33
24	Nigeria	4.29		1.69	76	Romania	2.74		1.41
25	Norway	4.27		1.42	77	Guatemala	2.69		1.40
26	Brazil	4.26		1.49	78	Estonia	2.68		1.33
27	Austria	4.19		1.57	79	Honduras	2.68		1.40
28	Morocco	4.11		1.54	80	Croatia	2.67		1.49
29	France	4.07		1.34	81	Dominican Republic	2.65		1.42
30	Netherlands	3.84		1.54	82	Bolivia	2.64		1.48
31	Belgium	3.83		1.44	83	Ecuador	2.64		1.54
32	China	3.80		1.45	83	El Salvador	2.64		1.26
33	Kenya	3.73		1.55	85	Mozambique	2.63		1.70
34	Portugal	3.64		1.39	86	Hungary	2.63		1.29
35	Zambia	3.61		1.99	87	Vietnam	2.62		1.52
36	Luxembourg	3.61		1.91	88	Chad	2.61		1.75
37	Australia	3.60		1.31	89	Ghana	2.57		1.51
38	South Africa	3.58		1.47	90	Peru	2.53		1.31
39	Spain	3.56		1.33	91	Bulgaria	2.52		1.49
40	Tanzania	3.56		1.83	92	Ethiopia	2.50		1.69
41	Tunisia	3.52		1.70	93	Georgia	2.47		1.33
42	Bangladesh	3.52		1.93	94	Serbia and Montenegro	2.45		1.55
43	Iceland	3.45		1.10	95	Malta	2.37		1.19
44	Cyprus	3.44		1.57	96	Paraguay	2.36		1.45
45	Turkey	3.43		1.51	97	Nicaragua	2.30		1.48
46	Uganda	3.39		1.88	98	Algeria	2.29		1.46
47	New Zealand	3.39		1.22	99	Bosnia and Herzegovina	2.26		1.38
48	Panama	3.38		1.76	100	Venezuela	2.25		1.00
49	Mexico	3.37		1.51	101	Uruguay	2.20		1.22
50	Philippines	3.35		1.48	102	Macedonia	2.10		1.15
51	Bahrain	3.34		1.81	103	Mali	2.06		1.33
52	Lithuania	3.33		1.28	104	Angola	1.80		1.44

0 1 2 3 4 5 6 7

Source: World Economic Forum, Executive Opinion Survey 2004

I. Environment Component – Market Environment
1.06 Collaboration in clusters, 2004

Collaboration in your clusters with suppliers, service providers and partners in your country is (1 = almost nonexistent, 7 = extensive)

RANK	COUNTRY	SCORE		SD	RANK	COUNTRY	SCORE		SD
1	Japan	6.00		0.74	53	Latvia	3.61		1.46
2	Finland	5.85		0.94	54	Tanzania	3.60		1.83
3	United States	5.71		1.02	55	Uganda	3.60		1.57
4	Taiwan	5.49		0.80	56	Jamaica	3.57		1.32
5	Denmark	5.45		0.83	57	Slovenia	3.51		1.43
6	Sweden	5.24		0.90	58	Mexico	3.48		1.32
7	Germany	5.22		1.45	59	Estonia	3.48		1.30
8	Belgium	5.15		1.13	60	Pakistan	3.47		1.25
9	Switzerland	5.01		1.47	61	Romania	3.46		1.39
10	Netherlands	4.96		1.22	62	Poland	3.46		1.38
11	Singapore	4.94		1.13	63	Jordan	3.45		1.29
12	United Kingdom	4.93		1.04	64	Greece	3.42		1.33
13	Canada	4.82		1.43	65	Chile	3.40		1.45
14	France	4.79		1.21	66	Bahrain	3.38		1.80
15	China	4.78		1.33	67	Panama	3.32		1.45
16	Italy	4.76		1.56	68	Sri Lanka	3.32		1.48
17	Austria	4.73		1.58	69	Botswana	3.26		1.52
18	Brazil	4.73		1.05	70	Philippines	3.26		1.28
19	Israel	4.67		1.78	71	Costa Rica	3.25		1.31
20	Korea	4.61		1.06	72	Bulgaria	3.25		1.58
21	Luxembourg	4.59		1.34	73	Serbia and Montenegro	3.22		1.41
22	Hong Kong	4.50		1.52	74	Hungary	3.21		1.24
23	Malaysia	4.43		1.19	75	Malawi	3.16		1.44
24	India	4.40		1.24	76	Ghana	3.15		1.45
25	Norway	4.38		1.20	77	Vietnam	3.11		1.46
26	Croatia	4.33		1.35	78	Colombia	3.11		1.06
27	Ireland	4.31		1.37	79	Gambia	3.05		1.54
28	United Arab Emirates	4.23		1.46	80	Trinidad and Tobago	3.01		1.19
29	Egypt	4.22		1.58	81	Algeria	3.00		1.31
30	South Africa	4.21		1.47	81	Zimbabwe	3.00		1.39
31	Australia	4.20		1.37	83	Georgia	2.96		1.45
32	Indonesia	4.18		1.04	84	Dominican Republic	2.94		1.38
33	Thailand	4.18		0.97	85	Bosnia and Herzegovina	2.93		1.44
34	New Zealand	4.17		1.28	86	Mozambique	2.90		1.55
35	Zambia	4.08		1.61	87	Namibia	2.90		1.32
36	Morocco	4.07		1.43	88	Chad	2.88		1.84
37	Russia	4.01		1.35	89	Argentina	2.81		1.23
38	Portugal	4.00		1.13	90	El Salvador	2.81		1.35
39	Ukraine	3.99		1.05	91	Bangladesh	2.78		1.37
40	Nigeria	3.95		1.58	92	Peru	2.77		1.29
41	Czech Republic	3.84		1.28	93	Guatemala	2.72		1.22
42	Lithuania	3.84		1.19	94	Malta	2.67		1.37
43	Spain	3.81		1.26	95	Honduras	2.66		1.12
44	Slovak Republic	3.81		1.26	96	Macedonia	2.66		1.49
45	Tunisia	3.81		1.47	97	Uruguay	2.64		1.22
46	Iceland	3.80		1.32	98	Ethiopia	2.63		1.31
47	Cyprus	3.80		1.31	99	Ecuador	2.47		1.21
48	Turkey	3.74		1.32	100	Bolivia	2.46		1.20
49	Madagascar	3.68		1.55	101	Paraguay	2.43		1.21
50	Mali	3.68		1.66	102	Nicaragua	2.34		1.26
51	Mauritius	3.66		1.40	103	Venezuela	2.29		1.07
52	Kenya	3.65		1.40	104	Angola	2.23		1.31

0 1 2 3 4 5 6 7

0 1 2 3 4 5 6 7

Source: World Economic Forum, Executive Opinion Survey 2004

I. Environment Component – Market Environment
1.07 University-industry collaboration, 2004

In its research and development (R&D) activity, business collaboration with universities is (1 = minimal or nonexistent, 7 = intensive and ongoing)

RANK	COUNTRY	SCORE	SD	RANK	COUNTRY	SCORE	SD
1	Finland	5.79	1.06	53	Nigeria	3.04	1.54
2	United States	5.45	1.30	54	Trinidad and Tobago	3.04	1.26
3	Sweden	5.32	0.82	55	Morocco	3.02	1.53
4	Germany	5.20	1.13	56	Luxembourg	3.00	1.68
5	Singapore	5.12	1.20	56	Madagascar	3.00	1.57
6	Taiwan	5.07	1.24	58	Egypt	2.97	1.73
7	Switzerland	5.04	1.25	59	Mauritius	2.94	1.39
8	Japan	5.00	1.10	60	Ukraine	2.93	1.27
8	United Kingdom	5.00	1.10	61	Serbia and Montenegro	2.92	1.24
10	Netherlands	4.87	1.25	62	Turkey	2.91	1.36
11	Denmark	4.83	0.96	63	Kenya	2.90	1.56
12	Israel	4.81	1.60	64	Italy	2.89	1.29
13	Canada	4.65	1.32	65	United Arab Emirates	2.88	1.40
14	Austria	4.63	1.36	66	Namibia	2.87	1.53
15	Malaysia	4.56	1.19	67	Venezuela	2.86	1.44
16	Belgium	4.47	1.21	68	Philippines	2.86	1.31
17	Hong Kong	4.36	1.40	69	Bahrain	2.80	1.54
18	Australia	4.35	1.23	70	Hungary	2.76	1.59
19	South Africa	4.30	1.30	71	Romania	2.74	1.26
20	Iceland	4.29	1.49	72	Uruguay	2.72	0.93
21	Ireland	4.29	1.23	73	Bosnia and Herzegovina	2.69	1.42
22	China	4.23	1.41	74	Macedonia	2.67	1.54
23	Norway	4.17	1.37	75	Mozambique	2.66	1.42
24	Korea	4.15	1.32	76	Malta	2.65	1.14
25	France	4.04	1.43	77	Latvia	2.65	1.28
26	New Zealand	4.02	1.32	78	Zimbabwe	2.64	1.13
27	Indonesia	3.97	1.06	79	Gambia	2.60	1.89
28	Brazil	3.82	1.39	80	Sri Lanka	2.59	1.32
29	Czech Republic	3.79	1.45	81	Malawi	2.58	1.18
30	Slovenia	3.77	1.40	82	Vietnam	2.57	1.48
31	Thailand	3.71	1.50	83	Panama	2.55	1.40
32	Tunisia	3.65	1.64	84	Guatemala	2.53	1.13
33	Spain	3.63	1.38	85	Ghana	2.52	1.33
34	India	3.61	1.22	86	Cyprus	2.51	1.23
35	Uganda	3.57	1.76	87	Zambia	2.43	1.47
36	Lithuania	3.50	1.30	88	Argentina	2.42	1.18
37	Estonia	3.46	1.48	89	Bulgaria	2.40	1.29
38	Mali	3.46	1.90	90	Ecuador	2.37	1.17
39	Slovak Republic	3.40	1.28	91	Peru	2.33	0.97
40	Russia	3.36	1.51	92	Algeria	2.31	1.35
41	Portugal	3.32	1.25	93	Dominican Republic	2.29	1.24
42	Greece	3.30	1.35	94	Bolivia	2.22	1.10
43	Jamaica	3.24	1.35	95	Nicaragua	2.19	1.25
44	Chile	3.21	1.28	96	Georgia	2.19	1.11
45	Mexico	3.13	1.34	97	Bangladesh	2.12	1.32
46	Colombia	3.11	1.35	98	Pakistan	2.03	1.40
47	Jordan	3.11	1.42	99	Paraguay	2.03	1.17
48	Botswana	3.11	1.40	100	Honduras	2.02	1.19
49	Poland	3.09	1.49	101	Ethiopia	2.00	1.10
50	Tanzania	3.08	1.66	102	El Salvador	1.96	0.95
51	Croatia	3.06	1.47	103	Chad	1.95	1.36
52	Costa Rica	3.06	1.31	104	Angola	1.64	0.86

Source: World Economic Forum, Executive Opinion Survey 2004

I. Environment Component – Market Environment
1.08 Quality of scientific research institutions, 2004

Scientific research institutions in your country (e.g., university laboratories, government laboratories) are (1 = nonexistent, 7 = the best in their fields globally)

RANK	COUNTRY	SCORE	SD	RANK	COUNTRY	SCORE	SD
1	United States	6.29	1.07	53	Thailand	3.71	1.27
2	Sweden	6.00	0.67	54	Malawi	3.69	1.14
3	Israel	5.76	0.44	55	Nigeria	3.69	1.52
4	Finland	5.76	0.78	56	Egypt	3.69	1.58
5	United Kingdom	5.76	0.99	57	Mauritius	3.68	1.20
6	Switzerland	5.74	0.90	58	Mexico	3.67	1.26
7	Japan	5.60	0.95	59	Chile	3.67	1.16
8	Germany	5.54	0.73	60	Zimbabwe	3.66	1.47
9	Australia	5.47	1.04	61	Bulgaria	3.65	1.34
10	Denmark	5.38	0.65	62	Trinidad and Tobago	3.59	1.21
11	Netherlands	5.37	0.97	63	Slovak Republic	3.58	1.30
12	Canada	5.31	0.83	64	Zambia	3.57	1.36
13	Singapore	5.23	0.89	65	Georgia	3.57	1.53
14	Belgium	5.22	1.18	66	Colombia	3.57	1.13
15	France	5.19	1.00	67	Sri Lanka	3.56	1.23
16	Norway	5.09	1.24	67	United Arab Emirates	3.56	1.41
17	India	5.05	1.06	69	Romania	3.55	1.42
17	New Zealand	5.05	1.06	70	Turkey	3.52	1.25
19	Russia	4.87	1.32	71	Namibia	3.48	1.52
20	Czech Republic	4.82	1.35	72	Malta	3.48	1.31
21	Taiwan	4.78	1.00	73	Uruguay	3.47	1.24
22	Ireland	4.77	1.11	74	Gambia	3.45	1.61
23	Austria	4.74	1.35	75	Philippines	3.45	1.31
24	Hungary	4.72	1.06	76	Ethiopia	3.45	1.48
25	Malaysia	4.71	1.14	77	Latvia	3.40	1.20
26	Iceland	4.68	1.11	78	Venezuela	3.36	1.33
27	South Africa	4.50	1.16	79	Italy	3.32	1.45
28	Hong Kong	4.48	1.20	80	Mozambique	3.31	1.54
29	Lithuania	4.47	1.21	81	Morocco	3.27	1.20
30	Slovenia	4.45	1.11	82	Madagascar	3.26	1.22
31	Kenya	4.44	1.36	83	Argentina	3.25	1.19
32	Korea	4.43	1.25	84	Macedonia	3.23	1.60
33	Uganda	4.40	1.22	85	Mali	3.20	1.35
34	Ghana	4.37	1.29	86	Bahrain	3.18	1.42
35	Jamaica	4.30	1.24	87	Panama	3.14	1.27
36	Estonia	4.26	1.26	88	Algeria	3.10	1.02
37	Brazil	4.26	1.28	89	Cyprus	3.05	1.37
38	Costa Rica	4.23	1.03	90	Dominican Republic	2.92	1.06
39	Poland	4.20	1.21	91	Guatemala	2.91	1.13
40	China	4.08	1.25	92	Bangladesh	2.91	1.30
41	Portugal	4.05	1.07	93	Vietnam	2.87	1.04
42	Ukraine	4.00	1.25	94	Pakistan	2.84	1.48
43	Jordan	3.93	1.22	95	Peru	2.81	0.99
44	Tunisia	3.92	1.13	96	Ecuador	2.77	1.04
45	Tanzania	3.90	1.56	97	Bosnia and Herzegovina	2.67	1.19
46	Croatia	3.83	1.40	98	Bolivia	2.48	0.91
47	Botswana	3.82	1.37	99	Nicaragua	2.44	1.04
48	Luxembourg	3.80	1.40	100	Angola	2.37	1.36
49	Indonesia	3.79	1.24	101	El Salvador	2.36	1.11
50	Serbia and Montenegro	3.78	1.36	102	Honduras	2.23	0.98
51	Spain	3.75	1.15	103	Paraguay	2.17	0.93
52	Greece	3.74	1.10	104	Chad	2.07	1.22

0 1 2 3 4 5 6 7   0 1 2 3 4 5 6 7

Source: World Economic Forum, Executive Opinion Survey 2004

I. Environment Component – Market Environment
1.09 Subsidies for firm-level research and development, 2004

For firms conducting R&D in your country, direct government subsidies to individual companies or R&D tax credits (1 = never occur, 7 = are widespread and large)

RANK	COUNTRY	SCORE	SD	RANK	COUNTRY	SCORE	SD
1	Singapore	5.55	1.14	53	Romania	2.96	1.25
2	Luxembourg	5.44	1.12	54	Mauritius	2.94	1.24
3	Taiwan	5.27	1.17	55	Poland	2.93	1.12
4	Canada	5.02	1.20	56	Croatia	2.92	1.36
5	Finland	4.90	1.46	57	Malta	2.88	1.43
6	Malaysia	4.87	0.96	58	Ukraine	2.86	1.28
7	Israel	4.69	1.35	59	Cyprus	2.84	1.29
8	Ireland	4.68	1.16	60	Trinidad and Tobago	2.78	1.29
9	France	4.66	1.23	61	Gambia	2.75	1.69
10	Japan	4.66	1.30	62	Mexico	2.73	1.14
11	United Kingdom	4.64	1.12	63	Slovak Republic	2.72	1.12
12	Austria	4.64	1.38	64	Philippines	2.69	1.26
13	Tunisia	4.62	1.41	65	Serbia and Montenegro	2.69	1.08
14	Germany	4.59	1.23	66	Colombia	2.68	1.29
15	Australia	4.58	1.24	67	Kenya	2.67	1.34
16	United States	4.50	1.30	68	Jamaica	2.66	1.45
17	India	4.46	1.20	69	Ghana	2.61	1.42
18	Netherlands	4.39	1.14	69	Madagascar	2.61	1.62
19	Norway	4.38	1.28	71	Sri Lanka	2.61	1.23
20	Belgium	4.33	1.26	72	Zambia	2.61	1.69
21	Korea	4.32	1.20	73	Chile	2.60	1.17
22	Indonesia	4.10	1.19	74	New Zealand	2.57	1.22
23	Denmark	4.08	1.35	75	United Arab Emirates	2.57	1.52
24	China	3.99	1.26	76	Vietnam	2.52	1.43
25	Spain	3.76	1.25	77	Bulgaria	2.45	1.00
26	Hungary	3.70	1.29	78	Bahrain	2.36	1.40
27	Portugal	3.68	1.23	79	Zimbabwe	2.33	1.36
28	Slovenia	3.65	1.38	80	Macedonia	2.31	1.24
29	Czech Republic	3.51	1.30	81	Latvia	2.29	1.13
30	Iceland	3.47	1.54	81	Malawi	2.29	1.15
31	Greece	3.45	1.17	83	Uruguay	2.25	1.00
32	Thailand	3.43	1.59	84	Mozambique	2.23	1.30
33	Hong Kong	3.42	1.41	85	Costa Rica	2.21	1.17
34	Botswana	3.41	1.58	86	Panama	2.16	1.23
35	Egypt	3.38	1.93	87	Ethiopia	2.08	1.28
36	Lithuania	3.37	1.33	88	Bosnia and Herzegovina	2.07	1.04
37	Turkey	3.35	1.46	89	Chad	2.04	1.52
38	Mali	3.32	1.81	90	Bangladesh	2.00	1.38
39	South Africa	3.31	1.28	91	Venezuela	1.98	0.97
40	Brazil	3.31	1.45	92	Argentina	1.90	1.01
41	Uganda	3.31	1.70	93	Pakistan	1.77	1.15
42	Morocco	3.30	1.38	94	Angola	1.75	1.14
43	Switzerland	3.29	1.45	95	Honduras	1.75	1.14
44	Russia	3.26	1.36	96	Guatemala	1.74	1.02
45	Algeria	3.21	1.67	97	Georgia	1.70	1.08
46	Tanzania	3.19	1.78	98	Nicaragua	1.62	1.13
47	Namibia	3.17	1.71	99	Ecuador	1.61	0.83
48	Italy	3.11	1.20	100	Dominican Republic	1.52	0.85
49	Jordan	3.06	1.43	101	Peru	1.50	0.72
50	Sweden	3.06	1.35	102	El Salvador	1.48	0.67
51	Estonia	3.04	1.21	103	Bolivia	1.45	0.82
52	Nigeria	2.99	1.54	104	Paraguay	1.45	0.78

0 1 2 3 4 5 6 7   0 1 2 3 4 5 6 7

Source: World Economic Forum, Executive Opinion Survey 2004

I. Environment Component – Market Environment
1.10 Brain drain, 2004

Your country's talented people (1 = normally leave to pursue opportunities in other countries, 7 = almost always remain in the country)

RANK	COUNTRY	SCORE	SD	RANK	COUNTRY	SCORE	SD
1	United States	6.25	0.96	53	Russia	3.26	1.34
2	Norway	6.00	0.74	54	Guatemala	3.25	1.39
3	United Arab Emirates	5.73	1.40	55	Trinidad and Tobago	3.24	1.13
4	Switzerland	5.37	0.97	56	Mauritius	3.24	1.07
5	Iceland	5.36	0.95	57	Namibia	3.23	1.61
6	Japan	5.32	1.34	58	Pakistan	3.19	1.68
7	Chile	5.29	1.21	59	Turkey	3.18	1.40
8	Netherlands	5.16	1.09	60	El Salvador	3.17	1.57
9	United Kingdom	5.13	1.06	61	Vietnam	3.09	1.59
10	Hong Kong	5.13	1.34	62	South Africa	3.03	1.14
11	Finland	5.02	1.23	63	Slovak Republic	2.97	1.27
12	Bahrain	4.98	1.41	64	Jordan	2.89	1.27
13	Israel	4.94	1.18	65	Mozambique	2.82	1.59
14	Luxembourg	4.93	1.31	66	Uruguay	2.81	1.17
15	Taiwan	4.85	1.08	67	Paraguay	2.81	1.42
16	Singapore	4.80	1.23	68	Lithuania	2.81	1.27
17	Belgium	4.69	1.37	69	Ecuador	2.79	1.35
18	Malaysia	4.69	1.29	70	Argentina	2.76	1.22
19	Ireland	4.60	1.17	71	Nicaragua	2.76	1.39
20	Costa Rica	4.54	1.49	72	Morocco	2.73	1.69
21	Sweden	4.50	1.19	73	Colombia	2.72	1.31
21	Thailand	4.50	1.31	74	Kenya	2.71	1.51
23	Denmark	4.44	1.29	75	Madagascar	2.68	1.48
24	Canada	4.37	1.42	76	Nigeria	2.67	1.60
25	Spain	4.32	1.53	77	Angola	2.66	1.58
26	Austria	4.28	1.34	78	Jamaica	2.61	1.27
27	Estonia	4.27	1.57	78	Malawi	2.61	1.02
28	Botswana	4.22	1.45	80	Croatia	2.61	1.32
29	Brazil	4.22	1.26	81	Uganda	2.60	1.43
30	Australia	4.22	1.27	82	Ukraine	2.58	1.36
31	Germany	4.18	1.56	83	Ghana	2.56	1.34
32	Panama	4.16	1.34	84	Georgia	2.55	1.06
33	Slovenia	4.12	1.17	85	Sri Lanka	2.55	1.18
34	Portugal	4.12	1.15	86	Algeria	2.54	1.35
35	Indonesia	4.05	1.33	87	Poland	2.52	1.46
36	Malta	4.00	1.20	88	Mali	2.51	1.25
37	Greece	3.97	1.69	89	Egypt	2.51	1.53
38	Czech Republic	3.94	1.57	90	Peru	2.46	1.28
39	Tunisia	3.84	1.36	91	Venezuela	2.45	1.38
40	Korea	3.82	1.43	92	Gambia	2.45	1.33
41	China	3.77	1.26	93	Serbia and Montenegro	2.36	1.44
42	Hungary	3.70	1.62	94	Bolivia	2.33	1.32
43	Cyprus	3.57	1.66	95	Bosnia and Herzegovina	2.26	1.27
44	Dominican Republic	3.48	1.54	96	Ethiopia	2.16	1.17
45	Italy	3.47	1.54	97	Philippines	2.15	0.98
46	France	3.47	1.03	98	Bulgaria	2.12	1.47
47	New Zealand	3.45	1.76	99	Chad	2.11	1.35
48	Tanzania	3.39	1.78	100	Bangladesh	2.09	0.93
49	Honduras	3.38	1.37	101	Romania	2.06	1.13
50	India	3.31	1.29	102	Macedonia	2.06	1.31
51	Latvia	3.29	1.37	103	Zambia	2.05	1.18
52	Mexico	3.27		104	Zimbabwe	1.63	0.76

Source: World Economic Forum, Executive Opinion Survey 2004

I. Environment Component – Market Environment
1.11 Ease of access to loans, 2004

How easy is it to obtain a bank loan in your country with only a good business plan and no collateral? (1 = impossible, 7 = easy)

RANK	COUNTRY	SCORE		SD	RANK	COUNTRY	SCORE		SD
1	United Kingdom	5.48		1.26	53	Morocco	3.23		1.86
2	Hong Kong	5.24		1.67	54	Tanzania	3.20		1.79
3	Finland	5.23		1.43	55	Romania	3.18		1.83
4	Sweden	5.11		1.41	56	Japan	3.12		1.42
5	Denmark	5.08		1.14	57	Uganda	3.11		1.90
6	Lithuania	5.01		1.41	58	Costa Rica	3.06		1.67
7	New Zealand	4.96		1.41	59	Sri Lanka	3.04		1.70
8	Iceland	4.96		1.49	60	Bulgaria	3.00		1.78
9	Ireland	4.87		1.22	61	Gambia	2.97		1.68
10	Netherlands	4.86		1.44	62	Bosnia and Herzegovina	2.97		1.60
11	United States	4.86		1.72	63	Israel	2.94		1.57
12	Luxembourg	4.73		1.78	64	Peru	2.94		1.29
13	Taiwan	4.59		1.42	65	Dominican Republic	2.90		1.47
14	Norway	4.52		1.47	66	Poland	2.80		1.31
15	Singapore	4.49		1.73	67	Croatia	2.78		1.78
16	Bahrain	4.49		1.83	67	Russia	2.78		1.47
17	Estonia	4.44		1.72	69	Ghana	2.75		1.61
18	United Arab Emirates	4.43		1.91	70	Serbia and Montenegro	2.70		1.54
19	Australia	4.42		1.59	71	Paraguay	2.69		1.59
20	France	4.38		1.60	72	Guatemala	2.65		1.33
21	Portugal	4.24		1.26	73	Pakistan	2.62		1.52
22	Canada	4.17		1.90	74	Italy	2.61		1.40
23	Malaysia	4.16		1.35	75	Zimbabwe	2.60		1.35
24	Belgium	4.11		1.61	76	Georgia	2.58		1.57
25	Panama	4.05		1.82	77	Jamaica	2.57		1.33
26	Chile	3.99		1.75	78	Zambia	2.53		1.70
27	Indonesia	3.97		1.38	79	China	2.52		1.34
28	Switzerland	3.94		1.54	79	Nicaragua	2.52		1.70
29	Greece	3.93		1.68	81	Korea	2.50		1.19
30	Kenya	3.92		1.98	82	Philippines	2.48		1.26
31	Slovenia	3.91		1.71	83	Mexico	2.44		1.35
32	Mauritius	3.84		1.27	84	Algeria	2.43		1.48
33	India	3.76		1.63	85	Venezuela	2.41		1.24
34	Botswana	3.76		1.71	86	Bangladesh	2.38		1.66
35	Cyprus	3.74		1.79	87	Egypt	2.37		1.69
36	Austria	3.73		1.70	88	Turkey	2.36		1.65
37	Malta	3.62		1.76	89	Malawi	2.33		1.33
38	Slovak Republic	3.61		1.77	90	Macedonia	2.32		1.72
38	Spain	3.61		1.77	91	Mali	2.32		1.63
40	El Salvador	3.55		1.75	92	Nigeria	2.31		1.49
41	South Africa	3.54		1.38	93	Vietnam	2.31		1.29
42	Namibia	3.53		1.96	94	Ukraine	2.30		1.57
43	Hungary	3.52		1.65	95	Ecuador	2.29		1.31
44	Thailand	3.46		1.60	96	Madagascar	2.22		1.57
45	Germany	3.45		1.36	97	Uruguay	2.17		1.00
46	Jordan	3.42		1.61	98	Honduras	2.13		1.24
47	Trinidad and Tobago	3.36		1.49	99	Argentina	2.09		0.98
48	Latvia	3.30		1.64	100	Angola	1.98		1.19
49	Brazil	3.30		1.61	101	Bolivia	1.92		1.17
50	Czech Republic	3.29		1.44	102	Ethiopia	1.88		1.23
51	Tunisia	3.26		1.64	103	Chad	1.82		1.27
52	Colombia	3.24		1.57	104	Mozambique	1.76		1.23

0 1 2 3 4 5 6 7

Source: World Economic Forum, Executive Opinion Survey 2004

I. Environment Component – Market Environment
1.12 Administrative burden, 2004

Complying with administrative requirements (permits, regulations, reporting) issued by the central government in your country is (1 = burdensome, 7 = not burdensome)

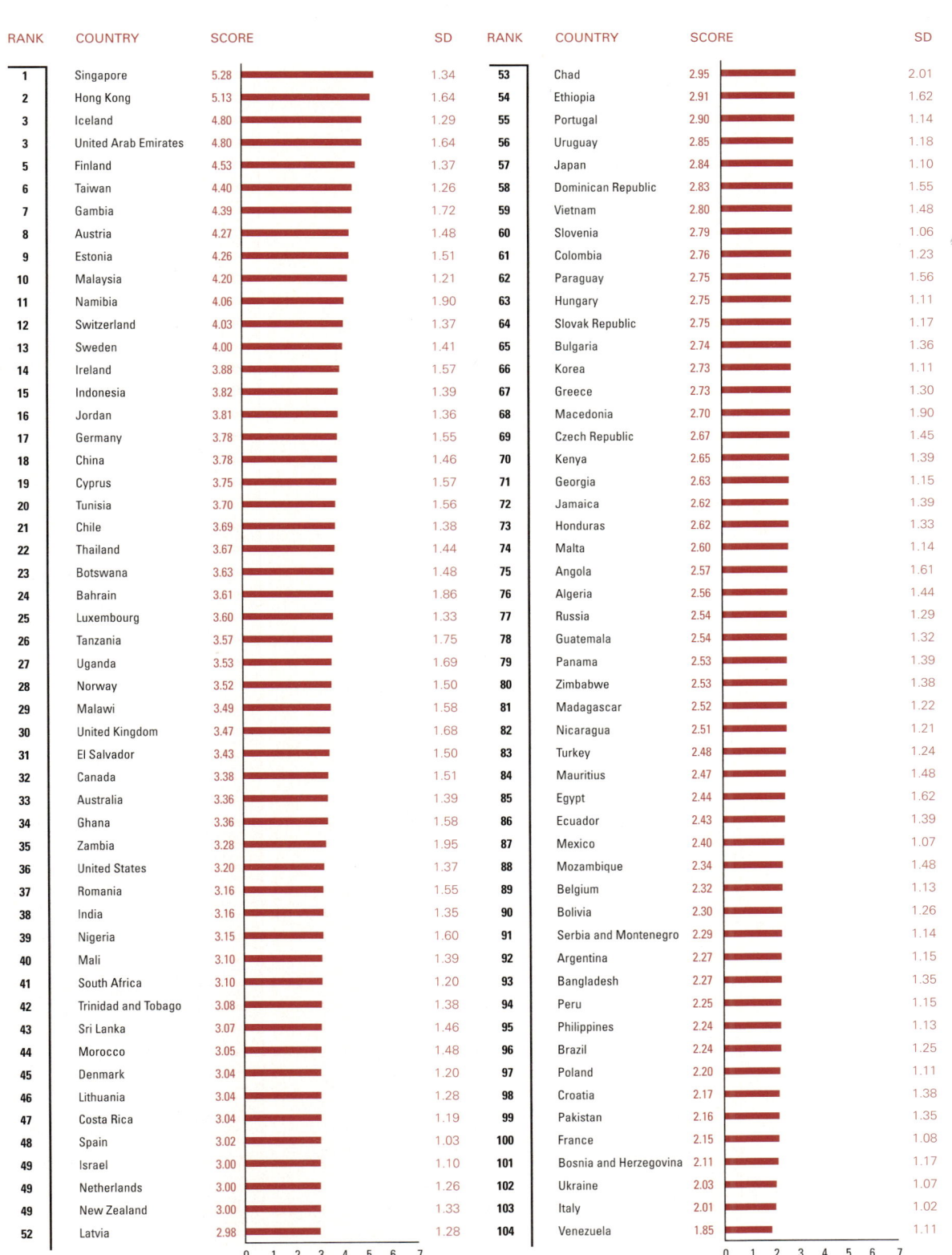

RANK	COUNTRY	SCORE	SD	RANK	COUNTRY	SCORE	SD
1	Singapore	5.28	1.34	53	Chad	2.95	2.01
2	Hong Kong	5.13	1.64	54	Ethiopia	2.91	1.62
3	Iceland	4.80	1.29	55	Portugal	2.90	1.14
3	United Arab Emirates	4.80	1.64	56	Uruguay	2.85	1.18
5	Finland	4.53	1.37	57	Japan	2.84	1.10
6	Taiwan	4.40	1.26	58	Dominican Republic	2.83	1.55
7	Gambia	4.39	1.72	59	Vietnam	2.80	1.48
8	Austria	4.27	1.48	60	Slovenia	2.79	1.06
9	Estonia	4.26	1.51	61	Colombia	2.76	1.23
10	Malaysia	4.20	1.21	62	Paraguay	2.75	1.56
11	Namibia	4.06	1.90	63	Hungary	2.75	1.11
12	Switzerland	4.03	1.37	64	Slovak Republic	2.75	1.17
13	Sweden	4.00	1.41	65	Bulgaria	2.74	1.36
14	Ireland	3.88	1.57	66	Korea	2.73	1.11
15	Indonesia	3.82	1.39	67	Greece	2.73	1.30
16	Jordan	3.81	1.36	68	Macedonia	2.70	1.90
17	Germany	3.78	1.55	69	Czech Republic	2.67	1.45
18	China	3.78	1.46	70	Kenya	2.65	1.39
19	Cyprus	3.75	1.57	71	Georgia	2.63	1.15
20	Tunisia	3.70	1.56	72	Jamaica	2.62	1.39
21	Chile	3.69	1.38	73	Honduras	2.62	1.33
22	Thailand	3.67	1.44	74	Malta	2.60	1.14
23	Botswana	3.63	1.48	75	Angola	2.57	1.61
24	Bahrain	3.61	1.86	76	Algeria	2.56	1.44
25	Luxembourg	3.60	1.33	77	Russia	2.54	1.29
26	Tanzania	3.57	1.75	78	Guatemala	2.54	1.32
27	Uganda	3.53	1.69	79	Panama	2.53	1.39
28	Norway	3.52	1.50	80	Zimbabwe	2.53	1.38
29	Malawi	3.49	1.58	81	Madagascar	2.52	1.22
30	United Kingdom	3.47	1.68	82	Nicaragua	2.51	1.21
31	El Salvador	3.43	1.50	83	Turkey	2.48	1.24
32	Canada	3.38	1.51	84	Mauritius	2.47	1.48
33	Australia	3.36	1.39	85	Egypt	2.44	1.62
34	Ghana	3.36	1.58	86	Ecuador	2.43	1.39
35	Zambia	3.28	1.95	87	Mexico	2.40	1.07
36	United States	3.20	1.37	88	Mozambique	2.34	1.48
37	Romania	3.16	1.55	89	Belgium	2.32	1.13
38	India	3.16	1.35	90	Bolivia	2.30	1.26
39	Nigeria	3.15	1.60	91	Serbia and Montenegro	2.29	1.14
40	Mali	3.10	1.39	92	Argentina	2.27	1.15
41	South Africa	3.10	1.20	93	Bangladesh	2.27	1.35
42	Trinidad and Tobago	3.08	1.38	94	Peru	2.25	1.15
43	Sri Lanka	3.07	1.46	95	Philippines	2.24	1.13
44	Morocco	3.05	1.48	96	Brazil	2.24	1.25
45	Denmark	3.04	1.20	97	Poland	2.20	1.11
46	Lithuania	3.04	1.28	98	Croatia	2.17	1.38
47	Costa Rica	3.04	1.19	99	Pakistan	2.16	1.35
48	Spain	3.02	1.03	100	France	2.15	1.08
49	Israel	3.00	1.10	101	Bosnia and Herzegovina	2.11	1.17
49	Netherlands	3.00	1.26	102	Ukraine	2.03	1.07
49	New Zealand	3.00	1.33	103	Italy	2.01	1.02
52	Latvia	2.98	1.28	104	Venezuela	1.85	1.11

Source: World Economic Forum, Executive Opinion Survey 2004

I. Environment Component – Market Environment
1.13 Ease to start a new business, 2004

Starting a new business in your country is generally (1 = extremely difficult and time consuming, 7 = easy)

RANK	COUNTRY	SCORE		SD	RANK	COUNTRY	SCORE		SD
1	Hong Kong	6.34		0.94	53	Malawi	3.89		1.68
2	Iceland	5.96		1.22	54	Botswana	3.85		1.65
3	New Zealand	5.78		0.94	55	Chad	3.78		2.14
4	Singapore	5.71		1.25	56	Tanzania	3.77		1.59
5	Finland	5.62		1.19	57	Jamaica	3.75		1.67
6	United Kingdom	5.60		1.19	58	Kenya	3.71		1.75
7	United States	5.55		1.33	59	Croatia	3.71		1.66
8	Estonia	5.55		1.34	60	Mauritius	3.70		1.42
9	United Arab Emirates	5.46		1.28	61	Belgium	3.69		1.55
10	Sweden	5.42		0.96	62	Panama	3.69		1.47
11	Switzerland	5.35		1.44	63	France	3.68		1.52
12	Tunisia	5.26		1.42	64	Ethiopia	3.66		1.90
13	Taiwan	5.24		1.38	65	Portugal	3.63		1.32
14	Luxembourg	5.20		1.24	66	Guatemala	3.60		1.54
15	Canada	5.12		1.46	67	Dominican Republic	3.60		1.54
16	Hungary	5.03		1.54	68	Mali	3.58		1.81
17	Malaysia	5.02		1.34	69	Egypt	3.57		1.91
18	Australia	5.02		1.40	70	Costa Rica	3.54		1.41
19	Bahrain	4.95		1.60	71	Spain	3.53		1.29
20	Cyprus	4.90		1.46	72	Lithuania	3.47		1.47
21	Norway	4.83		1.27	73	El Salvador	3.47		1.67
22	Israel	4.75		1.44	74	Ghana	3.42		1.75
23	Netherlands	4.75		1.48	75	Romania	3.41		1.58
24	Namibia	4.74		1.79	76	Italy	3.33		1.55
25	Thailand	4.69		1.41	76	Zimbabwe	3.33		1.40
26	Denmark	4.67		1.34	78	Czech Republic	3.30		1.54
27	Sri Lanka	4.62		1.54	79	Nigeria	3.28		1.80
28	Ireland	4.60		1.35	80	Bangladesh	3.26		1.72
29	Slovak Republic	4.59		1.49	81	Poland	3.26		1.45
30	Jordan	4.58		1.54	82	Paraguay	3.24		1.61
31	Gambia	4.54		1.73	83	Georgia	3.23		1.62
32	China	4.47		1.47	84	Greece	3.22		1.30
33	Indonesia	4.46		1.07	85	Colombia	3.22		1.47
34	Uganda	4.44		1.85	86	Algeria	3.21		1.70
35	Austria	4.34		1.65	87	Russia	3.18		1.58
36	Slovenia	4.33		1.47	88	Nicaragua	3.01		1.72
37	Japan	4.27		1.52	89	Bosnia and Herzegovina	2.97		1.64
38	South Africa	4.25		1.59	90	Uruguay	2.97		1.35
39	Latvia	4.24		1.60	91	Mexico	2.94		1.44
40	Vietnam	4.22		1.74	92	Pakistan	2.88		1.58
41	Zambia	4.21		1.96	93	Honduras	2.87		1.54
42	Serbia and Montenegro	4.17		1.82	94	Peru	2.86		1.35
43	Turkey	4.13		1.74	95	Argentina	2.82		1.43
44	Korea	4.09		1.52	96	Brazil	2.75		1.44
45	Germany	4.08		1.67	97	Ukraine	2.75		1.56
46	Malta	4.08		1.47	98	Ecuador	2.72		1.27
47	Madagascar	4.05		1.62	99	Angola	2.65		1.81
48	India	4.04		1.50	100	Bolivia	2.63		1.26
49	Trinidad and Tobago	4.01		1.35	101	Venezuela	2.58		1.37
50	Morocco	3.98		1.65	102	Bulgaria	2.47		1.37
51	Philippines	3.97		1.47	103	Macedonia	2.36		1.48
52	Chile	3.90		1.52	104	Mozambique	2.35		1.55

0 1 2 3 4 5 6 7

Source: World Economic Forum, Executive Opinion Survey 2004

Political and Regulatory Environment

I. Environment Component – Political and Regulatory Environment
2.01 Effectiveness of lawmaking, 2004

How effective is your national parliament/congress as a lawmaking and oversight institution? (1 = very ineffective, 7 = very effective, the best in the world)

RANK	COUNTRY	SCORE	SD	RANK	COUNTRY	SCORE	SD
1	Singapore	6.02	0.98	53	Belgium	3.54	1.54
2	United Kingdom	5.58	1.14	54	Gambia	3.53	1.72
3	Denmark	5.52	1.00	55	Bahrain	3.49	1.50
4	Iceland	5.29	1.30	56	Jamaica	3.32	1.49
5	Australia	5.26	1.38	57	Malawi	3.25	1.42
6	Malaysia	5.09	1.06	58	Hungary	3.20	1.26
7	Luxembourg	5.07	1.22	59	Slovenia	3.18	1.12
8	Sweden	5.00	1.03	60	Sri Lanka	3.16	1.37
8	United States	5.00	1.47	61	Brazil	3.13	1.24
10	Ghana	4.95	1.11	62	Nigeria	3.06	1.54
11	New Zealand	4.76	1.47	63	Russia	3.05	1.31
12	Norway	4.74	1.42	64	Slovak Republic	2.98	1.31
13	France	4.73	1.31	65	Czech Republic	2.97	1.50
14	Finland	4.71	1.42	66	Latvia	2.97	1.19
15	Namibia	4.68	1.35	67	Trinidad and Tobago	2.96	1.35
16	South Africa	4.67	1.42	68	Madagascar	2.89	1.20
17	Netherlands	4.63	1.39	69	Croatia	2.85	1.48
18	Estonia	4.62	1.19	70	Romania	2.84	1.40
19	Canada	4.54	1.41	71	Italy	2.82	1.36
20	Switzerland	4.40	1.49	72	Algeria	2.80	1.45
21	India	4.39	1.45	73	Mozambique	2.73	1.25
22	Ireland	4.38	1.15	74	Uruguay	2.71	1.08
23	Malta	4.36	1.18	75	Colombia	2.69	1.12
24	China	4.34	1.59	76	Angola	2.67	1.00
25	Spain	4.34	1.32	77	Bangladesh	2.61	1.35
26	Tunisia	4.25	1.46	78	Honduras	2.58	1.12
27	Botswana	4.22	1.55	79	Georgia	2.53	1.15
27	Japan	4.22	1.27	80	Macedonia	2.49	1.48
29	Indonesia	4.13	1.23	81	Korea	2.47	1.23
30	Austria	4.09	1.45	82	Bulgaria	2.44	1.30
31	Tanzania	4.08	1.46	83	Ukraine	2.31	1.17
32	Mauritius	4.06	1.18	84	Poland	2.29	1.06
33	United Arab Emirates	3.96	1.73	85	Serbia and Montenegro	2.28	1.29
34	Israel	3.94	1.53	86	Philippines	2.25	1.14
34	Turkey	3.94	1.35	87	Ethiopia	2.24	1.39
36	Thailand	3.88	1.25	88	Costa Rica	2.23	1.15
37	Cyprus	3.86	1.54	89	Paraguay	2.18	1.08
38	Uganda	3.83	1.39	90	Pakistan	2.15	0.97
39	Greece	3.81	1.22	91	Chad	2.11	1.23
40	Jordan	3.76	1.38	92	Bosnia and Herzegovina	2.11	1.23
41	Taiwan	3.75	1.37	93	El Salvador	2.09	1.10
42	Chile	3.73	1.29	94	Zimbabwe	2.07	1.31
43	Vietnam	3.72	1.56	95	Dominican Republic	2.03	1.06
44	Germany	3.69	1.55	96	Mexico	2.00	1.10
45	Portugal	3.66	1.26	97	Bolivia	1.85	0.84
46	Mali	3.65	1.41	98	Guatemala	1.81	0.97
47	Egypt	3.62	1.75	99	Panama	1.77	1.04
48	Morocco	3.62	1.43	100	Peru	1.72	0.77
49	Kenya	3.60	1.45	101	Ecuador	1.66	0.87
50	Lithuania	3.58	1.32	102	Argentina	1.62	0.74
51	Zambia	3.57	1.84	103	Nicaragua	1.59	0.97
52	Hong Kong	3.55	1.66	104	Venezuela	1.37	0.60

Source: World Economic Forum, Executive Opinion Survey 2004

I. Environment Component – Political and Regulatory Environment
2.02 Laws relating to ICT, 2004

Laws relating to information and communication technology (electronic commerce, digital signatures, consumer protection) are (1 = nonexistent, 7 = well-developed and enforced)

RANK	COUNTRY	SCORE	SD	RANK	COUNTRY	SCORE	SD
1	Singapore	5.53	1.02	53	Jamaica	3.69	1.43
2	Norway	5.52	1.44	54	Colombia	3.68	1.39
3	Estonia	5.47	1.23	55	Uganda	3.63	1.61
4	Finland	5.45	1.05	56	Poland	3.58	1.11
5	United Kingdom	5.43	1.19	57	Mexico	3.58	1.27
6	Denmark	5.38	0.97	58	Croatia	3.57	1.49
7	Germany	5.29	1.29	59	Italy	3.57	1.54
8	Australia	5.24	1.02	60	Dominican Republic	3.56	1.62
9	Hong Kong	5.23	1.27	61	Bulgaria	3.50	1.52
10	New Zealand	5.19	1.21	62	Cyprus	3.48	1.57
11	United States	5.16	1.30	63	Egypt	3.46	1.78
12	Sweden	5.10	1.25	64	Greece	3.44	1.15
13	Iceland	5.09	1.35	65	Ghana	3.38	1.70
14	Korea	5.05	0.99	66	Botswana	3.37	1.44
15	France	5.05	1.28	67	Peru	3.36	1.34
16	Malaysia	4.96	1.00	68	Latvia	3.35	1.17
17	Japan	4.89	1.06	69	Tanzania	3.35	1.61
18	Austria	4.89	1.28	70	Nigeria	3.34	1.54
19	Canada	4.81	1.35	71	Costa Rica	3.28	1.53
20	South Africa	4.78	1.26	72	Venezuela	3.27	1.61
21	Taiwan	4.78	1.11	73	Kenya	3.25	1.47
22	Malta	4.71	1.11	74	Zambia	3.25	1.83
23	Switzerland	4.66	1.20	75	Russia	3.21	1.34
24	Ireland	4.64	1.25	76	Gambia	3.21	1.69
25	United Arab Emirates	4.62	1.27	77	Turkey	3.17	1.36
26	Luxembourg	4.61	1.37	78	Sri Lanka	3.13	1.26
27	Slovenia	4.56	1.24	79	Trinidad and Tobago	3.12	1.40
28	Chile	4.56	1.26	80	El Salvador	3.09	1.41
29	Netherlands	4.48	1.27	81	Ukraine	3.06	1.16
30	Belgium	4.46	1.09	82	Mozambique	3.03	1.63
31	Israel	4.40	0.83	83	Uruguay	3.02	1.37
32	Tunisia	4.36	1.49	84	Zimbabwe	3.00	1.49
33	India	4.35	1.25	85	Argentina	2.97	1.33
34	Bahrain	4.26	1.66	86	Ecuador	2.94	1.28
35	Jordan	4.25	1.44	87	Malawi	2.89	1.32
36	Portugal	4.22	1.24	88	Mali	2.85	1.33
37	Indonesia	4.18	1.05	89	Macedonia	2.81	1.58
38	Czech Republic	4.14	1.31	90	Madagascar	2.70	1.35
39	Spain	4.10	1.29	91	Bosnia and Herzegovina	2.61	1.24
40	Romania	4.06	1.30	92	Nicaragua	2.60	1.38
41	Brazil	4.04	1.44	93	Serbia and Montenegro	2.58	1.17
42	Mauritius	4.00	1.41	94	Vietnam	2.53	1.51
43	Slovak Republic	3.98	1.38	95	Honduras	2.50	1.26
44	Philippines	3.95	1.19	96	Guatemala	2.47	1.30
45	Thailand	3.94	1.33	97	Algeria	2.43	1.35
46	Lithuania	3.93	1.23	98	Angola	2.22	1.46
47	Panama	3.88	1.40	99	Ethiopia	2.21	1.31
48	Hungary	3.82	1.30	100	Georgia	2.18	1.31
49	Pakistan	3.81	1.24	101	Bangladesh	2.17	1.16
50	Morocco	3.77	1.58	102	Paraguay	2.14	1.02
51	China	3.74	1.37	103	Bolivia	2.04	1.14
52	Namibia	3.73	1.72	104	Chad	1.82	1.18

Source: World Economic Forum, Executive Opinion Survey 2004

I. Environment Component – Political and Regulatory Environment
2.03 Effectiveness of judiciary, 2004

Is the judiciary in your country independent of political influence by members of government, citizens or firms? (1 = no, heavily influenced, 7 = yes, entirely independent)

RANK	COUNTRY	SCORE	SD	RANK	COUNTRY	SCORE	SD
1	Denmark	6.56	1.04	53	Uganda	4.04	1.74
2	Finland	6.56	0.72	54	Slovenia	3.94	1.63
3	Germany	6.41	0.85	55	Mauritius	3.91	1.52
4	Norway	6.36	1.05	56	Zambia	3.74	2.02
5	Netherlands	6.34	0.89	57	Tanzania	3.71	1.78
6	Iceland	6.16	1.49	58	Indonesia	3.70	1.35
7	Switzerland	6.14	1.04	59	Vietnam	3.69	1.72
8	Sweden	6.11	1.20	60	Turkey	3.67	1.70
9	United Kingdom	6.09	1.30	61	China	3.65	1.64
10	Australia	6.06	1.32	62	Slovak Republic	3.58	1.52
11	Luxembourg	6.03	1.15	63	Italy	3.56	1.91
12	New Zealand	6.00	1.31	64	Sri Lanka	3.49	1.65
13	Canada	5.88	1.32	65	Mexico	3.45	1.60
14	Hong Kong	5.85	1.51	66	Latvia	3.44	1.46
15	South Africa	5.78	1.37	67	Dominican Republic	3.40	1.42
16	Austria	5.74	1.43	68	Lithuania	3.27	1.41
17	Ireland	5.68	1.58	69	Kenya	3.17	1.73
18	Portugal	5.63	1.32	70	Nigeria	3.16	1.74
19	Estonia	5.54	1.27	71	El Salvador	3.13	1.78
20	Israel	5.50	1.59	72	Poland	3.09	1.46
21	Belgium	5.47	1.77	73	Colombia	3.02	1.37
22	Japan	5.38	1.58	74	Philippines	3.02	1.55
23	Namibia	5.32	1.76	75	Romania	2.99	1.51
24	Singapore	5.28	1.39	76	Mali	2.97	1.81
25	Malta	5.27	1.42	77	Algeria	2.92	1.55
26	United States	5.22	1.55	78	Bosnia and Herzegovina	2.85	1.43
27	Cyprus	5.20	1.67	79	Guatemala	2.82	1.48
28	Botswana	5.15	1.48	80	Madagascar	2.71	1.54
29	Uruguay	5.02	1.44	81	Bulgaria	2.70	1.59
30	United Arab Emirates	4.99	1.96	82	Bangladesh	2.60	1.33
31	Malaysia	4.98	1.34	83	Mozambique	2.52	1.55
32	India	4.96	1.61	84	Russia	2.51	1.41
33	Costa Rica	4.89	1.58	85	Serbia and Montenegro	2.50	1.24
34	France	4.88	1.62	86	Ethiopia	2.41	1.45
35	Trinidad and Tobago	4.83	1.64	87	Georgia	2.40	1.12
36	Malawi	4.81	1.51	88	Croatia	2.39	1.30
37	Chile	4.80	1.60	89	Pakistan	2.38	0.99
38	Tunisia	4.77	1.66	90	Angola	2.30	1.29
39	Taiwan	4.76	1.68	91	Macedonia	2.28	1.51
40	Ghana	4.75	1.55	92	Honduras	2.23	1.43
41	Hungary	4.61	1.64	93	Bolivia	2.19	1.34
42	Greece	4.60	1.42	94	Peru	2.18	1.07
43	Jordan	4.59	1.66	95	Ukraine	2.15	1.23
44	Thailand	4.53	1.60	96	Panama	2.14	1.34
45	Czech Republic	4.51	1.54	97	Argentina	2.09	1.15
46	Bahrain	4.36	1.83	98	Egypt	2.00	1.00
47	Morocco	4.25	1.91	99	Paraguay	1.96	1.16
48	Korea	4.19	1.60	100	Chad	1.96	1.30
49	Jamaica	4.12	1.69	101	Zimbabwe	1.87	1.17
50	Spain	4.08	1.81	102	Ecuador	1.78	0.89
51	Brazil	4.07	1.51	103	Venezuela	1.49	1.17
52	Gambia	4.05	1.93	104	Nicaragua	1.42	0.99

Source: World Economic Forum, Executive Opinion Survey 2004

I. Environment Component – Political and Regulatory Environment
2.04 Intellectual property protection, 2004

Intellectual property protection in your country is (1 = weak or nonexistent, 7 = equal to the world's most stringent)

RANK	COUNTRY	SCORE		SD	RANK	COUNTRY	SCORE		SD
1	Sweden	6.32		0.75	53	Uruguay	3.61		1.26
2	Denmark	6.25		0.79	54	China	3.56		1.45
3	United States	6.23		1.16	55	Mauritius	3.55		1.30
4	Germany	6.19		0.76	56	Jamaica	3.54		1.49
5	Finland	6.14		1.01	57	Malawi	3.51		1.50
6	United Kingdom	6.09		1.35	58	Botswana	3.49		1.66
7	Norway	6.05		0.72	59	India	3.45		1.32
8	Switzerland	6.04		1.07	60	Dominican Republic	3.40		1.37
9	Iceland	6.04		0.79	61	Lithuania	3.36		1.39
10	Netherlands	6.01		0.93	62	Mexico	3.33		1.40
11	Australia	5.97		1.15	63	Sri Lanka	3.30		1.47
12	New Zealand	5.93		1.01	64	Romania	3.29		1.50
13	Singapore	5.73		1.07	65	Colombia	3.28		1.09
14	France	5.71		1.17	66	Mali	3.27		1.73
15	Austria	5.68		1.19	67	Gambia	3.22		1.82
16	Canada	5.66		1.21	68	Trinidad and Tobago	3.21		1.34
17	Luxembourg	5.63		1.00	69	Madagascar	3.15		1.29
18	Belgium	5.53		1.27	70	Zimbabwe	3.14		1.43
19	Hong Kong	5.40		1.32	71	Kenya	3.13		1.27
20	Japan	5.31		1.38	72	Latvia	3.11		1.32
21	Ireland	5.22		1.06	73	Nigeria	2.98		1.52
22	South Africa	5.04		1.44	74	Tanzania	2.96		1.56
23	Israel	5.00		1.30	75	Zambia	2.96		1.80
24	Taiwan	4.93		1.23	76	Croatia	2.94		1.53
25	Malaysia	4.84		1.23	77	Algeria	2.85		1.47
26	Tunisia	4.83		1.40	78	Guatemala	2.85		1.38
27	United Arab Emirates	4.77		1.48	79	Poland	2.83		1.36
28	Jordan	4.77		1.35	80	Turkey	2.81		1.26
29	Estonia	4.73		1.37	81	Bulgaria	2.79		1.51
30	Portugal	4.63		1.33	82	Philippines	2.71		0.98
31	Spain	4.49		1.61	83	Ukraine	2.70		1.33
32	Slovenia	4.48		1.44	84	Russia	2.68		1.31
33	Namibia	4.45		1.96	85	Uganda	2.65		1.31
34	Bahrain	4.39		1.53	86	Honduras	2.61		1.28
35	Greece	4.35		1.24	87	Macedonia	2.51		1.54
36	Korea	4.22		1.22	88	Argentina	2.50		1.08
37	Hungary	4.19		1.26	89	Mozambique	2.47		1.20
38	Egypt	4.15		1.73	90	Peru	2.42		1.00
39	Thailand	4.02		1.42	91	Nicaragua	2.41		1.22
40	Morocco	4.00		1.66	92	Venezuela	2.40		1.22
41	Cyprus	3.93		1.52	93	Vietnam	2.39		1.19
42	Chile	3.91		1.40	94	Georgia	2.35		1.20
43	Czech Republic	3.90		1.56	95	Ecuador	2.34		1.14
44	Ghana	3.88		1.93	96	Serbia and Montenegro	2.29		1.12
45	Italy	3.87		1.64	97	Paraguay	2.19		0.94
46	Panama	3.87		1.41	98	Pakistan	2.18		1.48
47	Indonesia	3.86		1.36	99	Ethiopia	2.17		1.27
48	Costa Rica	3.84		1.46	100	Bangladesh	2.10		1.27
49	Slovak Republic	3.83		1.56	101	Bosnia and Herzegovina	2.05		0.94
50	Malta	3.73		1.34	102	Chad	2.01		1.27
51	Brazil	3.71		1.41	103	Bolivia	2.01		1.07
52	El Salvador	3.70		1.52	104	Angola	1.75		0.81

0 1 2 3 4 5 6 7   0 1 2 3 4 5 6 7

Source: World Economic Forum, Executive Opinion Survey 2004

Infrastructure Environment

I. Environment Component – Infrastructure Environment
3.01 Telephone mainlines, 2002

Telephone mainlines per 1,000 people

* estimated data

RANK	COUNTRY	SCORE		RANK	COUNTRY	SCORE
1	Luxembourg	796.82		53	Brazil	223.20
2	Switzerland	744.24		54	Argentina	218.84
3	Sweden	735.68		55	Ukraine	216.07
4	Norway	734.37		56	Romania	194.42
5	Denmark	688.63		57	Malaysia	190.40
6	Cyprus	687.96		58	Colombia	179.39
7	Iceland	652.77		59	Jamaica	169.73
8	Germany	650.86		60	China	166.92
9	United States	645.81		61	Mexico	146.66
10	Canada	635.45		62	Georgia	131.43
11	Netherlands	617.72		63	Jordan	126.57
12	United Kingdom	590.61		64	Panama	122.00
13	France	568.92		65	Zimbabwe	* 120.92
14	Hong Kong	564.66		66	Zambia	* 118.27
15	Japan	558.32		67	Tunisia	117.37
16	Australia	538.60		68	Venezuela	112.75
17	Finland	523.45		69	Dominican Republic	110.43
18	Malta	523.41		70	Egypt	110.38
19	Spain	506.24		71	Ecuador	110.20
20	Slovenia	506.09		72	South Africa	106.57
21	Ireland	502.42		73	Thailand	105.03
22	Belgium	494.44		74	El Salvador	103.40
23	Greece	491.27		75	Botswana	87.21
24	Austria	488.79		76	Guatemala	70.51
25	Korea	488.59		77	Bolivia	67.61
26	Italy	480.70		78	Peru	66.03
27	Israel	467.18		79	Namibia	64.75
28	Singapore	462.86		80	Algeria	60.97
29	New Zealand	448.07		81	Vietnam	48.36
30	Portugal	421.33		82	Honduras	48.15
31	Croatia	417.24		83	Paraguay	47.25
32	Bulgaria	367.67		84	Sri Lanka	46.61
33	Czech Republic	362.33		85	Philippines	41.66
34	Hungary	361.15		86	India	39.76
35	Estonia	350.55		87	Morocco	38.03
36	United Arab Emirates	313.55		88	Indonesia	36.54
37	Taiwan	* 304.26		89	Nicaragua	31.96
38	Latvia	301.08		90	Gambia	27.95
39	Poland	295.12		91	Pakistan	25.04
40	Turkey	281.17		92	Ghana	12.66
41	Uruguay	279.63		93	Kenya	10.28
42	Macedonia	271.33		94	Malawi	7.00
43	Mauritius	270.32		95	Angola	6.10
44	Lithuania	270.30		96	Nigeria	5.85
45	Slovak Republic	268.24		97	Mali	5.33
46	Bahrain	263.08		98	Ethiopia	5.25
47	Costa Rica	250.54		99	Bangladesh	5.12
48	Trinidad and Tobago	249.75		100	Tanzania	4.69
49	Russia	242.18		101	Mozambique	4.59
50	Bosnia and Herzegovina	236.65		102	Madagascar	3.74
51	Serbia and Montenegro	232.55		103	Uganda	2.23
52	Chile	230.36		104	Chad	1.50

0 100 200 300 400 500 600 700 800   0 100 200 300 400 500 600 700 800

Source: World Bank. World Development Indicators 2002

I. Environment Component – Infrastructure Environment
3.02 Secure Internet servers, 2003

Secure Internet servers per 1,000,000 people

* estimated data

RANK	COUNTRY	SCORE		RANK	COUNTRY	SCORE
1	Iceland	634.48		53	Namibia	4.50
2	United States	471.14		54	Jamaica	4.44
3	Canada	342.38		55	Venezuela	4.12
4	New Zealand	327.18		56	Mexico	4.02
5	Australia	291.83		57	Netherlands	3.60
6	Switzerland	268.19		58	El Salvador	3.54
7	Luxembourg	245.03		59	Bulgaria	3.04
8	United Kingdom	228.33		60	Guatemala	2.93
9	Ireland	196.00		61	Thailand	2.85
10	Denmark	184.81		62	Peru	2.68
11	Finland	179.23		63	Dominican Republic	2.53
12	Sweden	179.21		64	Colombia	2.38
13	Singapore	170.23		65	Honduras	2.32
14	Norway	161.33		66	Macedonia	* 2.13
15	Malta	149.75		67	Ecuador	1.77
16	Austria	142.72		68	Jordan	1.64
17	Hong Kong	109.71		69	Russia	1.63
18	Germany	102.44		70	Nicaragua	1.45
19	Cyprus	97.26		71	Romania	1.35
20	Japan	93.01		72	Tunisia	1.33
21	Israel	87.81		73	Philippines	1.21
22	Estonia	68.46		74	Sri Lanka	1.20
23	Belgium	55.92		75	Bolivia	1.14
24	Slovenia	48.00		76	Bosnia and Herzegovina	0.95
25	Spain	47.79		77	Georgia	0.78
26	France	47.59		78	Paraguay	0.68
27	Costa Rica	34.29		79	Ukraine	0.58
28	Portugal	31.58		80	Serbia and Montenegro	0.57
29	Taiwan	* 28.34		81	Zimbabwe	0.54
30	United Arab Emirates	27.67		82	Morocco	0.49
31	Panama	27.42		83	Indonesia	0.27
32	Italy	24.91		84	India	0.26
33	Croatia	24.32		85	Ghana	0.24
34	Latvia	23.04		86	Egypt	0.24
35	Czech Republic	22.45		87	Pakistan	0.16
36	Bahrain	19.34		88	China	0.14
37	Greece	18.64		89	Algeria	0.13
38	Chile	14.75		90	Kenya	0.13
39	Korea	14.42		91	Mozambique	0.11
40	South Africa	14.40		92	Uganda	0.08
41	Mauritius	14.17		93	Mali	0.08
42	Hungary	14.04		94	Angola	0.07
43	Uruguay	11.47		95	Madagascar	0.06
44	Poland	10.08		96	Vietnam	0.04
45	Trinidad and Tobago	10.00		97	Ethiopia	0.03
46	Botswana	* 9.51		98	Nigeria	0.02
47	Slovak Republic	8.89		99	Bangladesh	0.01
48	Brazil	8.85		100	Chad	* 0.00
49	Lithuania	8.53		100	Gambia	* 0.00
50	Argentina	7.14		100	Malawi	* 0.00
51	Malaysia	7.13		100	Tanzania	* 0.00
52	Turkey	6.96		100	Zambia	* 0.00

0 100 200 300 400 500 600 700 800

0 100 200 300 400 500 600 700 800

Source: World Bank, World Development Indicators 2003

I. Environment Component – Infrastructure Environment
3.03 Internet hosts, 2003

Internet hosts per 10,000 inhabitants

* estimated data

RANK	COUNTRY	SCORE		RANK	COUNTRY	SCORE
1	Iceland	3776.59		53	Mauritius	33.21
2	Finland	2445.45		54	Colombia	26.05
3	Denmark	2311.66		55	Costa Rica	25.78
4	Netherlands	2187.54		56	Peru	24.22
5	Australia	1445.56		57	Panama	23.00
6	Norway	1268.24		58	Romania	21.27
7	New Zealand	1216.40		59	Serbia and Montenegro	18.89
8	Singapore	1127.50		60	Ukraine	18.83
9	Sweden	1059.71		61	Bahrain	18.43
10	Canada	1019.07		62	Bosnia and Herzegovina	17.22
11	Japan	1015.04		63	Macedonia	17.12
12	Hong Kong	845.70		64	Guatemala	16.55
13	Switzerland	761.17		65	Thailand	16.51
14	Austria	710.99		66	Namibia	15.82
15	Israel	680.66		67	Paraguay	15.67
16	Luxembourg	622.83		68	Venezuela	13.74
17	United Kingdom	534.45		69	Nicaragua	12.90
18	Estonia	492.68		70	Botswana	10.67
19	France	399.91		71	Georgia	9.66
20	Ireland	397.08		72	Bolivia	8.05
21	Hungary	373.45		73	El Salvador	6.28
22	Germany	315.52		74	Jordan	5.68
23	Czech Republic	270.77		75	Jamaica	5.48
24	Uruguay	257.74		76	Gambia	4.06
25	Portugal	224.75		77	Philippines	3.50
26	Spain	221.58		78	Zimbabwe	3.49
27	Slovenia	214.44		79	Indonesia	2.82
28	Slovak Republic	211.27		80	Honduras	2.82
29	Belgium	204.05		81	Kenya	2.60
30	Poland	203.76		82	Ecuador	2.45
31	Lithuania	195.21		83	Zambia	1.74
32	Argentina	193.32		84	Mozambique	1.72
33	United Arab Emirate	187.23		85	Tanzania	1.50
34	Malta	180.63		86	China	1.23
35	Latvia	179.40		87	Morocco	1.16
36	Greece	177.54		88	Uganda	0.99
37	Brazil	177.22		89	Sri Lanka	0.99
38	Mexico	128.83		90	Pakistan	0.98
39	Chile	128.12		91	India	0.82
40	Taiwan	* 127.67		92	Egypt	0.46
41	Italy	109.15		93	Madagascar	0.44
42	Dominican Republic	73.79		94	Tunisia	0.28
43	Cyprus	72.04		95	Algeria	0.27
44	Croatia	67.37		96	Ghana	0.19
45	South Africa	64.14		97	Mali	0.14
46	Bulgaria	63.25		98	Nigeria	0.09
47	Trinidad and Tobago	61.56		99	Vietnam	0.04
48	United States	59.78		100	Malawi	0.01
49	Korea	53.09		101	Chad	0.01
50	Turkey	50.38		102	Angola	0.01
51	Malaysia	44.25		103	Ethiopia	0.00
52	Russia	43.14		104	Bangladesh	0.00

Source: International Telecommunication Union, June 2004

Individual Readiness

II. Readiness Component – Individual Readiness
4.01 Quality of math and science education, 2004

Math and science education in your country's schools (1 = lag far behind most other countries, 7 = are among the best in the world)

RANK	COUNTRY	SCORE		SD	RANK	COUNTRY	SCORE		SD
1	Singapore	6.23		0.87	53	Thailand	4.15		1.30
2	Finland	6.07		0.81	54	Georgia	4.13		1.53
3	France	6.02		1.17	55	China	4.09		1.51
4	Belgium	5.97		1.15	56	Turkey	4.09		1.48
5	Romania	5.72		1.42	57	Indonesia	4.00		1.34
6	Denmark	5.63		1.01	58	Zimbabwe	3.90		1.52
7	Taiwan	5.58		0.99	59	Botswana	3.89		1.41
8	Switzerland	5.56		0.96	60	Vietnam	3.86		1.48
9	Tunisia	5.55		1.34	61	Trinidad and Tobago	3.85		1.48
10	Czech Republic	5.52		1.12	62	Bahrain	3.83		1.66
11	Austria	5.50		1.01	63	Sri Lanka	3.80		1.52
11	India	5.50		1.36	64	Madagascar	3.73		1.57
13	Slovak Republic	5.48		1.01	65	Ghana	3.68		1.40
14	Japan	5.47		1.21	66	Mauritius	3.68		1.43
15	Hong Kong	5.43		1.06	67	Algeria	3.67		1.32
16	Hungary	5.39		1.06	68	Kenya	3.63		1.60
17	Bulgaria	5.33		1.62	69	Uruguay	3.60		1.40
18	Australia	5.29		1.20	70	Uganda	3.57		1.61
19	Lithuania	5.19		1.26	71	Zambia	3.57		1.83
20	Estonia	5.16		1.05	72	Nigeria	3.52		1.64
21	Sweden	5.15		1.18	73	Egypt	3.31		1.60
22	Canada	5.15		1.32	74	Panama	3.27		1.58
23	Russia	5.08		1.47	75	Tanzania	3.25		1.66
24	Netherlands	5.07		1.26	76	Namibia	3.23		1.38
25	Ireland	5.05		1.12	77	Mali	3.18		1.54
26	Slovenia	5.02		1.32	78	Colombia	3.16		1.24
27	Malaysia	4.94		1.18	79	Brazil	3.13		1.29
28	Jordan	4.93		1.03	80	Argentina	3.12		1.46
29	Iceland	4.92		1.04	81	Malawi	3.11		1.53
30	Greece	4.92		1.19	82	Venezuela	3.10		1.52
31	Cyprus	4.88		1.26	83	Gambia	3.07		1.50
32	Latvia	4.80		1.28	84	Philippines	3.05		1.39
33	Serbia and Montenegro	4.80		1.31	85	Jamaica	3.05		1.51
34	Ukraine	4.79		1.30	86	Portugal	3.00		1.07
35	Malta	4.78		1.18	87	Chile	2.99		1.24
36	New Zealand	4.76		1.40	88	Mexico	2.97		1.28
37	Croatia	4.73		1.56	89	Ethiopia	2.95		1.37
38	Spain	4.71		1.30	90	El Salvador	2.91		1.25
39	Poland	4.69		1.80	91	Mozambique	2.83		1.38
40	United Arab Emirates	4.58		1.45	92	Bangladesh	2.82		1.42
41	Korea	4.52		1.40	93	Ecuador	2.81		1.35
42	Israel	4.50		1.21	94	Bolivia	2.75		1.23
43	Macedonia	4.48		1.78	95	Nicaragua	2.57		1.27
44	Luxembourg	4.47		1.17	96	South Africa	2.57		1.40
45	Norway	4.43		1.34	97	Dominican Republic	2.56		1.43
46	United States	4.43		1.43	98	Paraguay	2.38		1.20
47	Bosnia and Herzegovina	4.38		1.67	99	Pakistan	2.36		0.93
48	Morocco	4.35		1.73	100	Chad	2.36		1.54
49	Germany	4.34		1.25	101	Peru	2.33		1.17
50	Italy	4.34		1.62	102	Honduras	2.02		0.90
51	United Kingdom	4.31		1.22	103	Guatemala	2.00		1.06
52	Costa Rica	4.30		1.29	104	Angola	1.82		1.02

0 1 2 3 4 5 6 7   0 1 2 3 4 5 6 7

Source: World Economic Forum, Executive Opinion Survey 2004

II. Readiness Component – Individual Readiness
4.02 Quality of educational system, 2004

The educational system in your country (1 = does not meet the needs of a competitive economy, 7 = meets the needs of a competitive economy)

RANK	COUNTRY	SCORE		SD	RANK	COUNTRY	SCORE		SD
1	Finland	6.00		1.01	53	Serbia and Montenegro	3.51		1.70
2	Singapore	5.81		1.02	54	Ghana	3.51		1.60
3	Iceland	5.72		1.17	54	Uruguay	3.51		1.43
4	Belgium	5.71		1.06	56	Poland	3.50		1.70
5	Denmark	5.71		0.81	57	Mauritius	3.47		1.58
6	Switzerland	5.57		1.05	58	Namibia	3.43		1.63
7	Australia	5.35		1.25	59	Bosnia and Herzegovina	3.39		1.73
8	Sweden	5.25		1.16	60	Korea	3.36		1.44
9	Ireland	5.23		1.23	61	Uganda	3.35		1.88
10	Austria	5.21		1.33	62	Croatia	3.32		1.70
11	Taiwan	5.19		1.35	63	Zimbabwe	3.30		1.64
12	Canada	5.19		1.37	64	Trinidad and Tobago	3.29		1.57
13	Norway	5.13		1.22	65	China	3.22		1.41
14	France	5.12		1.42	66	Thailand	3.19		1.36
15	Tunisia	5.09		1.56	67	Philippines	3.12		1.63
16	Netherlands	5.00		1.36	68	Nigeria	3.06		1.65
17	Hong Kong	4.85		1.41	69	Colombia	3.02		1.27
18	Malaysia	4.72		1.16	70	Zambia	2.98		1.95
19	Luxembourg	4.57		1.30	71	Chile	2.97		1.32
20	Malta	4.54		1.37	72	South Africa	2.96		1.37
21	New Zealand	4.49		1.67	73	Portugal	2.90		1.00
22	United Kingdom	4.47		1.30	74	Jamaica	2.88		1.48
23	United Arab Emirates	4.46		1.63	75	Malawi	2.86		1.61
24	Costa Rica	4.46		1.42	76	Georgia	2.71		1.68
25	Slovenia	4.42		1.49	77	Mexico	2.71		1.25
26	Cyprus	4.41		1.39	78	Algeria	2.67		1.53
27	United States	4.40		1.64	79	Egypt	2.66		1.62
28	Estonia	4.35		1.48	80	Ethiopia	2.64		1.49
29	Israel	4.31		1.20	81	Turkey	2.64		1.44
30	Spain	4.31		1.48	82	Tanzania	2.60		1.40
31	Japan	4.25		1.62	83	Mozambique	2.60		1.46
32	Romania	4.22		1.67	84	Argentina	2.59		1.27
33	Lithuania	4.16		1.50	85	Brazil	2.58		1.19
34	Slovak Republic	4.14		1.37	86	Mali	2.51		1.49
35	Indonesia	4.13		1.20	87	El Salvador	2.51		1.28
36	Hungary	4.13		1.67	88	Sri Lanka	2.50		1.30
37	Czech Republic	4.12		1.44	89	Vietnam	2.42		1.33
38	Latvia	4.11		1.52	90	Panama	2.36		1.38
39	India	4.11		1.69	91	Dominican Republic	2.25		1.39
40	Bulgaria	4.10		1.81	92	Madagascar	2.18		1.15
41	Jordan	4.08		1.37	93	Venezuela	2.10		1.12
42	Kenya	3.93		1.67	94	Bolivia	2.09		1.06
43	Gambia	3.91		1.62	95	Bangladesh	2.06		1.21
44	Botswana	3.78		1.51	96	Peru	2.05		0.96
45	Macedonia	3.76		1.97	97	Nicaragua	2.03		1.19
46	Germany	3.72		1.61	98	Ecuador	2.02		1.12
47	Ukraine	3.68		1.46	99	Pakistan	2.00		1.00
48	Bahrain	3.64		1.83	100	Paraguay	1.98		0.99
49	Russia	3.57		1.60	101	Chad	1.85		1.26
50	Greece	3.56		1.29	102	Honduras	1.66		0.80
51	Morocco	3.55		1.70	103	Guatemala	1.63		0.87
52	Italy	3.52		1.66	104	Angola	1.59		0.86

0 1 2 3 4 5 6 7   0 1 2 3 4 5 6 7

Source: World Economic Forum, Executive Opinion Survey 2004

II. Readiness Component – Individual Readiness
4.03 Quality of public schools, 2004

The public (free) schools in your country are (1 = of poor quality, 7 = equal to the best in the world)

RANK	COUNTRY	SCORE	SD	RANK	COUNTRY	SCORE	SD
1	Finland	6.56	0.69	53	United Arab Emirates	3.88	1.67
2	Iceland	6.44	0.65	54	Macedonia	3.87	1.91
3	Belgium	6.34	0.91	55	Uruguay	3.81	1.27
4	Netherlands	6.16	0.77	56	Thailand	3.71	1.29
5	Switzerland	6.15	0.76	57	Mauritius	3.62	1.30
6	Denmark	6.12	0.73	58	Gambia	3.61	1.52
7	Ireland	6.05	0.80	59	Namibia	3.58	1.26
8	Singapore	6.04	0.99	60	China	3.54	1.52
9	France	5.98	1.08	61	Trinidad and Tobago	3.52	1.46
10	Austria	5.91	1.05	62	South Africa	3.44	1.19
11	Canada	5.76	1.11	63	Vietnam	3.42	1.33
12	Australia	5.70	1.32	64	Sri Lanka	3.35	1.50
13	New Zealand	5.64	1.27	65	Bosnia and Herzegovina	3.32	1.53
14	Norway	5.61	0.78	66	Algeria	3.14	1.29
15	Sweden	5.55	1.05	67	Kenya	3.09	1.34
16	Estonia	5.51	1.10	68	Jamaica	3.05	1.54
17	Japan	5.51	1.43	69	Zimbabwe	3.03	1.52
18	Czech Republic	5.48	1.26	70	Turkey	3.00	1.27
19	Malaysia	5.33	0.95	71	Panama	2.87	1.49
20	Taiwan	5.29	1.15	72	Argentina	2.87	1.36
21	Luxembourg	5.20	1.06	73	Egypt	2.83	1.66
22	Hungary	5.18	1.05	74	Mexico	2.82	1.13
23	Slovenia	5.15	1.30	75	Chile	2.80	1.22
24	Tunisia	5.14	1.37	76	Colombia	2.78	1.11
25	Hong Kong	5.08	1.16	77	Ghana	2.74	1.36
26	Slovak Republic	5.07	1.17	78	Uganda	2.68	1.43
27	United States	4.99	1.41	79	El Salvador	2.57	1.23
28	Germany	4.98	1.24	80	Georgia	2.57	1.15
29	Israel	4.94	0.90	81	Brazil	2.57	0.99
30	Cyprus	4.83	1.41	82	Zambia	2.46	1.91
31	United Kingdom	4.83	1.42	83	India	2.45	1.33
32	Romania	4.72	1.67	84	Tanzania	2.37	1.16
33	Portugal	4.59	1.14	85	Philippines	2.36	1.23
34	Italy	4.55	1.82	86	Paraguay	2.34	1.12
35	Lithuania	4.55	1.45	87	Pakistan	2.33	1.09
36	Latvia	4.52	1.22	88	Mozambique	2.26	1.21
37	Malta	4.51	1.42	89	Ethiopia	2.21	1.30
38	Spain	4.50	1.59	90	Dominican Republic	2.21	1.08
39	Korea	4.34	1.30	91	Mali	2.21	1.20
40	Bahrain	4.29	1.42	92	Nicaragua	2.17	1.12
41	Croatia	4.27	1.62	93	Madagascar	2.16	1.11
42	Bulgaria	4.27	1.88	94	Nigeria	2.13	1.28
43	Poland	4.22	1.65	95	Bolivia	2.09	0.95
44	Botswana	4.22	1.46	96	Venezuela	2.06	1.02
45	Russia	4.21	1.73	97	Bangladesh	2.01	1.10
46	Greece	4.08	1.15	98	Ecuador	2.00	1.02
47	Morocco	4.06	1.64	99	Honduras	1.92	0.98
48	Indonesia	4.03	1.37	100	Chad	1.89	1.09
49	Costa Rica	4.02	1.30	101	Malawi	1.86	1.00
50	Serbia and Montenegro	4.01	1.49	102	Peru	1.84	0.84
51	Jordan	3.95	1.34	103	Guatemala	1.70	0.83
52	Ukraine	3.88	1.43	104	Angola	1.44	0.81

Source: World Economic Forum, Executive Opinion Survey 2004

II. Readiness Component – Individual Readiness
4.04 Internet access in schools, 2004

Internet access in schools is (1 = very limited, 7 = pervasive; most children have frequent access)

RANK	COUNTRY	SCORE		SD	RANK	COUNTRY	SCORE		SD
1	Iceland	6.72		0.61	53	Italy	3.60		1.69
2	Finland	6.57		0.59	54	Pakistan	3.60		1.48
3	Korea	6.56		0.69	55	Vietnam	3.59		1.69
4	Hong Kong	6.41		0.79	56	Mauritius	3.58		1.54
5	Sweden	6.40		0.75	57	Philippines	3.54		1.76
6	Singapore	6.39		0.89	58	Morocco	3.52		1.86
7	Denmark	6.33		0.87	59	Peru	3.50		1.58
8	Taiwan	6.12		0.97	60	Costa Rica	3.49		1.54
9	Canada	6.03		1.05	61	Poland	3.41		1.29
10	Estonia	6.00		1.05	62	Mexico	3.40		1.65
10	Switzerland	6.00		0.89	63	Egypt	3.40		1.89
12	Australia	5.97		1.10	64	Argentina	3.39		1.59
13	Austria	5.97		1.03	65	Botswana	3.38		1.68
14	United States	5.94		1.06	66	Bulgaria	3.36		1.72
15	New Zealand	5.92		0.73	67	Turkey	3.34		1.50
16	Netherlands	5.92		1.04	68	South Africa	3.32		1.51
17	Malta	5.88		0.90	69	Panama	3.28		1.63
18	Norway	5.87		1.06	70	El Salvador	3.26		1.57
19	Luxembourg	5.82		1.22	71	Venezuela	3.21		1.27
20	United Kingdom	5.80		1.04	72	Dominican Republic	3.16		1.64
21	Slovenia	5.78		0.93	73	Jamaica	3.15		1.50
22	Japan	5.78		1.00	74	Macedonia	3.06		1.89
23	Israel	5.73		0.96	75	Serbia and Montenegro	3.01		1.54
24	United Arab Emirates	5.60		1.25	76	Bosnia and Herzegovina	2.99		1.72
25	Germany	5.33		1.18	77	Colombia	2.98		1.47
26	Czech Republic	5.27		1.42	78	Trinidad and Tobago	2.92		1.43
27	Belgium	5.24		0.97	79	Gambia	2.83		1.95
28	Malaysia	5.19		1.11	80	Sri Lanka	2.71		1.51
29	Hungary	5.05		1.58	81	Guatemala	2.68		1.50
30	Bahrain	4.98		1.42	82	Nicaragua	2.67		1.69
31	France	4.94		1.10	83	Bolivia	2.54		1.44
32	Portugal	4.93		1.34	84	Ukraine	2.53		1.27
33	Chile	4.75		1.42	85	Uganda	2.52		1.47
34	Ireland	4.67		1.51	86	Zambia	2.51		1.73
35	Lithuania	4.66		1.57	87	Ecuador	2.46		1.41
36	Cyprus	4.60		1.57	88	Tanzania	2.43		1.41
37	Tunisia	4.59		1.58	89	Ghana	2.40		1.31
38	Spain	4.41		1.45	90	Nigeria	2.31		1.34
39	Latvia	4.39		1.54	91	Georgia	2.27		1.26
40	Jordan	4.38		1.52	92	Paraguay	2.17		1.30
41	Slovak Republic	4.24		1.47	93	Zimbabwe	2.13		1.01
42	Thailand	4.12		1.44	94	Honduras	2.08		1.10
43	Greece	4.03		1.32	95	Madagascar	2.05		1.39
44	Romania	4.01		1.66	96	Mali	1.97		1.42
45	Indonesia	4.00		1.07	97	Mozambique	1.96		1.34
46	India	3.89		1.38	98	Kenya	1.94		1.19
47	Croatia	3.85		1.74	99	Algeria	1.87		1.34
48	Uruguay	3.81		1.34	100	Malawi	1.78		1.12
49	China	3.80		1.58	101	Bangladesh	1.69		1.13
50	Brazil	3.79		1.53	102	Ethiopia	1.64		1.11
51	Namibia	3.77		1.82	103	Angola	1.52		0.96
52	Russia	3.65		1.63	104	Chad	1.20		0.62

Source: World Economic Forum, Executive Opinion Survey 2004

II. Readiness Component – Individual Readiness
4.05 Buyer sophistication, 2004

Buyers in your country are (1 = unsophisticated and make choices based on the lowest price, 7 = knowledgeable and buy based on superior performance attributes)

RANK	COUNTRY	SCORE	SD	RANK	COUNTRY	SCORE	SD
1	United States	5.94	1.06	53	Malta	3.76	1.37
2	Japan	5.94	1.08	54	Panama	3.72	1.65
3	United Kingdom	5.91	0.80	55	Czech Republic	3.68	1.53
4	Switzerland	5.90	0.97	56	Ghana	3.67	1.67
5	Luxembourg	5.87	1.17	57	Sri Lanka	3.63	1.45
6	Hong Kong	5.79	1.22	58	Zambia	3.61	2.03
7	France	5.78	1.04	59	Jordan	3.58	1.58
8	Australia	5.67	1.20	59	Tanzania	3.58	1.74
9	Canada	5.64	1.14	61	Ukraine	3.56	1.65
10	New Zealand	5.64	0.91	62	Argentina	3.56	1.34
11	Korea	5.55	1.09	63	El Salvador	3.54	1.47
12	Finland	5.54	1.16	64	Turkey	3.54	1.48
13	Belgium	5.53	1.23	65	Vietnam	3.48	1.62
14	Sweden	5.53	1.26	66	Botswana	3.46	1.44
15	Singapore	5.52	1.23	67	Colombia	3.46	1.29
16	Netherlands	5.50	1.15	68	Zimbabwe	3.45	1.18
17	Germany	5.43	1.36	69	Nigeria	3.45	1.73
18	Norway	5.39	0.84	70	Mexico	3.45	1.32
19	Iceland	5.38	1.10	71	Lithuania	3.45	1.32
20	Denmark	5.29	1.30	72	Romania	3.44	1.64
21	Taiwan	5.27	1.20	73	Hungary	3.43	1.37
22	Ireland	5.08	1.37	74	Kenya	3.40	1.66
23	Austria	5.03	1.30	75	Poland	3.35	1.35
24	Malaysia	5.03	1.21	76	Uruguay	3.34	1.17
25	Tunisia	4.78	1.55	77	Slovak Republic	3.31	1.50
26	Israel	4.75	1.34	78	Bulgaria	3.28	1.76
27	South Africa	4.59	1.22	79	Uganda	3.26	1.72
28	Slovenia	4.56	1.36	80	Dominican Republic	3.16	1.45
29	Pakistan	4.55	1.99	81	Bangladesh	3.12	1.48
30	Estonia	4.53	1.32	82	Venezuela	3.02	1.35
31	Italy	4.51	1.60	83	Gambia	2.92	1.70
32	United Arab Emirates	4.48	1.66	84	Serbia and Montenegro	2.92	1.64
33	Spain	4.44	1.50	85	Algeria	2.91	1.59
34	India	4.44	1.34	86	Macedonia	2.91	1.75
35	Bahrain	4.40	1.37	87	Malawi	2.89	1.23
36	Chile	4.27	1.45	88	Egypt	2.87	1.74
37	Cyprus	4.27	1.34	89	Croatia	2.78	1.38
38	Namibia	4.17	1.64	90	Peru	2.75	1.19
39	Indonesia	4.15	1.29	91	Guatemala	2.71	1.22
40	Portugal	4.13	1.30	92	Georgia	2.70	1.26
41	Brazil	4.09	1.45	93	Bosnia and Herzegovina	2.65	1.55
42	China	4.06	1.40	94	Mali	2.63	1.56
43	Greece	4.04	1.30	95	Ecuador	2.46	1.10
44	Thailand	4.04	1.44	96	Ethiopia	2.42	1.35
45	Mauritius	4.03	1.59	97	Chad	2.37	1.74
46	Costa Rica	4.02	1.40	98	Honduras	2.37	1.07
47	Morocco	4.01	1.69	99	Paraguay	2.36	1.13
48	Russia	3.92	1.58	100	Nicaragua	2.33	1.09
49	Philippines	3.85	1.65	101	Madagascar	2.15	1.15
50	Trinidad and Tobago	3.82	1.34	102	Bolivia	2.09	0.96
51	Jamaica	3.82	1.44	103	Mozambique	2.05	1.18
52	Latvia	3.81	1.48	104	Angola	1.84	0.83

0 1 2 3 4 5 6 7

Source: World Economic Forum, Executive Opinion Survey 2004

II. Readiness Component – Individual Readiness
4.06 Buyer dynamism, 2004

Buyers in your country are (1 = slow to adopt new products and processes, 7 = actively seeking the latest products, technologies and processes)

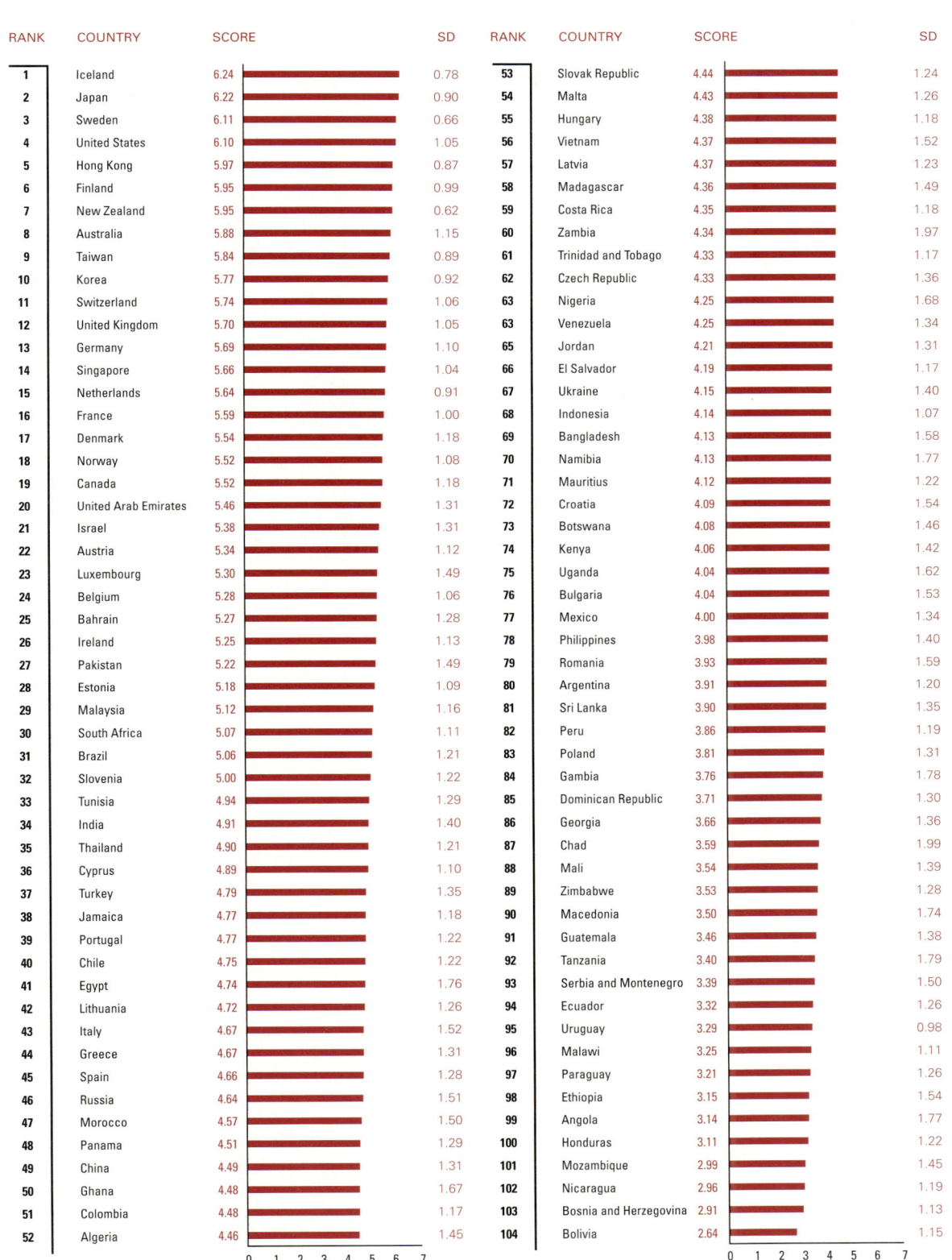

RANK	COUNTRY	SCORE	SD	RANK	COUNTRY	SCORE	SD
1	Iceland	6.24	0.78	53	Slovak Republic	4.44	1.24
2	Japan	6.22	0.90	54	Malta	4.43	1.26
3	Sweden	6.11	0.66	55	Hungary	4.38	1.18
4	United States	6.10	1.05	56	Vietnam	4.37	1.52
5	Hong Kong	5.97	0.87	57	Latvia	4.37	1.23
6	Finland	5.95	0.99	58	Madagascar	4.36	1.49
7	New Zealand	5.95	0.62	59	Costa Rica	4.35	1.18
8	Australia	5.88	1.15	60	Zambia	4.34	1.97
9	Taiwan	5.84	0.89	61	Trinidad and Tobago	4.33	1.17
10	Korea	5.77	0.92	62	Czech Republic	4.33	1.36
11	Switzerland	5.74	1.06	63	Nigeria	4.25	1.68
12	United Kingdom	5.70	1.05	63	Venezuela	4.25	1.34
13	Germany	5.69	1.10	65	Jordan	4.21	1.31
14	Singapore	5.66	1.04	66	El Salvador	4.19	1.17
15	Netherlands	5.64	0.91	67	Ukraine	4.15	1.40
16	France	5.59	1.00	68	Indonesia	4.14	1.07
17	Denmark	5.54	1.18	69	Bangladesh	4.13	1.58
18	Norway	5.52	1.08	70	Namibia	4.13	1.77
19	Canada	5.52	1.18	71	Mauritius	4.12	1.22
20	United Arab Emirates	5.46	1.31	72	Croatia	4.09	1.54
21	Israel	5.38	1.31	73	Botswana	4.08	1.46
22	Austria	5.34	1.12	74	Kenya	4.06	1.42
23	Luxembourg	5.30	1.49	75	Uganda	4.04	1.62
24	Belgium	5.28	1.06	76	Bulgaria	4.04	1.53
25	Bahrain	5.27	1.28	77	Mexico	4.00	1.34
26	Ireland	5.25	1.13	78	Philippines	3.98	1.40
27	Pakistan	5.22	1.49	79	Romania	3.93	1.59
28	Estonia	5.18	1.09	80	Argentina	3.91	1.20
29	Malaysia	5.12	1.16	81	Sri Lanka	3.90	1.35
30	South Africa	5.07	1.11	82	Peru	3.86	1.19
31	Brazil	5.06	1.21	83	Poland	3.81	1.31
32	Slovenia	5.00	1.22	84	Gambia	3.76	1.78
33	Tunisia	4.94	1.29	85	Dominican Republic	3.71	1.30
34	India	4.91	1.40	86	Georgia	3.66	1.36
35	Thailand	4.90	1.21	87	Chad	3.59	1.99
36	Cyprus	4.89	1.10	88	Mali	3.54	1.39
37	Turkey	4.79	1.35	89	Zimbabwe	3.53	1.28
38	Jamaica	4.77	1.18	90	Macedonia	3.50	1.74
39	Portugal	4.77	1.22	91	Guatemala	3.46	1.38
40	Chile	4.75	1.22	92	Tanzania	3.40	1.79
41	Egypt	4.74	1.76	93	Serbia and Montenegro	3.39	1.50
42	Lithuania	4.72	1.26	94	Ecuador	3.32	1.26
43	Italy	4.67	1.52	95	Uruguay	3.29	0.98
44	Greece	4.67	1.31	96	Malawi	3.25	1.11
45	Spain	4.66	1.28	97	Paraguay	3.21	1.26
46	Russia	4.64	1.51	98	Ethiopia	3.15	1.54
47	Morocco	4.57	1.50	99	Angola	3.14	1.77
48	Panama	4.51	1.29	100	Honduras	3.11	1.22
49	China	4.49	1.31	101	Mozambique	2.99	1.45
50	Ghana	4.48	1.67	102	Nicaragua	2.96	1.19
51	Colombia	4.48	1.17	103	Bosnia and Herzegovina	2.91	1.13
52	Algeria	4.46	1.45	104	Bolivia	2.64	1.15

0 1 2 3 4 5 6 7   0 1 2 3 4 5 6 7

Source: World Economic Forum, Executive Opinion Survey 2004

II. Readiness Component – Individual Readiness
4.07 Residential telephone connection charge, 2002

Residential telephone connection charge (US$) (adjusted by GDP per capita/1,000)

* estimated data

RANK	COUNTRY	SCORE
1	Singapore	0.79
2	Luxembourg	0.82
3	United States	1.13
4	Turkey	1.38
5	Germany	1.44
6	New Zealand	1.44
7	Trinidad and Tobago	1.45
8	France	1.49
9	Greece	1.75
10	Norway	1.94
11	United Arab Emirates	2.03
12	Belgium	2.12
13	Switzerland	2.27
14	Iceland	2.38
15	Italy	* 2.56
16	Canada	* 2.65
17	Hong Kong	2.69
18	Austria	2.77
19	Malaysia	3.11
20	Finland	3.24
21	Ireland	3.29
22	Israel	3.40
23	Cyprus	3.46
24	United Kingdom	3.69
25	Netherlands	* 3.75
26	Korea	3.78
27	Sweden	* 3.86
28	Bahrain	4.11
29	Romania	4.30
30	Spain	4.38
31	Australia	4.40
32	Taiwan	* 4.48
33	Portugal	4.64
34	Jamaica	4.66
35	Malta	4.67
36	Denmark	* 5.32
37	Slovenia	5.39
38	Zimbabwe	6.86
39	Estonia	7.68
40	Mauritius	7.87
41	Chile	7.95
42	Mexico	* 8.50
43	Panama	* 9.10
44	South Africa	9.77
45	Poland	* 10.13
45	Slovak Republic	* 10.13
47	Costa Rica	10.96
48	Botswana	11.03
49	Croatia	12.22
50	Lithuania	12.66
51	Czech Republic	12.76
52	Argentina	14.51

RANK	COUNTRY	SCORE
53	Hungary	15.65
54	Venezuela	15.89
55	Japan	17.05
56	Philippines	17.49
57	Uruguay	* 18.25
58	Honduras	21.51
59	Bulgaria	21.78
60	Namibia	23.44
61	Latvia	23.54
62	Tunisia	24.20
63	Brazil	* 24.42
64	Algeria	26.23
65	Indonesia	28.35
66	Ecuador	28.75
67	Zambia	28.96
68	Dominican Republic	29.28
69	India	30.48
70	Ukraine	30.72
71	Macedonia	* 32.56
72	Thailand	34.21
73	Morocco	37.35
74	Colombia	* 38.29
75	Serbia and Montenegro	39.77
76	Guatemala	* 40.00
77	El Salvador	* 40.91
78	Jordan	44.00
79	Angola	47.40
80	China	* 58.00
81	Bolivia	* 58.17
81	Paraguay	* 58.17
83	Russia	63.28
84	Pakistan	65.33
85	Peru	66.71
86	Kenya	67.35
87	Bosnia and Herzegovina	77.48
88	Mozambique	89.65
89	Madagascar	96.57
90	Egypt	99.92
91	Malawi	108.88
92	Georgia	117.76
93	Ghana	140.03
94	Vietnam	145.26
95	Nigeria	154.86
96	Gambia	156.78
97	Tanzania	159.08
98	Sri Lanka	161.80
99	Chad	247.43
100	Uganda	254.73
101	Mali	260.13
102	Ethiopia	379.05
103	Nicaragua	389.08
104	Bangladesh	486.83

Source: International Telecommunication Union, June 2004

II. Readiness Component – Individual Readiness
4.08 Affordability of Internet access, 2003

Total Internet price for 20 hours of usage as a % of GNI

RANK	COUNTRY	SCORE
1	Hong Kong	0.2
2	United States	0.5
3	Singapore	0.6
4	Canada	0.7
4	Denmark	0.7
4	Germany	0.7
4	Switzerland	0.7
4	Taiwan	0.7
9	France	0.8
9	Japan	0.8
9	Norway	0.8
9	United Arab Emirates	0.8
13	Iceland	0.9
13	Luxembourg	0.9
15	Italy	1
16	Australia	1.1
16	New Zealand	1.1
16	Sweden	1.1
16	United Kingdom	1.1
20	Finland	1.2
20	Netherlands	1.2
22	Ireland	1.4
23	Belgium	1.5
24	Austria	1.7
24	Cyprus	1.7
24	Spain	1.7
27	Israel	2.1
28	Hungary	2.3
28	Malta	2.3
28	Portugal	2.3
31	Trinidad and Tobago	2.5
32	Malaysia	2.9
33	Slovenia	3.1
34	Argentina	3.9
34	Estonia	3.9
34	Greece	3.9
37	Bahrain	4.1
37	Poland	4.1
39	Thailand	4.2
40	Croatia	4.4
41	Czech Republic	4.5
41	Egypt	4.5
43	Mexico	4.6
44	Mauritius	4.7
45	Russia	5.6
46	Venezuela	5.7
47	Chile	6.1
48	Slovak Republic	6.3
49	Bosnia and Herzegovina	6.9
50	Uruguay	7.3
51	Costa Rica	7.6
52	Bulgaria	8.3

RANK	COUNTRY	SCORE
53	Turkey	9.5
54	Tunisia	10.4
55	Panama	10.7
56	Botswana	10.9
57	Lithuania	11.2
58	Serbia and Montenegro	11.3
59	Brazil	11.8
60	Colombia	12.2
61	Algeria	12.4
62	China	13
63	Macedonia	13.3
64	South Africa	15.4
65	Dominican Republic	17.1
65	Romania	17.1
67	Jordan	18
68	Jamaica	18.5
69	Peru	19.2
70	Latvia	20
71	Philippines	20.1
72	Guatemala	21.4
73	Sri Lanka	21.5
74	India	21.9
75	Namibia	22.5
76	Morocco	25.5
77	Ukraine	26
78	Ecuador	26.3
79	El Salvador	27.8
80	Bolivia	29.8
81	Paraguay	37.3
82	Indonesia	37.6
83	Pakistan	45.7
84	Georgia	48.4
85	Honduras	52.9
86	Vietnam	55.4
87	Zimbabwe	58.3
88	Bangladesh	66.8
89	Gambia	116.2
90	Zambia	118.7
91	Nicaragua	138.6
92	Angola	143.3
93	Kenya	152.4
93	Korea	152.4
95	Ghana	194.8
96	Mali	289.8
97	Mozambique	290.2
98	Ethiopia	329.1
99	Madagascar	336.7
100	Nigeria	353.7
101	Chad	375.6
102	Uganda	464.4
103	Malawi	465
104	Tanzania	501.4

Source: International Telecommunication Union (WSIS 2003)

Business Readiness

II. Readiness Component – Business Readiness
5.01 Investment in training, 2004

The general approach of companies in your country to human resources is (1 = to invest little in training and employee development, 7 = to invest heavily to attract, train and retain employees)

RANK	COUNTRY	SCORE	SD	RANK	COUNTRY	SCORE	SD
1	Switzerland	5.90	0.93	53	Jamaica	3.74	1.16
2	Germany	5.89	0.89	54	Turkey	3.70	1.38
3	Sweden	5.89	0.58	55	Portugal	3.68	1.09
4	Japan	5.86	0.87	56	Kenya	3.66	1.34
5	Denmark	5.83	0.92	57	Romania	3.64	1.61
6	Netherlands	5.70	0.72	58	Cyprus	3.58	1.32
7	Finland	5.66	0.79	59	Nigeria	3.58	1.68
8	United States	5.66	0.96	60	El Salvador	3.56	1.22
9	Belgium	5.38	0.89	61	Italy	3.55	1.46
10	Norway	5.30	0.76	62	Colombia	3.52	1.26
11	Australia	5.26	1.21	63	Jordan	3.51	1.15
12	Luxembourg	5.24	1.12	64	Macedonia	3.49	1.84
13	Iceland	5.22	1.04	65	Sri Lanka	3.48	1.29
14	Singapore	5.18	1.10	66	Argentina	3.46	1.05
15	Taiwan	5.14	0.92	67	Mexico	3.44	1.29
16	United Kingdom	5.13	1.04	68	Hungary	3.42	1.18
17	Austria	5.10	1.22	69	Poland	3.40	1.21
18	Korea	5.08	1.03	70	Botswana	3.33	1.65
19	France	5.04	1.33	71	Gambia	3.28	1.71
20	Ireland	5.03	0.80	72	Ghana	3.22	1.66
21	Canada	4.99	1.22	73	Zambia	3.22	1.59
22	New Zealand	4.95	1.03	74	Dominican Republic	3.21	1.20
23	Hong Kong	4.93	1.37	75	Malawi	3.20	1.39
24	South Africa	4.90	1.12	76	Panama	3.19	1.30
25	Israel	4.81	1.22	77	Uganda	3.18	1.59
26	Malaysia	4.39	1.19	78	Vietnam	3.16	1.48
27	Brazil	4.26	1.22	79	Guatemala	3.15	1.13
28	Estonia	4.25	1.28	80	Tanzania	3.12	1.46
29	India	4.23	1.04	81	Russia	3.09	1.33
30	Spain	4.22	1.15	82	Peru	3.03	0.96
31	Costa Rica	4.22	1.32	83	Venezuela	3.02	0.90
32	Slovenia	4.18	1.20	84	Uruguay	3.02	1.03
33	Thailand	4.17	1.22	85	Ukraine	3.00	1.20
34	Namibia	4.17	1.58	86	Bulgaria	2.98	1.51
35	Slovak Republic	4.16	1.24	87	Mozambique	2.92	1.52
36	Philippines	4.12	1.20	88	Croatia	2.92	1.45
37	Mauritius	4.09	1.07	89	Georgia	2.88	1.28
38	Zimbabwe	4.07	1.26	90	Algeria	2.77	1.47
39	Indonesia	4.03	0.90	91	Madagascar	2.75	1.44
40	Chile	4.01	1.21	92	Serbia and Montenegro	2.73	1.20
41	Morocco	3.99	1.37	93	Honduras	2.68	0.98
42	Tunisia	3.97	1.56	94	Ecuador	2.66	1.02
43	United Arab Emirates	3.94	1.72	95	Bosnia and Herzegovina	2.60	1.56
44	Trinidad and Tobago	3.93	1.00	96	Nicaragua	2.47	1.28
45	Greece	3.93	1.06	97	Paraguay	2.45	0.79
46	Czech Republic	3.91	1.31	98	Angola	2.44	1.25
47	Malta	3.89	1.22	99	Bolivia	2.42	1.09
48	Lithuania	3.86	1.48	100	Mali	2.39	1.26
49	Egypt	3.85	1.74	101	Bangladesh	2.39	1.09
50	Bahrain	3.83	1.71	102	Ethiopia	2.29	1.10
51	Latvia	3.81	1.33	103	Chad	2.06	1.38
52	China	3.78	1.27	104	Pakistan	2.03	1.30

Source: World Economic Forum, Executive Opinion Survey 2004

II. Readiness Component – Business Readiness
5.02 Availability of training services, 2004

In your country, specialized research and training services are (1 = not available in the country, 7 = available from world-class local institutions)

RANK	COUNTRY	SCORE		SD	RANK	COUNTRY	SCORE		SD
1	United States	6.35		0.73	53	Jordan	3.97		1.34
2	Germany	6.17		0.87	54	Jamaica	3.97		1.35
3	Japan	6.03		0.80	55	Panama	3.94		1.55
4	Finland	6.00		0.97	56	Romania	3.93		1.36
4	United Kingdom	6.00		0.75	57	Kenya	3.93		1.30
6	Switzerland	5.97		1.01	58	Indonesia	3.92		1.02
7	Netherlands	5.77		0.85	59	Ukraine	3.92		0.89
8	Sweden	5.72		0.83	60	United Arab Emirates	3.91		1.72
9	Denmark	5.71		0.69	61	Greece	3.90		1.24
10	Belgium	5.69		0.82	62	Bulgaria	3.89		1.33
11	Austria	5.63		1.04	63	Cyprus	3.89		1.51
12	Australia	5.62		0.97	64	Argentina	3.88		1.44
13	Israel	5.60		1.45	65	Ghana	3.82		1.63
14	Canada	5.60		0.94	66	Hungary	3.80		1.36
15	France	5.58		1.11	67	Serbia and Montenegro	3.78		1.21
16	Taiwan	5.29		1.11	68	Bahrain	3.75		1.72
17	Brazil	5.14		1.23	69	Botswana	3.63		1.52
18	Norway	5.09		1.04	70	Uruguay	3.60		1.22
19	New Zealand	5.06		1.00	71	Guatemala	3.60		1.57
20	Singapore	5.04		1.31	72	Tanzania	3.56		1.96
21	South Africa	5.01		1.35	73	El Salvador	3.56		1.57
22	Ireland	4.92		1.05	74	Peru	3.55		1.44
23	Hong Kong	4.92		1.36	75	Dominican Republic	3.52		1.57
24	Tunisia	4.81		1.33	76	Madagascar	3.51		1.79
25	Italy	4.78		1.40	77	Mauritius	3.48		1.43
26	Korea	4.76		1.05	78	Mali	3.44		1.68
27	India	4.73		1.31	79	Trinidad and Tobago	3.42		1.27
28	Iceland	4.67		1.24	80	Bosnia and Herzegovina	3.39		1.60
29	Slovenia	4.60		1.40	81	Philippines	3.36		1.58
30	Estonia	4.59		1.36	82	Colombia	3.36		1.28
31	Spain	4.49		1.27	83	Sri Lanka	3.32		1.33
32	Czech Republic	4.47		1.39	84	Vietnam	3.32		1.47
33	Lithuania	4.47		0.97	85	Malawi	3.28		1.35
34	China	4.41		1.13	86	Ecuador	3.27		1.53
35	Malaysia	4.40		1.22	87	Zimbabwe	3.27		1.31
36	Costa Rica	4.36		1.33	88	Macedonia	3.25		1.65
37	Nigeria	4.27		1.56	89	Georgia	3.20		1.42
38	Slovak Republic	4.22		1.30	90	Pakistan	3.19		0.98
39	Chile	4.21		1.40	91	Algeria	3.12		1.50
40	Poland	4.21		1.53	92	Malta	3.09		1.33
41	Russia	4.20		1.30	93	Gambia	3.03		1.65
42	Croatia	4.20		1.47	94	Venezuela	3.00		1.19
43	Portugal	4.18		1.14	95	Nicaragua	2.96		1.56
44	Latvia	4.18		1.39	96	Ethiopia	2.93		1.49
45	Mexico	4.17		1.42	97	Paraguay	2.87		1.53
46	Uganda	4.16		1.60	98	Namibia	2.84		1.59
47	Thailand	4.15		1.24	99	Bolivia	2.82		1.46
48	Luxembourg	4.11		1.42	100	Mozambique	2.76		1.55
49	Egypt	4.05		1.77	101	Honduras	2.72		1.31
50	Turkey	4.01		1.37	102	Chad	2.61		2.01
51	Zambia	4.00		1.73	103	Bangladesh	2.48		1.33
52	Morocco	3.99		1.48	104	Angola	2.21		1.37

0 1 2 3 4 5 6 7   0 1 2 3 4 5 6 7

Source: World Economic Forum, Executive Opinion Survey 2004

II. Readiness Component – Business Readiness
5.03 Quality of business schools, 2004

Management or business schools in your country are (1 = limited or of poor quality, 7 = the best in the world)

RANK	COUNTRY	SCORE		SD	RANK	COUNTRY	SCORE		SD
1	United States	6.69		0.80	53	Turkey	4.12		1.28
2	France	6.36		0.75	54	Poland	4.11		1.17
3	United Kingdom	6.13		0.81	55	Nigeria	4.08		1.50
4	Switzerland	6.01		0.75	56	Greece	4.05		1.21
5	Canada	5.96		0.90	57	Indonesia	3.97		1.11
6	India	5.86		0.88	58	Korea	3.92		1.29
7	Australia	5.76		0.93	59	El Salvador	3.85		1.32
8	Spain	5.75		0.94	60	Guatemala	3.78		1.23
9	Netherlands	5.71		0.69	61	Panama	3.77		1.31
10	South Africa	5.67		0.96	62	Cyprus	3.76		1.29
11	Austria	5.63		0.99	63	Malta	3.76		1.33
12	Sweden	5.63		0.60	64	Madagascar	3.75		1.24
13	Finland	5.55		0.98	65	Dominican Republic	3.74		1.14
14	Denmark	5.50		0.98	66	Ukraine	3.74		1.08
15	Belgium	5.49		0.99	67	Jordan	3.72		1.28
16	Singapore	5.47		0.94	68	Uganda	3.70		1.49
17	Chile	5.46		0.75	69	Russia	3.66		1.15
18	Germany	5.45		0.98	70	Egypt	3.66		1.43
19	Ireland	5.35		0.98	71	China	3.63		1.25
20	Norway	5.27		1.03	72	Serbia and Montenegro	3.62		1.15
21	Tunisia	5.26		0.95	73	Romania	3.58		1.45
22	New Zealand	5.22		1.05	74	Gambia	3.55		1.53
23	Israel	5.19		0.66	75	Nicaragua	3.53		1.57
24	Costa Rica	5.17		0.95	76	Ecuador	3.53		1.17
25	Hong Kong	5.10		1.34	77	Zambia	3.50		1.55
26	Iceland	5.08		1.06	78	Bulgaria	3.46		1.25
27	Estonia	4.96		1.02	79	Croatia	3.45		1.36
28	Brazil	4.87		1.06	80	Mauritius	3.44		1.46
29	Argentina	4.87		1.00	81	Botswana	3.43		1.51
30	Philippines	4.83		1.13	82	Paraguay	3.38		1.13
31	Malaysia	4.81		1.12	83	Zimbabwe	3.34		1.52
32	Taiwan	4.80		1.11	84	Pakistan	3.34		1.20
33	Italy	4.67		1.23	85	Mali	3.32		1.45
34	Morocco	4.67		1.54	86	United Arab Emirates	3.31		1.47
35	Slovenia	4.64		1.06	87	Macedonia	3.25		1.54
36	Mexico	4.59		1.07	88	Bosnia and Herzegovina	3.23		1.13
37	Japan	4.54		1.32	89	Algeria	3.22		1.27
38	Thailand	4.50		0.98	90	Sri Lanka	3.19		1.24
39	Jamaica	4.49		1.20	91	Luxembourg	3.17		1.50
40	Peru	4.49		0.96	92	Bolivia	3.15		1.00
41	Lithuania	4.49		1.06	93	Bahrain	3.12		1.44
42	Colombia	4.46		1.03	94	Malawi	3.06		1.37
43	Portugal	4.39		1.00	95	Tanzania	3.03		1.48
44	Czech Republic	4.36		1.09	96	Bangladesh	2.99		1.36
45	Hungary	4.35		1.07	97	Ethiopia	2.95		1.14
46	Latvia	4.34		1.12	98	Georgia	2.65		1.05
47	Uruguay	4.25		1.04	99	Mozambique	2.65		1.27
48	Venezuela	4.24		1.34	100	Vietnam	2.64		1.05
49	Trinidad and Tobago	4.21		1.09	101	Honduras	2.62		1.21
50	Kenya	4.19		1.23	102	Namibia	2.48		1.55
51	Slovak Republic	4.19		1.10	103	Chad	2.22		1.39
52	Ghana	4.16		1.51	104	Angola	1.69		1.02

Source: World Economic Forum, Executive Opinion Survey 2004

II. Readiness Component – Business Readiness
5.04 Business investment in R&D, 2004

Companies in your country (1 = do not spend money on research and development, 7 = spend heavily on research and development relative to international peers)

RANK	COUNTRY	SCORE	SD	RANK	COUNTRY	SCORE	SD
1	Japan	5.82	0.85	53	Slovak Republic	3.05	1.07
2	United States	5.76	1.19	54	Croatia	3.05	1.36
3	Germany	5.75	0.83	55	Madagascar	3.03	1.32
4	Sweden	5.74	0.93	56	Greece	3.03	0.96
5	Switzerland	5.65	1.10	57	Mexico	3.01	1.09
6	Finland	5.62	0.89	58	Colombia	3.00	1.12
7	Israel	5.06	1.29	58	Poland	3.00	0.94
8	Denmark	5.04	0.91	60	Latvia	2.99	1.16
9	Singapore	4.77	1.19	61	Trinidad and Tobago	2.99	1.12
10	Netherlands	4.71	1.11	62	Zimbabwe	2.97	1.19
11	United Kingdom	4.67	1.08	63	Portugal	2.95	0.80
12	Taiwan	4.64	1.11	64	Sri Lanka	2.95	1.20
13	Iceland	4.60	0.96	65	Jordan	2.95	1.16
14	Korea	4.59	1.25	66	Morocco	2.94	1.37
15	France	4.56	1.15	67	Ghana	2.92	1.32
16	Belgium	4.54	1.02	68	Venezuela	2.91	1.10
17	Luxembourg	4.37	1.35	69	Tanzania	2.90	1.75
18	Canada	4.31	1.19	70	Guatemala	2.89	1.15
19	Australia	4.26	1.23	71	Vietnam	2.88	1.13
20	Norway	4.22	1.24	72	Egypt	2.85	1.67
21	Slovenia	4.11	1.20	73	Gambia	2.85	1.74
22	Austria	4.10	1.34	74	Panama	2.85	1.28
23	Malaysia	4.10	1.12	75	Argentina	2.84	1.18
24	South Africa	4.03	1.07	76	Turkey	2.81	1.14
25	Ireland	3.98	1.35	77	Malawi	2.81	1.24
26	India	3.95	1.19	78	Italy	2.80	1.18
27	China	3.89	1.26	79	Serbia and Montenegro	2.77	1.15
28	Indonesia	3.87	1.03	80	Zambia	2.73	1.41
29	Hong Kong	3.85	1.11	81	Mozambique	2.72	1.22
30	New Zealand	3.80	1.14	82	Mali	2.68	1.19
31	Brazil	3.72	1.37	83	Georgia	2.64	1.35
32	Kenya	3.66	1.43	84	Malta	2.64	0.91
33	Costa Rica	3.60	1.23	85	Cyprus	2.63	1.24
34	Lithuania	3.56	1.31	86	Uruguay	2.63	0.91
35	Czech Republic	3.49	1.22	87	Peru	2.58	0.90
36	Russia	3.48	1.46	88	Bahrain	2.53	1.24
37	Tunisia	3.46	1.38	89	Macedonia	2.51	1.46
38	Uganda	3.42	1.57	90	Bulgaria	2.49	1.23
39	Estonia	3.39	1.16	91	Ecuador	2.49	1.11
40	Jamaica	3.33	1.20	92	El Salvador	2.47	1.04
41	Spain	3.32	1.06	93	Dominican Republic	2.44	1.01
42	Namibia	3.29	1.66	94	Algeria	2.42	1.23
43	Thailand	3.21	1.53	95	Nicaragua	2.36	1.02
44	Botswana	3.21	1.45	96	Honduras	2.34	0.99
45	United Arab Emirates	3.20	1.44	97	Bosnia and Herzegovina	2.29	1.10
46	Chile	3.16	1.16	98	Bangladesh	2.23	1.03
47	Nigeria	3.15	1.43	99	Bolivia	2.19	0.91
48	Ukraine	3.13	1.35	100	Paraguay	2.17	0.88
49	Philippines	3.12	1.25	101	Pakistan	2.08	1.18
50	Mauritius	3.12	1.17	102	Angola	1.93	1.18
51	Romania	3.09	1.28	103	Chad	1.91	1.21
52	Hungary	3.06	1.03	104	Ethiopia	1.85	1.00

Source: World Economic Forum, Executive Opinion Survey 2004

II. Readiness Component – Business Readiness
5.05 Business monthly telephone subscription, 2002–3

Business monthly telephone subscription charge (adjusted against GDP per capita/1,000)

* estimated data

RANK	COUNTRY	SCORE
1	United Arab Emirates	0.15
2	Luxembourg	0.26
3	Serbia and Montenegro	0.31
4	Korea	0.33
5	Singapore	0.33
6	Finland	0.36
7	Iceland	0.36
8	Switzerland	0.38
9	Germany	0.38
10	Norway	0.41
11	Bahrain	0.44
12	Israel	0.48
13	France	0.49
14	Belgium	0.52
15	Canada	* 0.53
15	Taiwan	* 0.53
17	Spain	0.54
18	Ireland	0.57
19	Cyprus	0.58
20	Slovenia	0.59
21	Greece	0.60
22	Japan	0.62
23	Austria	0.63
24	Australia	0.72
25	Hong Kong	0.73
26	Portugal	0.77
27	United Kingdom	0.79
28	Tunisia	0.81
29	Botswana	0.93
30	Zimbabwe	0.95
31	Estonia	0.95
32	Malta	1.01
33	Thailand	1.02
34	United States	1.17
35	Turkey	1.24
36	Costa Rica	1.33
37	New Zealand	1.40
38	Lithuania	1.42
39	Czech Republic	1.45
40	Algeria	1.50
41	Egypt	1.54
42	Mauritius	1.65
43	Croatia	1.71
44	Hungary	1.92
45	Chile	2.02
46	Romania	2.22
47	Latvia	2.35
48	Bosnia and Herzegovina	2.61
49	Argentina	2.64
50	Denmark	* 2.6
51	Sweden	* 2.74
52	Netherlands	* 2.74

RANK	COUNTRY	SCORE
53	Italy	* 2.77
54	Malaysia	2.80
55	Russia	* 2.82
56	Macedonia	* 2.87
57	Bulgaria	2.94
58	Brazil	* 3.01
59	Ukraine	3.04
60	Poland	* 3.20
61	Colombia	* 3.27
62	Slovak Republic	* 3.42
63	Ghana	3.50
64	Georgia	3.53
65	China	* 3.59
66	Trinidad and Tobago	3.63
67	Mexico	* 3.66
68	South Africa	3.68
69	Panama	* 3.76
70	Vietnam	3.92
71	El Salvador	* 4.01
72	Namibia	4.20
73	Indonesia	4.50
74	Uruguay	* 4.69
75	Guatemala	* 4.73
76	Sri Lanka	4.85
77	Venezuela	5.11
78	Jamaica	5.22
79	Ecuador	5.75
80	Zambia	5.79
81	Honduras	6.14
82	Paraguay	* 6.20
83	Bolivia	* 6.21
84	Gambia	6.68
85	Jordan	6.91
86	Bangladesh	7.09
87	Peru	7.14
88	Morocco	7.47
89	Dominican Republic	8.05
90	Nigeria	8.60
91	Malawi	9.07
92	Pakistan	9.22
93	India	9.53
94	Angola	11.55
95	Madagascar	11.93
96	Tanzania	13.92
97	Kenya	14.64
98	Chad	16.50
99	Ethiopia	21.13
100	Uganda	23.16
101	Mali	23.48
102	Philippines	25.81
103	Nicaragua	39.47
104	Mozambique	41.48

Source: International Telecommunication Union, June 2004

II. Readiness Component – Business Readiness
5.06 Business telephone connection charge, 2002–3

Business telephone connection charge (US$) (adjusted by GDP per capita/1,000)

* estimated data

RANK	COUNTRY	SCORE		RANK	COUNTRY	SCORE
1	Mexico	0.00		53	Japan	17.05
2	Singapore	0.79		54	Venezuela	17.21
3	Luxembourg	0.82		55	Bulgaria	21.78
4	Turkey	1.38		56	Namibia	23.44
5	Germany	1.44		57	Latvia	23.54
6	France	1.49		58	Dominican Republic	23.66
7	Greece	1.75		59	Tunisia	24.20
8	United States	1.94		60	Philippines	25.99
9	Norway	1.94		61	Algeria	26.23
10	New Zealand	1.98		62	Uruguay	* 28.12
11	United Arab Emirates	2.03		63	India	30.48
12	Switzerland	* 2.07		64	Brazil	* 33.41
13	Belgium	2.12		65	Thailand	34.21
14	Iceland	2.38		66	Hungary	34.77
15	Italy	* 2.59		67	Indonesia	39.68
16	Hong Kong	2.69		68	Macedonia	* 49.61
17	Austria	2.77		69	Honduras	52.23
18	Canada	* 2.85		70	El Salvador	* 57.95
19	Trinidad and Tobago	2.90		71	Guatemala	* 59.44
20	Malaysia	3.11		72	Colombia	* 61.11
21	Finland	3.24		73	Pakistan	65.33
22	Ireland	3.29		74	Peru	66.71
23	Israel	3.40		75	Kenya	67.35
24	Cyprus	3.46		76	Morocco	74.69
25	Korea	3.78		77	Bosnia and Herzegovina	77.48
26	Netherlands	* 3.81		78	Serbia and Montenegro	79.54
27	Sweden	* 4.01		79	Zambia	86.88
28	Bahrain	4.11		80	Jordan	87.99
29	Denmark	* 4.14		81	Mozambique	89.65
30	Romania	4.30		82	Ecuador	95.85
31	Spain	4.38		83	Madagascar	96.57
32	Australia	4.40		84	Bolivia	* 102.57
33	Taiwan	* 4.48		84	China	* 102.57
34	Portugal	4.64		84	Paraguay	* 102.57
35	Slovenia	5.39		87	Malawi	108.88
36	United Kingdom	5.72		88	Russia	113.90
37	Jamaica	6.64		89	Angola	115.53
38	Estonia	7.68		90	Georgia	117.76
39	Chile	7.95		91	Ukraine	122.90
40	Malta	9.35		92	Ghana	140.03
41	South Africa	9.77		93	Vietnam	145.26
42	Poland	* 10.13		94	Nigeria	154.86
43	Slovak Republic	* 10.67		95	Gambia	156.78
44	Panama	* 10.89		96	Tanzania	159.08
45	Costa Rica	10.96		97	Sri Lanka	161.80
46	Botswana	11.03		98	Egypt	199.83
47	Croatia	12.22		99	Chad	247.43
48	Lithuania	12.66		100	Uganda	254.73
49	Czech Republic	12.76		101	Mali	260.13
50	Zimbabwe	13.71		102	Ethiopia	379.05
51	Argentina	14.51		103	Bangladesh	486.83
52	Mauritius	15.74		104	Nicaragua	591.87

Source: International Telecommunication Union, June 2004

Government Readiness

II. Readiness Component – Government Readiness
6.01 Government prioritization of ICT, 2004

Information and communication technologies are an overall priority for the government (1 = strongly disagree, 7 = strongly agree)

RANK	COUNTRY	SCORE	SD	RANK	COUNTRY	SCORE	SD
1	Singapore	6.06	1.01	53	New Zealand	4.43	1.25
2	Japan	6.05	0.86	54	Brazil	4.42	1.44
3	United Arab Emirates	5.91	1.06	55	Hungary	4.39	1.33
4	Mauritius	5.76	1.02	56	Bangladesh	4.39	1.69
5	Denmark	5.75	1.22	57	Namibia	4.39	1.61
6	Taiwan	5.75	1.14	58	Romania	4.34	1.51
7	Pakistan	5.71	1.30	59	Czech Republic	4.30	1.67
8	Finland	5.71	1.25	60	Greece	4.27	1.28
9	India	5.71	1.18	61	Slovenia	4.26	1.37
10	Ghana	5.66	1.30	62	Belgium	4.25	1.38
11	Malta	5.48	1.27	62	Morocco	4.25	1.74
12	Tunisia	5.47	1.40	64	Mexico	4.19	1.40
13	Egypt	5.46	1.47	65	Indonesia	4.18	1.30
14	Estonia	5.43	1.35	66	Slovak Republic	4.05	1.33
15	Korea	5.36	1.07	67	Croatia	4.03	1.67
16	Thailand	5.33	1.28	68	Russia	4.00	1.68
17	Jordan	5.29	1.38	69	Kenya	3.98	1.61
18	Gambia	5.29	1.56	70	Uruguay	3.98	1.30
19	Hong Kong	5.28	1.32	71	Colombia	3.98	1.39
20	United States	5.21	1.23	72	Trinidad and Tobago	3.96	1.40
21	Iceland	5.18	1.33	73	Italy	3.91	1.64
22	Norway	5.17	1.11	74	Nigeria	3.90	1.79
23	Malaysia	5.16	1.22	75	Costa Rica	3.86	1.44
24	Ireland	5.13	1.20	76	Latvia	3.84	1.46
25	Bahrain	5.12	1.33	77	Cyprus	3.84	1.52
26	Sweden	5.10	1.29	78	Algeria	3.83	1.88
27	Israel	5.06	1.57	79	Turkey	3.76	1.61
28	United Kingdom	5.04	1.30	80	Bosnia and Herzegovina	3.76	1.96
29	Mali	4.98	1.53	81	El Salvador	3.76	1.81
30	Netherlands	4.92	1.14	82	Ukraine	3.72	1.54
31	Germany	4.91	1.42	83	Ethiopia	3.71	1.74
32	Uganda	4.88	1.65	84	Zambia	3.70	1.78
33	Vietnam	4.88	1.82	85	Bulgaria	3.70	1.75
34	Austria	4.88	1.37	86	Nicaragua	3.58	1.58
35	Luxembourg	4.87	1.25	87	Macedonia	3.56	1.81
36	Australia	4.86	1.27	88	Malawi	3.53	1.69
37	Portugal	4.85	1.08	89	Angola	3.46	1.86
38	Botswana	4.85	1.32	90	Serbia and Montenegro	3.46	1.58
39	Chile	4.84	1.21	91	Venezuela	3.43	1.73
40	Sri Lanka	4.83	1.53	92	Dominican Republic	3.39	1.77
41	South Africa	4.82	1.37	93	Guatemala	3.32	1.43
42	Canada	4.80	1.52	94	Georgia	3.27	1.58
43	China	4.74	1.42	95	Bolivia	3.12	1.56
44	Madagascar	4.74	1.62	96	Panama	3.12	1.50
45	France	4.73	1.41	97	Chad	3.12	2.04
46	Jamaica	4.71	1.50	98	Zimbabwe	3.07	1.57
47	Switzerland	4.70	1.27	99	Poland	2.94	1.54
48	Lithuania	4.67	1.42	100	Honduras	2.89	1.63
49	Tanzania	4.54	1.66	101	Ecuador	2.87	1.40
50	Philippines	4.53	1.58	102	Peru	2.78	1.18
51	Mozambique	4.49	1.75	103	Argentina	2.67	1.18
52	Spain	4.47	1.43	104	Paraguay	2.61	1.41

Source: World Economic Forum, Executive Opinion Survey 2004

II. Readiness Component – Government Readiness
6.02 Government procurement of ICT, 2004

Government purchase decisions for the procurement of advanced technology products are (1 = based solely on price, 7 = based on technical performance and innovativeness)

RANK	COUNTRY	SCORE		SD	RANK	COUNTRY	SCORE		SD
1	Singapore	5.20		1.13	53	Mali	3.69		1.60
2	Luxembourg	5.08		0.95	54	Gambia	3.67		1.80
3	Taiwan	4.97		1.18	55	Czech Republic	3.67		1.38
4	Japan	4.87		0.94	56	Romania	3.67		1.47
5	Malaysia	4.78		1.21	57	Russia	3.66		1.47
6	United Arab Emirates	4.78		1.47	58	Mauritius	3.63		1.35
7	Finland	4.77		1.10	59	Hungary	3.61		1.10
8	United States	4.75		1.14	60	Portugal	3.59		1.04
9	Germany	4.63		1.19	61	Malta	3.59		1.23
10	France	4.62		1.30	62	Sri Lanka	3.54		1.28
11	Switzerland	4.59		1.24	63	Mexico	3.54		1.33
12	China	4.57		1.30	64	India	3.51		1.28
13	Australia	4.46		1.16	65	Slovak Republic	3.50		1.33
14	Tunisia	4.46		1.40	66	Colombia	3.49		1.34
15	Hong Kong	4.42		1.36	67	Serbia and Montenegro	3.47		1.48
16	Bahrain	4.38		1.63	68	Greece	3.46		1.17
17	Korea	4.37		1.16	69	Uruguay	3.40		1.12
18	South Africa	4.31		1.07	70	Ethiopia	3.39		1.46
19	Spain	4.30		1.22	71	Malawi	3.38		1.13
20	Israel	4.29		1.82	72	El Salvador	3.35		1.45
21	Botswana	4.28		1.45	73	Turkey	3.35		1.53
22	Denmark	4.25		1.26	74	Kenya	3.28		1.36
22	Sweden	4.25		1.29	75	Egypt	3.27		1.79
24	United Kingdom	4.24		1.32	76	Costa Rica	3.27		1.46
25	Thailand	4.20		1.28	77	Italy	3.25		1.37
26	Indonesia	4.10		1.02	78	Macedonia	3.25		1.95
27	Netherlands	4.10		1.39	79	Bulgaria	3.22		1.52
28	Norway	4.10		1.33	80	Cyprus	3.16		1.39
29	Ireland	4.06		1.27	81	Mozambique	3.13		1.50
30	Iceland	4.05		1.51	82	Croatia	3.12		1.33
31	Austria	4.05		1.49	83	Angola	3.12		1.62
32	Estonia	4.00		1.24	83	Zambia	3.12		1.68
32	Uganda	4.00		1.62	85	Dominican Republic	3.11		1.42
34	Nigeria	3.97		1.65	86	Poland	3.09		1.24
35	Canada	3.97		1.28	87	Venezuela	3.09		1.33
36	Vietnam	3.96		1.60	88	Philippines	3.04		1.21
37	Madagascar	3.90		1.51	89	Argentina	3.02		1.27
38	Belgium	3.90		1.21	90	Guatemala	3.02		1.15
39	Jordan	3.88		1.50	91	Latvia	3.01		1.37
40	Trinidad and Tobago	3.86		1.18	92	Zimbabwe	2.96		1.10
41	Brazil	3.85		1.42	93	Nicaragua	2.95		1.52
42	Algeria	3.84		1.70	94	Ukraine	2.94		1.20
43	Pakistan	3.84		1.71	95	Bosnia and Herzegovina	2.93		1.47
44	Slovenia	3.83		1.18	96	Bangladesh	2.93		1.62
45	Lithuania	3.82		1.28	97	Georgia	2.88		1.39
46	New Zealand	3.80		1.22	98	Panama	2.87		1.55
47	Jamaica	3.79		1.47	99	Chad	2.74		1.61
48	Morocco	3.77		1.66	100	Honduras	2.67		1.22
49	Namibia	3.77		1.59	101	Ecuador	2.66		1.31
50	Chile	3.76		1.24	102	Paraguay	2.64		1.33
51	Tanzania	3.74		1.62	103	Peru	2.45		1.12
52	Ghana	3.73		1.48	104	Bolivia	2.37		1.34

0 1 2 3 4 5 6 7   0 1 2 3 4 5 6 7

Source: World Economic Forum, Executive Opinion Survey 2004

Individual Usage

III. Usage Component – Individual Usage
7.01 Cellular mobile subscribers, 2003

Cellular mobile telephone subscribers per 100 inhabitants

RANK	COUNTRY	SCORE
1	Luxembourg	106.05
2	Hong Kong	105.75
3	Italy	101.76
4	Iceland	96.56
5	Czech Republic	96.46
6	Israel	95.45
7	Spain	91.61
8	Norway	90.89
9	Portugal	90.38
10	Finland	90.06
11	Sweden	88.89
12	Denmark	88.72
13	Austria	87.88
14	Slovenia	87.09
15	Singapore	85.25
16	Ireland	84.47
17	Switzerland	84.34
18	United Kingdom	84.07
19	Belgium	78.56
20	Germany	78.54
21	Greece	78.00
22	Hungary	76.88
23	Netherlands	76.76
24	Cyprus	74.40
25	United Arab Emirates	73.57
26	Malta	72.50
27	Australia	71.95
28	France	69.59
29	Korea	69.37
30	Slovak Republic	68.42
31	Japan	67.96
32	Lithuania	66.62
33	Estonia	65.02
34	New Zealand	64.82
35	Bahrain	63.84
36	Croatia	58.37
37	United States	54.30
38	Jamaica	53.30
39	Latvia	52.86
40	Poland	45.09
41	Malaysia	44.20
42	Chile	42.83
43	Canada	41.68
44	Turkey	40.84
45	Mauritius	37.87
46	Bulgaria	37.06
47	South Africa	36.36
48	Serbia and Montenegro	33.78
49	Romania	32.87
50	Paraguay	29.85
51	Botswana	29.71
52	Mexico	29.11

RANK	COUNTRY	SCORE
53	Trinidad and Tobago	27.81
54	Bosnia and Herzegovina	27.40
55	Dominican Republic	27.14
56	Panama	26.76
57	Brazil	26.36
58	Thailand	26.04
59	Venezuela	25.64
60	Morocco	24.34
61	Jordan	24.18
62	China	21.40
63	Uruguay	19.26
64	Tunisia	19.21
65	Philippines	19.13
66	Ecuador	18.41
67	Argentina	17.76
68	Macedonia	17.70
69	El Salvador	17.65
70	Bolivia	16.67
71	Colombia	14.13
72	Guatemala	13.15
73	Russia	12.01
74	Namibia	11.63
75	Costa Rica	11.10
76	Georgia	10.68
77	Peru	10.61
78	Indonesia	8.74
79	Egypt	8.45
80	Ukraine	8.38
81	Gambia	7.53
82	Kenya	5.02
83	Sri Lanka	4.92
84	Honduras	4.87
85	Algeria	4.56
86	Nicaragua	3.78
87	Ghana	3.56
88	Vietnam	3.37
89	Zimbabwe	3.22
90	Uganda	3.03
91	Nigeria	2.55
92	Tanzania	2.52
93	India	2.47
94	Mali	2.30
95	Mozambique	2.28
96	Zambia	2.15
97	Pakistan	1.75
98	Madagascar	1.71
99	Malawi	1.29
100	Bangladesh	1.01
101	Angola	0.93
102	Chad	0.80
103	Ethiopia	0.14
104	Taiwan	0.00

Source: International Communication Union, June 2004

III. Usage Component – Individual Usage
7.02 Telephone subscribers, 2002

Telephone subscribers per 100 inhabitants

* estimated data

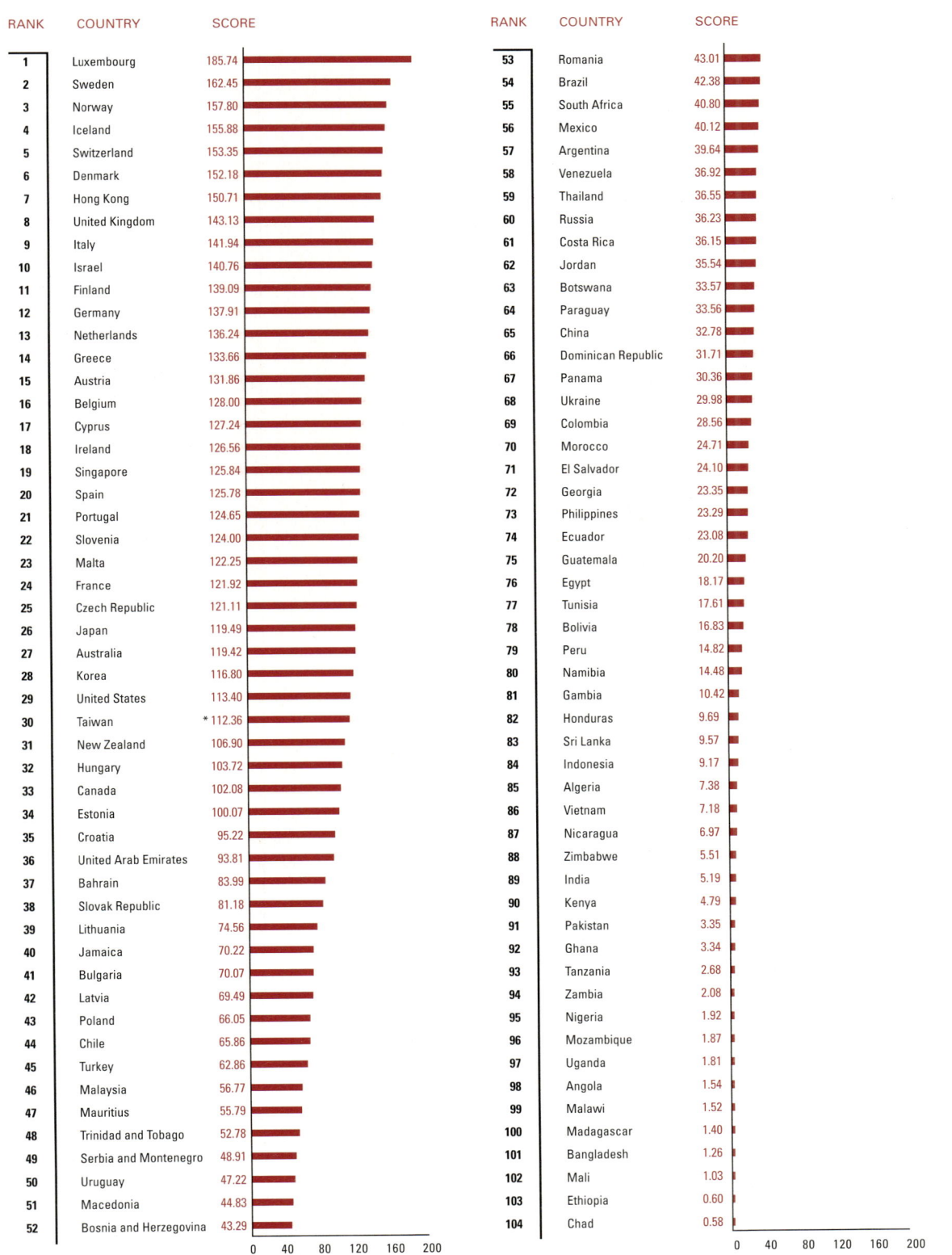

RANK	COUNTRY	SCORE		RANK	COUNTRY	SCORE
1	Luxembourg	185.74		53	Romania	43.01
2	Sweden	162.45		54	Brazil	42.38
3	Norway	157.80		55	South Africa	40.80
4	Iceland	155.88		56	Mexico	40.12
5	Switzerland	153.35		57	Argentina	39.64
6	Denmark	152.18		58	Venezuela	36.92
7	Hong Kong	150.71		59	Thailand	36.55
8	United Kingdom	143.13		60	Russia	36.23
9	Italy	141.94		61	Costa Rica	36.15
10	Israel	140.76		62	Jordan	35.54
11	Finland	139.09		63	Botswana	33.57
12	Germany	137.91		64	Paraguay	33.56
13	Netherlands	136.24		65	China	32.78
14	Greece	133.66		66	Dominican Republic	31.71
15	Austria	131.86		67	Panama	30.36
16	Belgium	128.00		68	Ukraine	29.98
17	Cyprus	127.24		69	Colombia	28.56
18	Ireland	126.56		70	Morocco	24.71
19	Singapore	125.84		71	El Salvador	24.10
20	Spain	125.78		72	Georgia	23.35
21	Portugal	124.65		73	Philippines	23.29
22	Slovenia	124.00		74	Ecuador	23.08
23	Malta	122.25		75	Guatemala	20.20
24	France	121.92		76	Egypt	18.17
25	Czech Republic	121.11		77	Tunisia	17.61
26	Japan	119.49		78	Bolivia	16.83
27	Australia	119.42		79	Peru	14.82
28	Korea	116.80		80	Namibia	14.48
29	United States	113.40		81	Gambia	10.42
30	Taiwan	* 112.36		82	Honduras	9.69
31	New Zealand	106.90		83	Sri Lanka	9.57
32	Hungary	103.72		84	Indonesia	9.17
33	Canada	102.08		85	Algeria	7.38
34	Estonia	100.07		86	Vietnam	7.18
35	Croatia	95.22		87	Nicaragua	6.97
36	United Arab Emirates	93.81		88	Zimbabwe	5.51
37	Bahrain	83.99		89	India	5.19
38	Slovak Republic	81.18		90	Kenya	4.79
39	Lithuania	74.56		91	Pakistan	3.35
40	Jamaica	70.22		92	Ghana	3.34
41	Bulgaria	70.07		93	Tanzania	2.68
42	Latvia	69.49		94	Zambia	2.08
43	Poland	66.05		95	Nigeria	1.92
44	Chile	65.86		96	Mozambique	1.87
45	Turkey	62.86		97	Uganda	1.81
46	Malaysia	56.77		98	Angola	1.54
47	Mauritius	55.79		99	Malawi	1.52
48	Trinidad and Tobago	52.78		100	Madagascar	1.40
49	Serbia and Montenegro	48.91		101	Bangladesh	1.26
50	Uruguay	47.22		102	Mali	1.03
51	Macedonia	44.83		103	Ethiopia	0.60
52	Bosnia and Herzegovina	43.29		104	Chad	0.58

Source: International Telecommunication Union, June 2004

III. Usage Component – Individual Usage
7.03 Public payphones, 2002

Public payphones per 1,000 inhabitants

* estimated data

RANK	COUNTRY	SCORE		RANK	COUNTRY	SCORE
1	Korea	10.81		53	Mexico	* 2.24
2	United Arab Emirates	9.43		54	Malta	2.19
3	China	7.56		55	Bosnia and Herzegovina	* 2.12
4	Brazil	7.38		56	United Kingdom	1.99
5	Greece	5.73		57	Estonia	1.90
6	Italy	* 5.26		58	India	1.88
7	Costa Rica	5.22		59	Lithuania	1.85
8	Canada	5.21		60	Indonesia	1.83
9	United States	5.09		61	Slovak Republic	* 1.79
10	Switzerland	5.00		62	Slovenia	1.78
11	Argentina	* 4.97		63	Algeria	* 1.78
12	Taiwan	* 4.91		64	Latvia	1.73
13	Malaysia	4.81		65	Trinidad and Tobago	* 1.70
14	New Zealand	4.80		66	Belgium	1.52
15	Chile	4.78		67	Namibia	* 1.44
16	Japan	4.57		68	Panama	* 1.40
17	Sweden	* 4.52		69	Jordan	1.40
18	Portugal	4.34		70	Ukraine	1.39
19	Hungary	4.09		71	Dominican Republic	1.35
20	Spain	* 4.09		72	Germany	1.33
21	Philippines	* 4.09		73	Russia	1.30
22	Venezuela	4.09		74	Hong Kong	1.29
23	Australia	4.06		75	Finland	1.25
24	Peru	4.03		76	Botswana	1.25
25	South Africa	3.98		77	Turkey	1.05
26	Guatemala	* 3.58		78	Luxembourg	0.96
27	Colombia	* 3.48		79	Serbia and Montenegro	0.94
28	Cyprus	3.45		80	Georgia	* 0.87
29	Thailand	* 3.38		81	Angola	* 0.85
30	Norway	* 3.38		82	Mali	* 0.84
31	Singapore	* 3.38		83	Pakistan	0.83
32	France	3.37		84	Gambia	* 0.80
33	Czech Republic	3.09		85	Egypt	0.68
34	Austria	3.04		86	Paraguay	* 0.67
35	Jamaica	* 2.98		87	Chad	* 0.55
36	Tunisia	2.91		88	Ecuador	0.38
37	Croatia	2.85		89	Honduras	0.37
38	Bolivia	* 2.72		90	Sri Lanka	0.35
39	Bahrain	2.72		91	Kenya	0.31
40	Denmark	* 2.71		92	Ghana	0.24
41	Netherlands	* 2.66		93	Mozambique	0.21
42	El Salvador	* 2.65		94	Zimbabwe	0.18
43	Uruguay	* 2.64		95	Uganda	0.13
44	Bulgaria	2.59		96	Vietnam	0.10
45	Morocco	2.54		97	Nicaragua	0.08
46	Poland	* 2.51		98	Zambia	0.08
47	Israel	2.50		99	Madagascar	0.06
48	Mauritius	2.43		100	Tanzania	0.05
49	Iceland	* 2.34		101	Ethiopia	0.05
50	Romania	2.31		102	Malawi	0.05
51	Macedonia	* 2.29		103	Nigeria	0.04
52	Ireland	2.27		104	Bangladesh	0.02

Source: International Telecommunication Union, June 2004

III. Usage Component – Individual Usage
7.04 Telephone lines, 2002

Telephone lines per 100 inhabitants

RANK	COUNTRY	SCORE		RANK	COUNTRY	SCORE
1	Luxembourg	79.68		53	Chile	23.04
2	Switzerland	74.42		54	Argentina	21.88
3	Sweden	73.57		55	Ukraine	21.61
4	Norway	73.44		56	China	20.92
5	Cyprus	67.50		57	Romania	20.48
6	Denmark	66.93		58	Colombia	20.03
7	Iceland	65.99		59	Malaysia	18.16
8	Germany	65.87		60	Jamaica	16.92
9	Canada	62.90		61	Mexico	15.77
10	United States	62.13		62	Georgia	13.30
11	Netherlands	61.43		63	Panama	12.87
12	United Kingdom	59.06		64	Egypt	12.73
13	Taiwan	59.00		65	Ecuador	11.91
14	France	56.60		66	Tunisia	11.77
15	Japan	55.83		67	El Salvador	11.55
16	Hong Kong	55.51		68	Dominican Republic	11.54
17	Australia	54.23		69	Jordan	11.36
18	Korea	53.83		70	Venezuela	11.27
19	Malta	52.07		71	South Africa	10.66
20	Belgium	49.44		72	Thailand	10.55
21	Finland	48.82		73	Botswana	7.49
22	Ireland	48.57		74	Bolivia	7.14
23	Italy	48.40		75	Guatemala	7.05
24	Austria	48.07		76	Algeria	6.93
25	Greece	45.43		77	Peru	6.71
26	Israel	45.30		78	Namibia	6.62
27	Singapore	45.03		79	Vietnam	5.41
28	New Zealand	44.77		80	Honduras	4.81
29	Spain	42.91		81	Sri Lanka	4.65
30	Croatia	41.72		82	India	4.63
31	Portugal	41.40		83	Paraguay	4.61
32	Slovenia	40.68		84	Philippines	4.17
33	Bulgaria	38.05		85	Morocco	4.05
34	Czech Republic	36.03		86	Indonesia	3.94
35	Estonia	35.06		87	Nicaragua	3.20
36	Hungary	34.86		88	Gambia	2.89
37	Poland	31.87		89	Pakistan	2.66
38	Mauritius	28.52		90	Zimbabwe	2.56
39	Latvia	28.34		91	Ghana	1.35
40	United Arab Emirates	28.11		92	Kenya	1.04
41	Uruguay	27.96		93	Malawi	0.81
42	Turkey	27.70		94	Zambia	0.79
43	Macedonia	27.13		95	Nigeria	0.69
44	Bahrain	26.76		96	Angola	0.67
45	Lithuania	25.31		97	Ethiopia	0.63
46	Costa Rica	25.05		98	Bangladesh	0.55
47	Trinidad and Tobago	24.98		99	Mali	0.53
48	Bosnia and Herzegovina	24.48		100	Mozambique	0.46
49	Serbia and Montenegro	24.27		101	Tanzania	0.42
50	Russia	24.22		102	Madagascar	0.36
51	Slovak Republic	24.08		103	Uganda	0.24
52	Brazil	24.05		104	Chad	0.15

0 10 20 30 40 50 60 70 80

0 10 20 30 40 50 60 70 80

International Telecommunication Union, June 2004

III. Usage Component – Individual Usage
7.05 Television sets, 2002

% of homes with a television

* estimated data

RANK	COUNTRY	SCORE		RANK	COUNTRY	SCORE
1	Turkey	108.45		53	Slovenia	90.51
2	Czech Republic	102.93		54	Bulgaria	* 90.01
3	Slovak Republic	100.00		54	Macedonia	* 90.01
4	Norway	99.97		56	Brazil	89.95
5	Switzerland	99.83		57	El Salvador	* 89.51
6	Japan	99.78		58	China	89.17
7	Hong Kong	99.56		59	Malaysia	88.91
8	Spain	99.49		60	Ecuador	88.80
9	Netherlands	99.42		61	Algeria	88.06
10	Belgium	99.33		62	Trinidad and Tobago	87.52
11	Singapore	98.60		63	Bosnia and Herzegovina	87.16
12	New Zealand	98.12		64	United Arab Emirates	85.91
13	Russia	98.03		65	Uruguay	* 83.97
14	United States	97.84		66	Argentina	* 83.87
15	Denmark	* 97.74		67	Venezuela	82.92
16	Austria	97.39		68	Dominican Republic	* 82.92
17	Ukraine	97.34		69	Vietnam	82.78
18	Cyprus	97.00		70	Guatemala	* 79.32
19	Ireland	96.91		71	Latvia	79.12
20	Iceland	96.80		72	Panama	77.42
21	Lithuania	96.70		73	Philippines	76.41
22	Australia	96.16		74	Costa Rica	* 76.37
23	Canada	* 95.49		75	Morocco	76.07
23	Italy	* 95.49		76	Georgia	75.76
25	Bahrain	95.42		77	Jamaica	70.04
26	United Kingdom	* 95.34		78	Bolivia	* 62.66
27	France	95.00		78	Paraguay	* 62.66
28	Chile	95.00		80	Indonesia	55.58
29	Egypt	94.80		81	South Africa	53.81
30	Hungary	* 94.54		82	Honduras	47.37
31	Sweden	93.91		83	Pakistan	39.34
32	Germany	93.88		84	Namibia	39.22
33	Luxembourg	93.79		85	Nicaragua	* 35.61
34	Mexico	93.56		86	India	32.00
35	Greece	* 93.52		87	Sri Lanka	31.64
36	Jordan	93.40		88	Bangladesh	29.00
37	Portugal	* 93.38		89	Zimbabwe	26.50
38	Malta	93.18		90	Zambia	25.99
39	Mauritius	93.00		91	Nigeria	25.60
40	Taiwan	* 92.67		92	Ghana	21.40
41	Israel	92.60		93	Kenya	17.08
42	Poland	92.33		94	Botswana	15.22
43	Croatia	* 92.12		95	Mali	14.80
44	Korea	92.07		96	Tanzania	14.24
45	Colombia	92.02		97	Gambia	12.37
46	Thailand	91.93		98	Angola	8.97
47	Serbia and Montenegro	91.81		99	Madagascar	7.86
48	Estonia	91.76		100	Mozambique	6.21
49	Tunisia	91.68		101	Uganda	6.19
50	Peru	* 91.52		102	Ethiopia	2.41
51	Finland	91.15		103	Malawi	2.31
52	Romania	* 90.83		104	Chad	2.29

Source: International Telecommunication Union, June 2004

III. Usage Component – Individual Usage
7.06 Broadband-DSL Internet subscribers, 2002–3

DSL Internet subscribers per 1,000 inhabitants

* estimated data

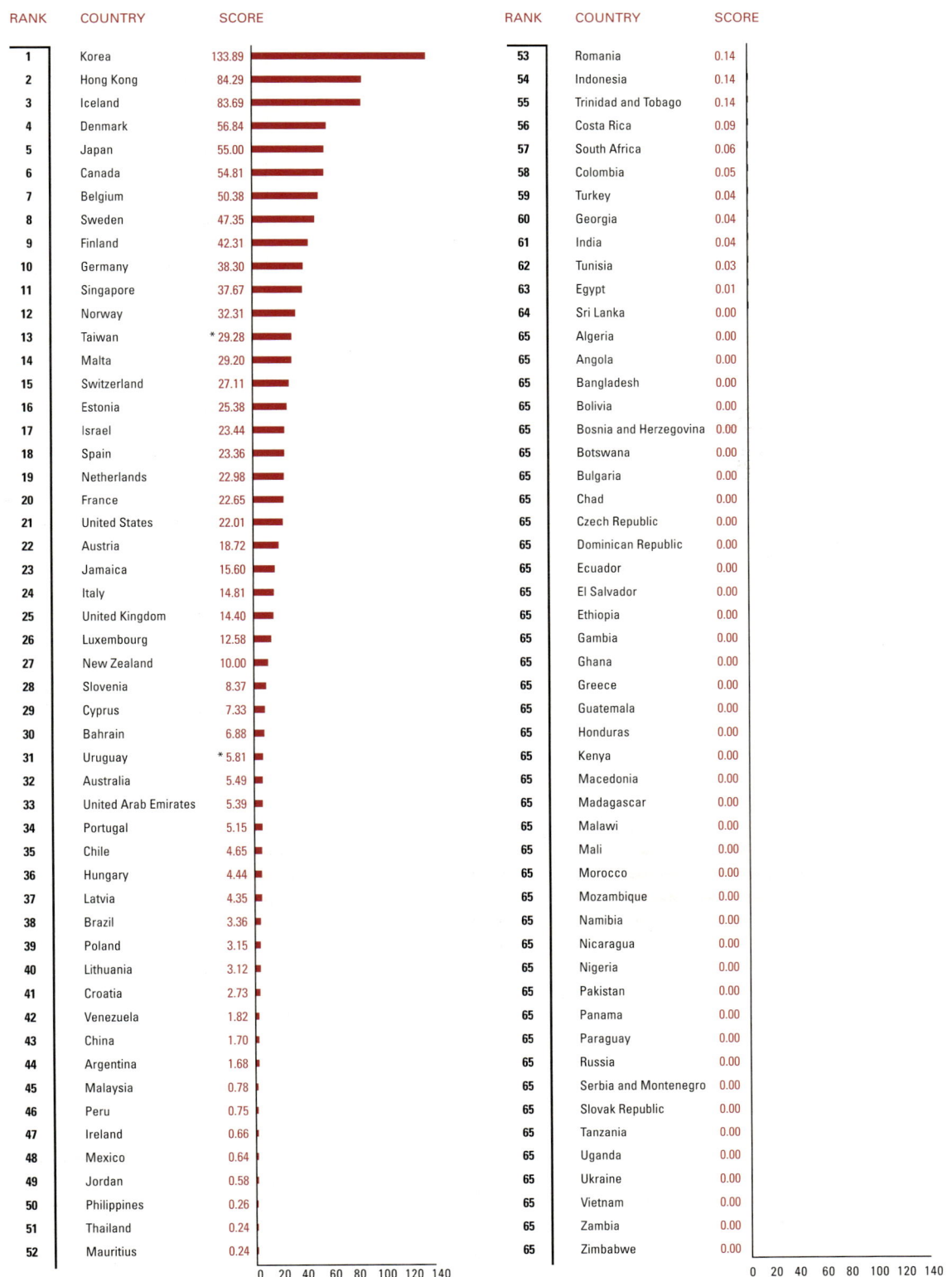

RANK	COUNTRY	SCORE		RANK	COUNTRY	SCORE
1	Korea	133.89		53	Romania	0.14
2	Hong Kong	84.29		54	Indonesia	0.14
3	Iceland	83.69		55	Trinidad and Tobago	0.14
4	Denmark	56.84		56	Costa Rica	0.09
5	Japan	55.00		57	South Africa	0.06
6	Canada	54.81		58	Colombia	0.05
7	Belgium	50.38		59	Turkey	0.04
8	Sweden	47.35		60	Georgia	0.04
9	Finland	42.31		61	India	0.04
10	Germany	38.30		62	Tunisia	0.03
11	Singapore	37.67		63	Egypt	0.01
12	Norway	32.31		64	Sri Lanka	0.00
13	Taiwan	* 29.28		65	Algeria	0.00
14	Malta	29.20		65	Angola	0.00
15	Switzerland	27.11		65	Bangladesh	0.00
16	Estonia	25.38		65	Bolivia	0.00
17	Israel	23.44		65	Bosnia and Herzegovina	0.00
18	Spain	23.36		65	Botswana	0.00
19	Netherlands	22.98		65	Bulgaria	0.00
20	France	22.65		65	Chad	0.00
21	United States	22.01		65	Czech Republic	0.00
22	Austria	18.72		65	Dominican Republic	0.00
23	Jamaica	15.60		65	Ecuador	0.00
24	Italy	14.81		65	El Salvador	0.00
25	United Kingdom	14.40		65	Ethiopia	0.00
26	Luxembourg	12.58		65	Gambia	0.00
27	New Zealand	10.00		65	Ghana	0.00
28	Slovenia	8.37		65	Greece	0.00
29	Cyprus	7.33		65	Guatemala	0.00
30	Bahrain	6.88		65	Honduras	0.00
31	Uruguay	* 5.81		65	Kenya	0.00
32	Australia	5.49		65	Macedonia	0.00
33	United Arab Emirates	5.39		65	Madagascar	0.00
34	Portugal	5.15		65	Malawi	0.00
35	Chile	4.65		65	Mali	0.00
36	Hungary	4.44		65	Morocco	0.00
37	Latvia	4.35		65	Mozambique	0.00
38	Brazil	3.36		65	Namibia	0.00
39	Poland	3.15		65	Nicaragua	0.00
40	Lithuania	3.12		65	Nigeria	0.00
41	Croatia	2.73		65	Pakistan	0.00
42	Venezuela	1.82		65	Panama	0.00
43	China	1.70		65	Paraguay	0.00
44	Argentina	1.68		65	Russia	0.00
45	Malaysia	0.78		65	Serbia and Montenegro	0.00
46	Peru	0.75		65	Slovak Republic	0.00
47	Ireland	0.66		65	Tanzania	0.00
48	Mexico	0.64		65	Uganda	0.00
49	Jordan	0.58		65	Ukraine	0.00
50	Philippines	0.26		65	Vietnam	0.00
51	Thailand	0.24		65	Zambia	0.00
52	Mauritius	0.24		65	Zimbabwe	0.00

International Telecommunication Union, June 2004

III. Usage Component – Individual Usage
7.07 Broadband-cable modem, 2002–3

Cable modem Internet subscribers per 1,000 inhabitants

* estimated data

RANK	COUNTRY	SCORE		RANK	COUNTRY	SCORE
1	Sweden	264.37		53	Zimbabwe	0.27
2	Korea	80.26		54	Georgia	0.20
3	Canada	78.83		55	Ecuador	0.14
4	Netherlands	57.76		56	Paraguay	0.08
5	United States	53.66		57	India	0.08
6	Switzerland	48.61		58	Sri Lanka	0.02
7	Denmark	45.11		59	Malawi	0.01
8	Austria	41.98		60	Bahrain	0.00
9	Singapore	38.53		60	Bangladesh	0.00
10	Hong Kong	36.86		60	Bolivia	0.00
11	Israel	34.38		60	Bosnia and Herzegovina	0.00
12	Belgium	34.32		60	Botswana	0.00
13	Portugal	31.25		60	Chad	0.00
14	Taiwan	* 25.23		60	Costa Rica	0.00
15	Malta	24.71		60	Croatia	0.00
16	Japan	20.19		60	Cyprus	0.00
17	Finland	16.42		60	Dominican Republic	0.00
18	United Kingdom	16.19		60	Egypt	0.00
19	Norway	15.50		60	Ethiopia	0.00
20	El Salvador	14.37		60	Gambia	0.00
21	Spain	13.91		60	Ghana	0.00
22	Australia	10.93		60	Greece	0.00
23	Slovenia	10.58		60	Guatemala	0.00
24	Estonia	9.77		60	Honduras	0.00
25	Hungary	7.80		60	Iceland	0.00
26	France	6.55		60	Italy	0.00
27	Chile	5.92		60	Jamaica	0.00
28	Lithuania	5.88		60	Jordan	0.00
29	Bulgaria	5.51		60	Kenya	0.00
30	Poland	* 4.44		60	Latvia	0.00
31	Mexico	* 4.07		60	Macedonia	0.00
32	Venezuela	2.62		60	Madagascar	0.00
33	Algeria	* 2.06		60	Malaysia	0.00
34	China	1.84		60	Mali	0.00
35	Thailand	1.79		60	Mauritius	0.00
36	Czech Republic	1.50		60	Morocco	0.00
37	Romania	1.46		60	Mozambique	0.00
38	Ireland	1.23		60	Namibia	0.00
39	New Zealand	1.15		60	Nigeria	0.00
40	Colombia	0.87		60	Pakistan	0.00
41	Uruguay	* 0.79		60	Panama	0.00
42	Brazil	0.73		60	Russia	0.00
43	Germany	0.73		60	Serbia and Montenegro	0.00
44	Argentina	* 0.67		60	South Africa	0.00
45	Slovak Republic	0.65		60	Tanzania	0.00
46	Turkey	0.58		60	Trinidad and Tobago	0.00
47	Peru	0.55		60	Tunisia	0.00
48	Nicaragua	0.42		60	Uganda	0.00
49	Angola	* 0.31		60	Ukraine	0.00
49	Indonesia	* 0.31		60	United Arab Emirates	0.00
51	Luxembourg	0.29		60	Vietnam	0.00
52	Philippines	* 0.28		60	Zambia	0.00

Source: International Telecommunication Union, June 2004

III. Usage Component – Individual Usage
7.08 Internet users, 2002

Internet users per 1,000 inhabitants, 2002

RANK	COUNTRY	SCORE	SD	RANK	COUNTRY	SCORE	SD
1	Iceland	64.79		53	Brazil	8.22	
2	Sweden	57.31		54	Bulgaria	8.08	
3	Korea	55.19		55	Thailand	7.76	
4	United States	55.14		56	Turkey	7.28	
5	Canada	51.28		57	South Africa	6.82	
6	Denmark	51.28		58	Dominican Republic	6.07	
7	Finland	50.89		59	Serbia and Montenegro	5.97	
8	Netherlands	50.63		60	Jordan	5.77	
9	Singapore	50.44		61	Tunisia	5.17	
10	Norway	50.26		62	Venezuela	5.06	
11	New Zealand	48.40		63	Macedonia	4.84	
12	Australia	48.17		64	El Salvador	4.65	
13	Japan	44.89		65	Colombia	4.62	
14	Germany	43.62		66	China	4.60	
15	Hong Kong	43.01		67	Philippines	4.40	
16	United Kingdom	42.31		68	Zimbabwe	4.30	
17	Austria	41.47		69	Ecuador	4.16	
18	Taiwan	38.14		70	Panama	4.14	
19	Slovenia	37.58		71	Russia	4.09	
20	Luxembourg	37.00		72	Botswana	3.49	
21	Italy	35.24		73	Guatemala	3.33	
22	Switzerland	35.10		74	Bolivia	3.24	
23	Belgium	32.83		75	Egypt	2.82	
24	Estonia	32.77		76	Namibia	2.67	
25	Malaysia	31.97		77	Bosnia and Herzegovina	2.62	
26	France	31.38		78	Honduras	2.52	
27	Malta	30.30		79	Morocco	2.36	
28	Israel	30.14		80	Indonesia	2.12	
29	Cyprus	29.37		81	Gambia	1.88	
30	Ireland	28.03		82	Vietnam	1.85	
31	United Arab Emirates	27.09		83	Ukraine	1.80	
32	Czech Republic	25.63		84	Paraguay	1.73	
33	Bahrain	24.56		85	Nicaragua	1.68	
34	Chile	23.75		86	Algeria	1.60	
35	Poland	23.00		87	India	1.59	
36	Jamaica	22.84		88	Georgia	1.49	
37	Portugal	19.35		89	Kenya	1.27	
38	Spain	19.31		90	Sri Lanka	1.06	
39	Costa Rica	19.31		91	Pakistan	1.03	
40	Croatia	18.04		92	Ghana	0.78	
41	Slovak Republic	16.04		93	Zambia	0.48	
42	Hungary	15.76		94	Uganda	0.40	
43	Lithuania	14.44		95	Nigeria	0.35	
44	Greece	13.48		96	Madagascar	0.35	
45	Latvia	13.31		97	Angola	0.29	
46	Uruguay	11.90		98	Mozambique	0.28	
47	Argentina	11.20		99	Malawi	0.26	
48	Trinidad and Tobago	10.60		100	Mali	0.24	
49	Mauritius	10.33		101	Tanzania	0.23	
50	Romania	10.15		102	Chad	0.19	
51	Mexico	9.85		103	Bangladesh	0.15	
52	Peru	8.97		104	Ethiopia	0.07	

Source: International Telecommunication Union (WSIS 2003)

Business Usage

III. Usage Component – Business Usage
8.01 Prevalence of foreign technology licensing, 2004

In your country, licensing foreign technology is (1 = uncommon, 7 = a common means of acquiring new technology)

RANK	COUNTRY	SCORE		SD	RANK	COUNTRY	SCORE		SD
1	Singapore	5.83		0.99	53	Romania	4.58		1.67
2	South Africa	5.83		1.02	54	Botswana	4.57		1.83
3	Hong Kong	5.74		1.19	55	Uganda	4.56		1.78
4	Taiwan	5.69		1.02	56	El Salvador	4.54		1.47
5	Australia	5.69		0.90	57	Austria	4.53		1.46
6	New Zealand	5.65		0.84	58	Egypt	4.49		1.73
7	Japan	5.62		1.37	59	China	4.49		1.50
8	India	5.54		1.08	60	Latvia	4.48		1.49
9	United Arab Emirates	5.49		1.48	61	France	4.46		1.42
10	Germany	5.45		1.25	62	Ghana	4.45		1.75
11	Switzerland	5.35		1.35	63	Korea	4.42		1.43
12	Thailand	5.33		1.19	64	Cyprus	4.33		1.54
13	Netherlands	5.31		1.24	65	Tanzania	4.32		1.92
14	Bahrain	5.28		1.67	66	Nigeria	4.24		1.73
15	United Kingdom	5.27		1.17	67	Pakistan	4.21		2.01
16	Malaysia	5.27		0.91	68	Costa Rica	4.13		1.60
17	Iceland	5.26		1.05	69	Venezuela	4.12		1.69
18	Jordan	5.22		1.38	70	Malawi	4.11		1.74
19	Denmark	5.21		1.35	71	Indonesia	4.10		1.25
20	Canada	5.17		1.34	72	Argentina	4.07		1.67
21	Czech Republic	5.16		1.24	73	Poland	4.06		1.51
22	Belgium	5.13		1.09	73	Zambia	4.06		1.72
23	Finland	5.13		1.24	75	Sri Lanka	4.05		1.54
24	Brazil	5.08		1.30	76	Zimbabwe	4.03		1.45
25	Portugal	5.07		1.17	77	Morocco	4.02		1.66
26	Sweden	5.05		1.64	78	Dominican Republic	3.88		1.37
27	Panama	5.05		1.34	79	Guatemala	3.86		1.54
28	United States	5.04		1.54	80	Bangladesh	3.80		1.86
29	Spain	5.02		1.40	81	Algeria	3.78		1.74
30	Tunisia	5.01		1.28	82	Colombia	3.76		1.37
31	Slovak Republic	4.92		1.36	83	Gambia	3.75		2.01
32	Norway	4.91		1.54	84	Peru	3.68		1.39
33	Turkey	4.88		1.49	85	Mozambique	3.67		1.87
34	Kenya	4.84		1.34	86	Uruguay	3.64		1.44
35	Ireland	4.82		1.34	87	Russia	3.60		1.40
36	Namibia	4.81		1.40	88	Ukraine	3.39		1.56
37	Chile	4.77		1.31	89	Bulgaria	3.38		1.46
38	Jamaica	4.76		1.78	90	Bosnia and Herzegovina	3.36		1.78
39	Mexico	4.76		1.45	91	Serbia and Montenegro	3.32		1.62
40	Israel	4.75		1.06	92	Ecuador	3.32		1.39
41	Greece	4.72		1.27	93	Macedonia	3.26		1.94
42	Hungary	4.72		1.30	94	Honduras	3.25		1.52
43	Croatia	4.72		1.50	95	Georgia	3.22		1.74
44	Lithuania	4.70		1.38	96	Madagascar	3.21		1.80
45	Slovenia	4.70		1.31	97	Mali	3.08		1.73
46	Philippines	4.70		1.54	98	Angola	2.98		1.74
47	Luxembourg	4.68		1.38	99	Vietnam	2.97		1.61
48	Mauritius	4.67		1.45	100	Paraguay	2.84		1.27
49	Estonia	4.65		1.53	101	Ethiopia	2.76		1.57
50	Italy	4.64		1.57	102	Nicaragua	2.66		1.34
51	Trinidad and Tobago	4.62		1.51	103	Bolivia	2.43		1.33
52	Malta	4.62		1.39	104	Chad	2.24		1.70

0 1 2 3 4 5 6 7

0 1 2 3 4 5 6 7

Source: World Economic Forum, Executive Opinion Survey 2004

III. Usage Component – Business Usage
8.02 Firm-level technology absorption, 2004

Companies in your country are (1 = not interested in absorbing new technology, 7 = aggressive in absorbing new technology)

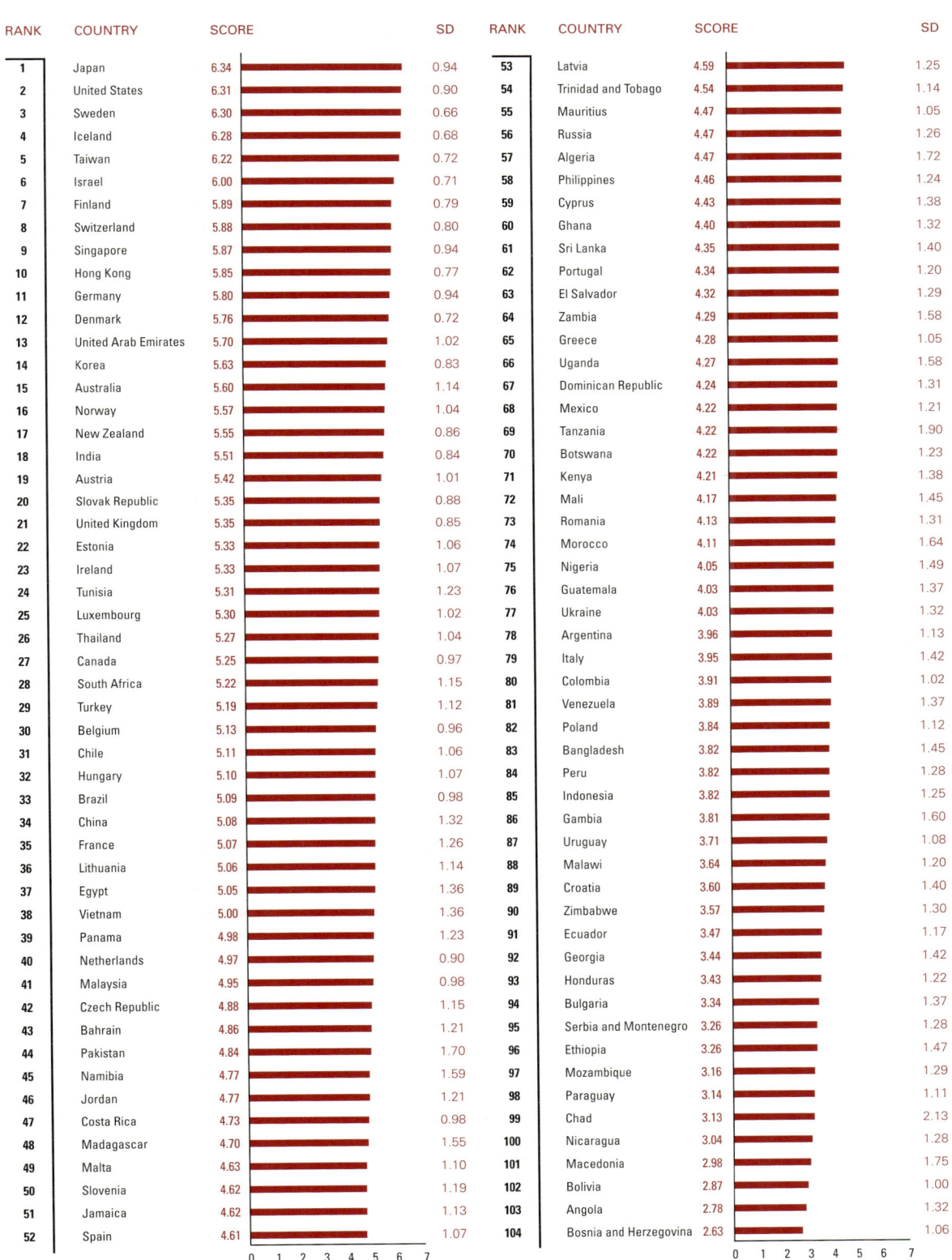

RANK	COUNTRY	SCORE	SD	RANK	COUNTRY	SCORE	SD
1	Japan	6.34	0.94	53	Latvia	4.59	1.25
2	United States	6.31	0.90	54	Trinidad and Tobago	4.54	1.14
3	Sweden	6.30	0.66	55	Mauritius	4.47	1.05
4	Iceland	6.28	0.68	56	Russia	4.47	1.26
5	Taiwan	6.22	0.72	57	Algeria	4.47	1.72
6	Israel	6.00	0.71	58	Philippines	4.46	1.24
7	Finland	5.89	0.79	59	Cyprus	4.43	1.38
8	Switzerland	5.88	0.80	60	Ghana	4.40	1.32
9	Singapore	5.87	0.94	61	Sri Lanka	4.35	1.40
10	Hong Kong	5.85	0.77	62	Portugal	4.34	1.20
11	Germany	5.80	0.94	63	El Salvador	4.32	1.29
12	Denmark	5.76	0.72	64	Zambia	4.29	1.58
13	United Arab Emirates	5.70	1.02	65	Greece	4.28	1.05
14	Korea	5.63	0.83	66	Uganda	4.27	1.58
15	Australia	5.60	1.14	67	Dominican Republic	4.24	1.31
16	Norway	5.57	1.04	68	Mexico	4.22	1.21
17	New Zealand	5.55	0.86	69	Tanzania	4.22	1.90
18	India	5.51	0.84	70	Botswana	4.22	1.23
19	Austria	5.42	1.01	71	Kenya	4.21	1.38
20	Slovak Republic	5.35	0.88	72	Mali	4.17	1.45
21	United Kingdom	5.35	0.85	73	Romania	4.13	1.31
22	Estonia	5.33	1.06	74	Morocco	4.11	1.64
23	Ireland	5.33	1.07	75	Nigeria	4.05	1.49
24	Tunisia	5.31	1.23	76	Guatemala	4.03	1.37
25	Luxembourg	5.30	1.02	77	Ukraine	4.03	1.32
26	Thailand	5.27	1.04	78	Argentina	3.96	1.13
27	Canada	5.25	0.97	79	Italy	3.95	1.42
28	South Africa	5.22	1.15	80	Colombia	3.91	1.02
29	Turkey	5.19	1.12	81	Venezuela	3.89	1.37
30	Belgium	5.13	0.96	82	Poland	3.84	1.12
31	Chile	5.11	1.06	83	Bangladesh	3.82	1.45
32	Hungary	5.10	1.07	84	Peru	3.82	1.28
33	Brazil	5.09	0.98	85	Indonesia	3.82	1.25
34	China	5.08	1.32	86	Gambia	3.81	1.60
35	France	5.07	1.26	87	Uruguay	3.71	1.08
36	Lithuania	5.06	1.14	88	Malawi	3.64	1.20
37	Egypt	5.05	1.36	89	Croatia	3.60	1.40
38	Vietnam	5.00	1.36	90	Zimbabwe	3.57	1.30
39	Panama	4.98	1.23	91	Ecuador	3.47	1.17
40	Netherlands	4.97	0.90	92	Georgia	3.44	1.42
41	Malaysia	4.95	0.98	93	Honduras	3.43	1.22
42	Czech Republic	4.88	1.15	94	Bulgaria	3.34	1.37
43	Bahrain	4.86	1.21	95	Serbia and Montenegro	3.26	1.28
44	Pakistan	4.84	1.70	96	Ethiopia	3.26	1.47
45	Namibia	4.77	1.59	97	Mozambique	3.16	1.29
46	Jordan	4.77	1.21	98	Paraguay	3.14	1.11
47	Costa Rica	4.73	0.98	99	Chad	3.13	2.13
48	Madagascar	4.70	1.55	100	Nicaragua	3.04	1.28
49	Malta	4.63	1.10	101	Macedonia	2.98	1.75
50	Slovenia	4.62	1.19	102	Bolivia	2.87	1.00
51	Jamaica	4.62	1.13	103	Angola	2.78	1.32
52	Spain	4.61	1.07	104	Bosnia and Herzegovina	2.63	1.06

Source: World Economic Forum, Executive Opinion Survey 2004

III. Usage Component – Business Usage
8.03 Capacity for innovation, 2004

Companies obtain technologies (1 = exclusively from licensing or imitating foreign companies, 7 = by conducting formal research and pioneering their own new products and processes)

RANK	COUNTRY	SCORE		SD	RANK	COUNTRY	SCORE		SD
1	Germany	6.28		0.60	53	Egypt	3.15		1.47
2	Sweden	6.18		0.73	54	Mexico	3.15		1.38
3	Japan	5.97		1.05	55	Greece	3.11		1.19
4	Finland	5.92		0.68	56	Chile	3.11		1.23
5	France	5.87		0.81	57	Turkey	3.10		1.29
6	Switzerland	5.87		0.69	58	Gambia	3.10		1.66
7	United States	5.80		1.03	59	Mauritius	3.06		1.14
8	Israel	5.69		1.20	60	Croatia	3.06		1.34
9	Denmark	5.50		0.66	61	Panama	3.01		1.41
10	United Kingdom	5.42		0.99	62	Uganda	3.01		1.62
11	Netherlands	5.39		0.74	63	Jordan	2.97		1.17
12	Belgium	5.19		0.78	64	Sri Lanka	2.95		1.29
13	Austria	5.11		0.91	65	Macedonia	2.94		1.50
14	Taiwan	4.95		1.11	66	Jamaica	2.92		1.16
15	Korea	4.76		1.06	67	Kenya	2.90		1.22
16	Iceland	4.61		1.62	68	Botswana	2.89		1.65
17	Luxembourg	4.55		1.40	69	Trinidad and Tobago	2.81		1.27
18	Singapore	4.55		1.32	70	Serbia and Montenegro	2.80		1.24
19	Canada	4.52		1.07	71	Zambia	2.79		1.49
20	Norway	4.48		0.85	72	El Salvador	2.79		1.26
21	Italy	4.43		1.32	73	Cyprus	2.79		1.20
22	Slovenia	4.39		1.01	74	Dominican Republic	2.79		1.18
23	Ireland	4.38		1.17	75	Georgia	2.78		1.20
24	Australia	4.29		1.10	76	Uruguay	2.78		1.10
25	New Zealand	4.16		1.10	77	United Arab Emirates	2.76		1.42
26	Hong Kong	4.10		1.50	78	Malta	2.76		0.95
27	Spain	4.07		1.03	79	Bosnia and Herzegovina	2.76		1.43
28	Czech Republic	4.00		1.16	80	Philippines	2.75		1.06
28	Indonesia	4.00		1.07	81	Madagascar	2.72		1.33
28	Ukraine	4.00		1.18	82	Peru	2.71		1.07
31	China	3.95		1.22	83	Mali	2.69		1.43
32	Malaysia	3.88		1.34	84	Ghana	2.69		1.46
33	India	3.84		1.20	85	Argentina	2.69		1.05
34	Pakistan	3.82		1.72	86	Venezuela	2.66		1.21
35	Russia	3.82		1.27	87	Guatemala	2.64		1.17
36	Hungary	3.71		1.16	88	Namibia	2.60		1.57
37	Brazil	3.69		1.21	89	Honduras	2.59		1.20
38	Costa Rica	3.63		1.20	90	Bulgaria	2.52		1.24
39	Estonia	3.62		1.08	91	Algeria	2.47		1.23
40	Morocco	3.52		1.47	92	Bahrain	2.46		1.38
41	Slovak Republic	3.52		1.11	93	Malawi	2.45		1.23
42	South Africa	3.51		1.17	94	Ecuador	2.45		1.04
43	Portugal	3.50		1.08	95	Tanzania	2.44		1.26
44	Lithuania	3.49		1.13	96	Bolivia	2.37		1.10
45	Romania	3.36		1.37	97	Ethiopia	2.33		1.16
46	Poland	3.36		1.28	98	Paraguay	2.23		0.97
47	Vietnam	3.34		1.24	99	Zimbabwe	2.17		0.99
48	Tunisia	3.34		1.45	100	Mozambique	2.11		1.30
49	Latvia	3.31		1.24	101	Chad	2.04		1.28
50	Nigeria	3.21		1.61	102	Bangladesh	2.01		0.94
51	Colombia	3.16		0.90	103	Nicaragua	1.88		0.85
52	Thailand	3.15		1.13	104	Angola	1.83		1.12

0 1 2 3 4 5 6 7   0 1 2 3 4 5 6 7

Source: World Economic Forum, Executive Opinion Survey 2004

III. Usage Component – Business Usage
8.04 Availability of new telephone lines, 2004

New telephone lines for your businesses are (1 = not available, 2 = as accessible and affordable as in the world's most technologically advanced countries)

RANK	COUNTRY	SCORE
1	Iceland	6.96
2	Hong Kong	6.90
3	Denmark	6.88
4	Switzerland	6.88
5	Finland	6.84
6	Singapore	6.84
7	Israel	6.82
8	France	6.81
9	Sweden	6.80
10	Netherlands	6.78
11	Norway	6.78
12	Japan	6.78
13	Germany	6.77
14	New Zealand	6.69
15	Chile	6.66
16	Canada	6.66
17	El Salvador	6.61
18	Austria	6.59
19	United Kingdom	6.59
20	United Arab Emirates	6.57
21	Cyprus	6.56
22	Korea	6.55
23	Taiwan	6.54
24	Belgium	6.53
25	Dominican Republic	6.52
26	Bahrain	6.50
27	United States	6.49
28	Luxembourg	6.43
29	Uruguay	6.41
30	Hungary	6.39
31	Portugal	6.39
32	Estonia	6.38
33	Greece	6.32
34	Australia	6.30
35	Brazil	6.29
36	Ireland	6.28
37	Jordan	6.25
38	Slovak Republic	6.24
39	Peru	6.21
40	Czech Republic	6.19
41	Tunisia	6.19
42	Slovenia	6.18
43	Lithuania	6.18
44	Malta	6.16
45	Morocco	6.14
46	India	6.00
47	Thailand	5.94
48	Latvia	5.88
49	Turkey	5.86
50	Spain	5.85
51	Croatia	5.85
52	Argentina	5.81

RANK	COUNTRY	SCORE
53	Namibia	5.74
54	Guatemala	5.73
55	Panama	5.72
56	Egypt	5.70
57	Macedonia	5.69
58	Sri Lanka	5.62
59	Vietnam	5.62
60	China	5.59
61	Italy	5.53
62	Colombia	5.51
63	Malaysia	5.49
64	Venezuela	5.42
65	Mauritius	5.41
66	Georgia	5.41
67	Mexico	5.35
68	Bulgaria	5.16
69	Bolivia	5.14
70	South Africa	5.10
71	Philippines	5.03
72	Uganda	5.03
73	Bosnia and Herzegovina	4.93
74	Russia	4.93
75	Jamaica	4.88
76	Romania	4.87
77	Botswana	4.86
78	Mozambique	4.77
79	Tanzania	4.70
80	Nigeria	4.48
81	Gambia	4.48
82	Zambia	4.47
83	Pakistan	4.47
84	Ukraine	4.43
85	Indonesia	4.41
86	Poland	4.32
87	Serbia and Montenegro	4.23
88	Trinidad and Tobago	4.06
89	Paraguay	3.84
90	Nicaragua	3.84
91	Madagascar	3.79
92	Mali	3.56
93	Chad	3.52
94	Algeria	3.52
95	Kenya	3.42
96	Ethiopia	3.36
97	Malawi	3.31
98	Costa Rica	3.30
99	Ecuador	3.30
100	Ghana	3.25
101	Angola	3.24
102	Zimbabwe	2.57
103	Bangladesh	2.37
104	Honduras	2.02

Source: World Economic Forum, Executive Opinion Survey 2004

III. Usage Component – Business Usage
8.05 Availability of cellular phones, 2004

Mobile or cellular telephones for your business are (1 = not available, 2 = as accessible and affordable as in the world's most technologically advanced countries)

RANK	COUNTRY	SCORE		RANK	COUNTRY	SCORE
1	Israel	7.00		53	Malta	6.33
1	Norway	7.00		54	Italy	6.29
3	Iceland	6.96		55	Morocco	6.28
4	Hong Kong	6.93		56	Paraguay	6.27
5	Denmark	6.92		57	Thailand	6.25
6	Austria	6.89		57	Vietnam	6.25
7	Finland	6.89		59	Uruguay	6.22
8	Germany	6.88		60	Argentina	6.19
9	Dominican Republic	6.86		61	Nicaragua	6.19
10	Czech Republic	6.85		62	Sri Lanka	6.16
11	France	6.84		63	Tunisia	6.09
12	Switzerland	6.81		64	Pakistan	6.08
13	Estonia	6.80		65	Botswana	6.08
13	Sweden	6.80		66	Mauritius	6.06
15	Netherlands	6.79		67	Bolivia	6.02
16	Korea	6.78		68	Romania	6.01
17	Singapore	6.78		69	Russia	6.01
18	Japan	6.77		70	Honduras	5.98
19	Belgium	6.76		71	Poland	5.98
20	Chile	6.76		72	Colombia	5.98
21	United Kingdom	6.74		73	Ireland	5.98
22	Portugal	6.73		74	Bulgaria	5.95
23	New Zealand	6.71		75	Namibia	5.94
24	Slovenia	6.70		76	Mexico	5.85
25	Bahrain	6.68		77	Uganda	5.85
26	Greece	6.66		78	Bangladesh	5.83
27	El Salvador	6.64		79	Serbia and Montenegro	5.79
28	Cyprus	6.62		80	Zambia	5.77
29	Lithuania	6.61		81	Macedonia	5.77
30	United Arab Emirates	6.59		82	Ecuador	5.73
31	Canada	6.59		83	Malaysia	5.71
32	Brazil	6.57		84	Gambia	5.69
33	Hungary	6.56		85	Ghana	5.63
34	Luxembourg	6.53		86	Mozambique	5.58
34	Slovak Republic	6.53		87	Egypt	5.54
36	Spain	6.53		88	Tanzania	5.45
37	Australia	6.52		89	China	5.42
38	South Africa	6.51		90	Kenya	5.28
39	Philippines	6.47		91	Malawi	5.28
40	Peru	6.47		92	Madagascar	5.25
41	Guatemala	6.47		93	Trinidad and Tobago	5.21
42	Turkey	6.43		94	Ukraine	5.18
43	Jordan	6.42		95	Nigeria	5.12
44	Georgia	6.42		96	Bosnia and Herzegovina	5.12
45	Panama	6.41		97	Algeria	4.72
46	India	6.41		98	Indonesia	4.68
47	Venezuela	6.40		99	Costa Rica	4.66
48	United States	6.40		100	Mali	4.66
49	Latvia	6.38		101	Zimbabwe	4.57
50	Croatia	6.38		102	Angola	4.40
51	Jamaica	6.38		103	Chad	4.29
52	Taiwan	6.34		104	Ethiopia	3.68

Source: World Economic Forum, Executive Opinion Survey 2004

Government Usage

III. Usage Component – Government Usage
9.01 Government success in ICT promotion, 2004

Government programs promoting the use of information and communication technologies are (1 = not very successful, 7 = highly successful)

RANK	COUNTRY	SCORE	SD	RANK	COUNTRY	SCORE	SD
1	Singapore	5.75	0.92	53	Slovenia	4.00	1.22
2	United Arab Emirates	5.58	0.96	54	Greece	3.97	1.20
3	Tunisia	5.28	1.39	55	Lithuania	3.95	1.22
4	Taiwan	5.24	1.04	56	Sri Lanka	3.92	1.42
5	Malta	5.21	1.09	57	Algeria	3.91	1.61
6	Estonia	5.04	1.14	58	Nigeria	3.90	1.68
7	Finland	5.02	1.21	59	Netherlands	3.90	1.28
8	Malaysia	4.96	0.97	60	Philippines	3.81	1.39
9	Jordan	4.93	1.22	61	Namibia	3.77	1.61
10	Denmark	4.92	1.06	62	Colombia	3.74	1.33
11	India	4.91	1.24	63	Belgium	3.74	1.12
12	Thailand	4.88	1.23	64	Croatia	3.70	1.38
13	Hong Kong	4.85	1.25	65	Kenya	3.66	1.45
14	Korea	4.84	1.17	66	Italy	3.62	1.52
15	Pakistan	4.79	1.31	67	Cyprus	3.58	1.37
16	Egypt	4.78	1.61	68	New Zealand	3.54	1.24
17	Japan	4.75	1.20	69	Spain	3.50	1.40
18	Ghana	4.73	1.34	70	Trinidad and Tobago	3.48	1.25
19	Ireland	4.59	1.21	71	Uruguay	3.47	1.15
20	Iceland	4.59	1.37	72	Slovak Republic	3.42	1.20
21	Israel	4.57	1.16	73	Malawi	3.40	1.61
22	Mauritius	4.56	1.33	74	Mexico	3.40	1.27
23	China	4.56	1.35	75	Czech Republic	3.38	1.53
24	Mali	4.54	1.46	76	Ukraine	3.33	1.25
25	Gambia	4.51	1.66	77	Costa Rica	3.33	1.25
26	France	4.49	1.40	78	Russia	3.30	1.33
27	Portugal	4.44	1.05	79	Angola	3.27	1.53
28	Vietnam	4.41	1.51	80	Zambia	3.27	1.68
29	Austria	4.39	1.44	81	Turkey	3.26	1.49
30	Uganda	4.39	1.54	82	El Salvador	3.25	1.51
31	Australia	4.39	1.10	83	Latvia	3.24	1.28
32	Madagascar	4.36	1.52	84	Bangladesh	3.19	1.44
33	Indonesia	4.36	1.09	85	Ethiopia	3.19	1.54
34	United States	4.35	1.20	86	Serbia and Montenegro	3.01	1.16
35	Bahrain	4.32	1.46	87	Bulgaria	2.99	1.42
36	Morocco	4.31	1.61	88	Panama	2.98	1.35
37	Hungary	4.30	1.31	89	Bosnia and Herzegovina	2.91	1.37
38	United Kingdom	4.29	1.20	90	Nicaragua	2.88	1.41
39	Luxembourg	4.28	1.16	91	Dominican Republic	2.87	1.49
40	Sweden	4.25	1.41	92	Venezuela	2.82	1.41
41	Mozambique	4.23	1.57	93	Macedonia	2.68	1.19
42	Tanzania	4.20	1.54	94	Poland	2.63	1.20
43	Switzerland	4.17	1.18	95	Zimbabwe	2.62	1.12
44	Botswana	4.14	1.33	96	Georgia	2.61	1.40
45	Norway	4.13	1.32	97	Guatemala	2.60	1.16
46	Jamaica	4.13	1.40	98	Honduras	2.55	1.23
47	Romania	4.11	1.44	99	Chad	2.55	1.74
48	Chile	4.08	1.29	100	Peru	2.51	1.11
49	Canada	4.07	1.35	101	Bolivia	2.47	1.32
50	Germany	4.06	1.26	102	Ecuador	2.31	1.11
51	South Africa	4.03	1.19	103	Paraguay	2.29	1.08
52	Brazil	4.02	1.27	104	Argentina	2.28	0.97

0 1 2 3 4 5 6 7   0 1 2 3 4 5 6 7

Source: World Economic Forum, Executive Opinion Survey 2004

III. Usage Component – Government Usage
9.02 Government online services, 2003

Sophistication of government online services (1–7 scale), 2003

* estimated data

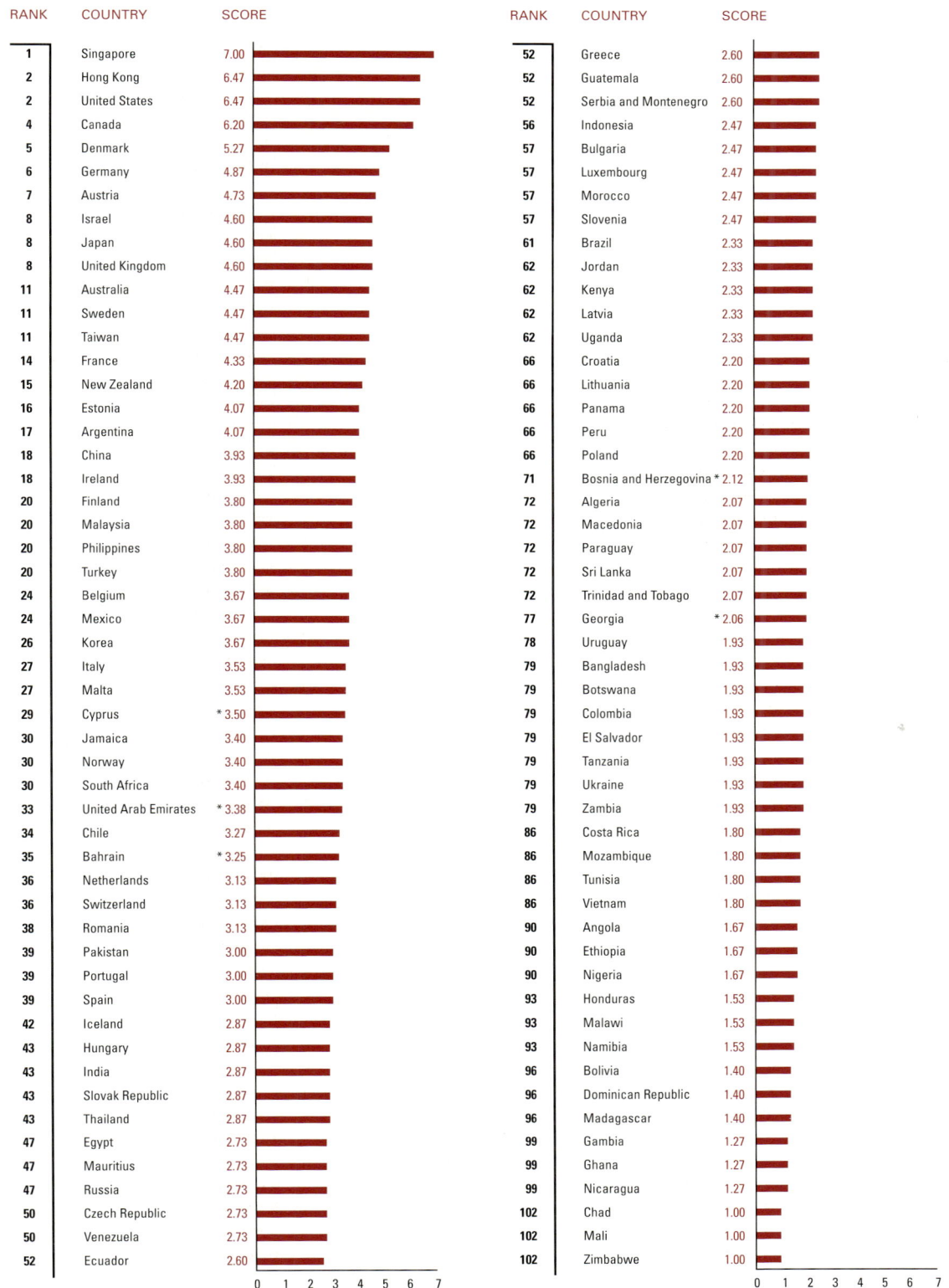

RANK	COUNTRY	SCORE		RANK	COUNTRY	SCORE
1	Singapore	7.00		52	Greece	2.60
2	Hong Kong	6.47		52	Guatemala	2.60
2	United States	6.47		52	Serbia and Montenegro	2.60
4	Canada	6.20		56	Indonesia	2.47
5	Denmark	5.27		57	Bulgaria	2.47
6	Germany	4.87		57	Luxembourg	2.47
7	Austria	4.73		57	Morocco	2.47
8	Israel	4.60		57	Slovenia	2.47
8	Japan	4.60		61	Brazil	2.33
8	United Kingdom	4.60		62	Jordan	2.33
11	Australia	4.47		62	Kenya	2.33
11	Sweden	4.47		62	Latvia	2.33
11	Taiwan	4.47		62	Uganda	2.33
14	France	4.33		66	Croatia	2.20
15	New Zealand	4.20		66	Lithuania	2.20
16	Estonia	4.07		66	Panama	2.20
17	Argentina	4.07		66	Peru	2.20
18	China	3.93		66	Poland	2.20
18	Ireland	3.93		71	Bosnia and Herzegovina *	2.12
20	Finland	3.80		72	Algeria	2.07
20	Malaysia	3.80		72	Macedonia	2.07
20	Philippines	3.80		72	Paraguay	2.07
20	Turkey	3.80		72	Sri Lanka	2.07
24	Belgium	3.67		72	Trinidad and Tobago	2.07
24	Mexico	3.67		77	Georgia *	2.06
26	Korea	3.67		78	Uruguay	1.93
27	Italy	3.53		79	Bangladesh	1.93
27	Malta	3.53		79	Botswana	1.93
29	Cyprus	* 3.50		79	Colombia	1.93
30	Jamaica	3.40		79	El Salvador	1.93
30	Norway	3.40		79	Tanzania	1.93
30	South Africa	3.40		79	Ukraine	1.93
33	United Arab Emirates	* 3.38		79	Zambia	1.93
34	Chile	3.27		86	Costa Rica	1.80
35	Bahrain	* 3.25		86	Mozambique	1.80
36	Netherlands	3.13		86	Tunisia	1.80
36	Switzerland	3.13		86	Vietnam	1.80
38	Romania	3.13		90	Angola	1.67
39	Pakistan	3.00		90	Ethiopia	1.67
39	Portugal	3.00		90	Nigeria	1.67
39	Spain	3.00		93	Honduras	1.53
42	Iceland	2.87		93	Malawi	1.53
43	Hungary	2.87		93	Namibia	1.53
43	India	2.87		96	Bolivia	1.40
43	Slovak Republic	2.87		96	Dominican Republic	1.40
43	Thailand	2.87		96	Madagascar	1.40
47	Egypt	2.73		99	Gambia	1.27
47	Mauritius	2.73		99	Ghana	1.27
47	Russia	2.73		99	Nicaragua	1.27
50	Czech Republic	2.73		102	Chad	1.00
50	Venezuela	2.73		102	Mali	1.00
52	Ecuador	2.60		102	Zimbabwe	1.00

Source: World Economic Forum, 2003

About the Authors

Scott Beardsley

Scott Beardsley is a Director in McKinsey & Company's Brussels Office. Since joining the firm in 1989, he has been particularly active in client service in the telecommunications, media, and technology sectors around the globe. He is a global leader of McKinsey's Telecommunications practice and is the leader of McKinsey's Strategy practice in Europe, the Middle East and Africa.

Mr Beardsley regularly writes articles for the *McKinsey Quarterly* and editorials for newspapers such as the *Wall Street Journal Europe* on the topics of telecommunications competition, liberalization and broadband. He has been quoted widely in publications such as *Forbes*, the *New York Times*, and *L'Echo*, and has written and presented his findings several times in the World Economic Forum's Competitiveness Report at the Annual Meeting in Davos. Mr Beardsley holds an MBA in Corporate Strategy and Marketing from the Massachusetts Institute of Technology (MIT) Sloan School of Management, where he graduated as a Henry S. Dupont III Scholar for outstanding academic performance. He obtained a B.S. in Electrical Engineering *magna cum laude* from Tufts University, where he was a Kodak Scholar.

Scott Brown

Scott Brown is the director of Market Insights with the Momentum Research Group. In this capacity, he facilitates the blending of research solutions with market knowledge to identify implications and opportunities for the development of marketing strategies.

Formerly, Mr Brown was a senior brand consultant with the brand consulting practice of Citigate Cunningham, responsible for developing corporate and executive communication platforms for clients such as Cisco Systems, Sun Microsystems, US Steel, Sybase and Adobe Systems. He has also worked in marketing for Principal Capital Management, the financial-services subsidiary of the Principal Financial Group. During his tenure with Principal, Mr Brown was responsible for managing media relations, advertising and employee communications activities for the US$19.3 billion commercial real estate operation. Mr Brown has also worked in public relations for a 630-bed acute-care facility and in investor relations for Meredith Corporation, publisher of *Better Homes & Gardens*.

Mr Brown earned his M.Sc. in Integrated Marketing Communications from Northwestern University, Illinois, in 1999. He also holds a B.Sc. in journalism from Drake University, Iowa.

Soumitra Dutta

Soumitra Dutta is the Roland Berger Professor of Business and Technology and Dean for Executive Education at INSEAD. He is also the faculty director of **Elab@INSEAD**, INSEAD's Center of Excellence in the digital economy (http://elab.insead.edu). Prior to joining the faculty of INSEAD, he was employed with Schlumberger in Japan and General Electric in the USA. Professor Dutta obtained his Ph.D. in computer science and his M.S. in business administration from the University of California at Berkeley.

His research and consulting have focused on breakthrough approaches to the interrelationships between innovation, technology and organizational design. He is co-editor of *The Global Information Technology Report 2003–2004: Towards an Equitable Information Society* (Oxford University Press, January 2004) and of *The Global Information Technology Report 2002–2003: Readiness for the Networked Future* (Oxford University Press, January 2003). His other publications include *The Bright Stuff: How Innovative People and Technology can make the Old Economy New* (Financial Times/Prentice Hall, 2002), *Embracing the Net: Get.Competitive"* (Financial Times/Prentice Hall, 2001) and *Process Reengineering, Organizational Change and Performance Improvement* (McGraw Hill, 1999). In addition, he has published over 50 articles in leading international journals.

A fellow of the World Economic Forum, he has won several awards for research and pedagogy. His research has been showcased in prominent international media and he has taught in and consulted with leading international corporations.
Personal URL: **http://www.insead.edu/faculty research/tom/dutta/index.htm.**

Andrew Elder

Andrew Elder is the primary methodological consultant for quantitative research projects for Momentum Research Group. His responsibilities include research design, sampling, statistical analysis, and interpretation of syndicated and custom research efforts.

Prior to joining Momentum Research Group in 1999, Mr Elder spent nearly five years designing and analysing technology research with IntelliQuest, Inc. He developed the analytic content of IntelliQuest's largest ongoing research projects, including the Computer Industry Media Study (CIMS(tm)) and the Worldwide Internet Tracking Service (WWITS(tm)). He is recognized as an expert among international research professionals, having presented research papers and tutorials across the globe. Mr Elder has extensive experience in the areas of market segmentation, conjoint analysis, and online survey methodology, frequently publishing on these subjects. He has also worked extensively in the area of IT-enabled productivity, helping to design and analyse several studies on this topic for Cisco Systems.

Mr Elder earned a B.S. in political science from Washington University in St. Louis, which he augmented with an MBA in marketing from the University of Texas at Austin.

Luis Enriquez

Luis Enriquez is a Principal in McKinsey & Company's Brussels office, where he has worked primarily in areas of corporate finance, strategy and telecommunications. He has had extensive experience in telecommunications, focusing on corporate finance, strategy, operations and regulation.

Prior to joining McKinsey, Dr Enriquez also worked extensively on telecommunication liberalization and regulation issues. In 1994 he assisted the Czech Ministry of Finance in developing price regulations to support the privatization of Cesky Telecom (then SPT Telecom), and taught courses and seminars for the ministry staff and other industry stakeholders. He has participated in proceedings on liberalization and privatization in Mexico, Argentina, Poland, and other eastern European and Latin American countries. He assisted the Chief Economist of the Federal Communications Commission in areas including interconnection, universal service subsidies and developing dispute-resolution mechanisms, and has worked with US incumbents and new entrants on various regulatory topics.

Dr Enriquez has a B.A. degree in Economics from Harvard University. He received a Ph.D. in Economics from the University of California at Berkeley, where he focused on the dynamics of interconnection among telecommunication networks.

Ferng-Ching Lin

Dr Ferng-Ching Lin has been the chairman of Institute for Information Industry since 2003. Before that, he was the President/CEO of the Institute for Information Industry (2000–03) and Science and Technology Advisor to the Prime Minister of Taiwan, R.O.C. (2002–04). He held the position of Director General of Bureau of Business Administration of Taipei City Government (1995–98). Dr Ferng-Ching Lin received his B.S. in Math from National Tsing-Hua University in Taiwan in 1970, and earned a Ph.D. in Math from the State University of New York at Buffalo in 1979. Before serving in the Taipei City Government, he was Chairman and Professor of the Department of Computer Science & Information Engineering at the National Taiwan University. He has

received awards for teaching and research excellence, including the Distinguished Research Award from the National Science Council and the Distinguished Teaching Award from the Ministry of Education.

Douglas Frosst

Douglas Frosst is the General Manager of Cisco Systems' Global Net Impact Research program. In this capacity he is responsible for Cisco's business research agenda, developing and delivering research, analysis and insight in areas of interest to Cisco's customers and partners. Recent research has focused on people, process and technology and the impact of this formula on organizational productivity, in both the public and private sectors. In forthcoming projects he proposes to explore the productivity characteristics of mobility and voice communications, and to undertake a detailed study of productivity best practices in small and medium enterprises around the world.

Mr Frosst has held several roles at Cisco since joining in 1993. He was responsible for key initiatives in the development of Cisco's access server product line. While exploring telephony services, he was an integral part of several acquisitions and post-closing integrations. Throughout he has been a significant contributor to Cisco's Service Provider marketing and strategy. Mr Frosst studied computer science and actuarial science at the University of Waterloo in Canada, and is currently a Fellow of the Wharton School at the University of Pennsylvania.

Victoria Gerus

Victoria Gerus is a regulatory specialist in European telecommunications, working from McKinsey's Brussels office. She supports McKinsey's client service and knowledge development in the domain for both the fixed and mobile subsectors.

Ms Gerus recently left McKinsey for a two-year period during which she worked at Ovum Ltd in London as a Senior Analyst in the Telecommunications Regulation and Competition Research Group and as the Service Manager for Interconnect@Ovum. Victoria then worked as head of Regulatory Policy at eircom Ltd. in Ireland.

Ms Gerus holds an M.A. in Political Science from Harvard University. She earned her B.A. in Political Science at Central Michigan University, United States.

Markus Haacker

Markus Haacker is an economist at the International Monetary Fund (IMF). His research interests are spread widely across the areas of economic growth and development. Most recently, he edited and co-authored *The Macroeconomics of HIV/AIDS* (IMF, 2004), a comprehensive resource on the economic and fiscal repercussions of the epidemic, with contributions from numerous international organizations and other institutions. He is chairman of the German publisher ARCO, and has worked as a management consultant with several arts organizations in Washington D.C.

Amit Jain

Amit Jain is the Research Program Manager for the Global Information Technology Project with the World Economic Forum at INSEAD. He has worked at INSEAD on various research projects in information and communication technology and knowledge management, and has been the Technical Director of INSEAD Online, INSEAD's e-Learning effort. Prior to joining INSEAD, he worked with Schlumberger in the Middle East, and has pursued several entrepreneurial opportunities in the information technology space. Amit Jain obtained his D.E.A. from the Université de Paris, Dauphine, MBA from INSEAD, and Bachelor of Technology in Mechanical Engineering from the Indian Institute of Technology.

Augusto Lopez-Claros

Augusto Lopez-Claros is Chief Economist and Director of the Global Competitiveness Program at the World Economic Forum in Geneva. Before joining the Forum he was Executive Director and Senior International Economist with Lehman Brothers International (London). He served as the resident representative for the International Monetary Fund in the Russian Federation (Moscow) from 1992 to 1995. Dr Lopez-Claros was educated in the United Kingdom and the United States, receiving a diploma in Mathematical Statistics from Cambridge University, and a Ph.D. in Economics from Duke University. Before joining the IMF he was professor of economics at the University of Chile in Santiago. He has written and lectured extensively in South America, the United States, and Europe on a broad range of subjects, including aspects of economic reform in transition economies, economic integration, interdependence and cooperation, governance, peace, and the role of international organizations.

Andreas Marschner

Andreas Marschner is an Engagement Manager in McKinsey & Company's Vienna Office, where he has worked primarily on regulatory strategy across various industries and on strategy and operations in travel and logistics.

Prior to joining McKinsey, Mr Marschner worked on macroeconomic performance differentials and country risk assessments for the International Finance Corporation (World Bank Group) in Washington, D.C.

Mr Marschner has an M.A. degree in Economics from Humboldt University Berlin, Germany. He received an M.A. in International Relations from the School of Advanced International Studies (SAIS), Johns Hopkins University, United States.

Elliot Maxwell

Elliot Maxwell advises corporate clients on Internet and e-commerce issues. From 1998 to 2001 he served as Special Advisor for the Digital Economy to US Secretary of Commerce William Daley, and prior to that worked on Internet and e-commerce policy at the Federal Communications Commission.

Mark Melford

Mark Melford is a Principal in Booz Allen's European Government and Transformation team, based in London. He is particularly interested in questions of strategy and industry structure in the Information and Communications Technology sectors. Mark advises the government of the UK and other leading nations on policy issues surrounding the adoption of technology by businesses.

Mark has an MBA from INSEAD in France, and an M.A. in General Engineering from the University of Cambridge, England.

Michael R. Nelson

Michael R. Nelson serves as Vice-President for Policy at the Internet Society and during the mid-1990s worked on Internet and telecommunications issues at the White House and the Federal Communications Commission in Washington, D.C.

Suvojoy Sengupta

Suvojoy Sengupta is a Principal in the IT strategy practice of Booz Allen Hamilton based in London. He advises clients on the use of technology in achieving step changes in business performance and in enhancing efficiency and effectiveness of the IT function. He has over ten years' experience working with clients in energy/utilities, telecoms and the public sector in Europe, the Middle East and North America.

Prior to joining Booz Allen, Mr Sengupta worked with a major technology and system integration company on large-scale IT system implementation programmes. He received an MBA from the Indian Institute of Management, Ahmedabad, and a Bachelor of Technology from the Indian Institute of Technology, Delhi.

Myles Wright

Myles Wright is a Principal in the London Office of Booz Allen Hamilton. Mr Wright is focused on helping clients define and implement their outsourcing and offshoring strategies a variety of industries, particularly within the financial services industry. His expertise spans both IT and

business process outsourcing and offshoring.
Prior to joining Booz Allen, Mr Wright was a manager at Mitchell Madison Group in London for four years; previously he worked for Xerox Corporation for five years in their solutions and outsourcing business.

Mr Wright received a B.Sc. in Economics from the University of Western Ontario, and an MBA from the Ivey Business School at the University of Western Ontario.

List of Partner Institutes

Algeria
Centre de Recherche en Economie Appliquée pour le Développement (CREAD)
Professor Yassine Ferfera

Angola
Serviços de Organização e Finanças (SOF)
Marcolino Meireles, Manager

Argentina
IAE—Universidad Austral
Marcelo Paladino, Research Director
Ariel A. Casarin, Assistant Professor

Australia
Australian Industry Group
Heather Ridout, Chief Executive, Australian Industry Group
Tony Pensabene, National Manager, Economics & Business

Austria
WIFO—Austrian Institute of Economic Research
Professor Karl Aiginger, Deputy Director
Gerhard Schwarz, Coordinator, Survey Department

Bahrain
Economic Development Board
Sulaf Zakharia, Manager, Research Services Unit

Bangladesh
Centre for Policy Dialogue (CPD)
Dr Debapriya Bhattacharya, Executive Director
Professor Mustafizur Rahman, Research Director
Dr Uttam Kumar Deb, Research Fellow

Belgium
Vlerick Leuven Gent Management School
Professor Dr Lutgart Van den Berghe, Executive Director, Chairman Competence Centre, Entrepreneurship, Governance & Strategy
Harry Bowen, Professor, Economics and International Business Lucy Amez, Research Assistant

Bolivia
Universidad Catolica Boliviana Lic. Marcela A. De Guzman, Directora, Depto. Economia

Bosnia and Hercegovina
The Center for Management and Information Technologies, MIT Center of the Faculty of Economics, University of Sarajevo
Professor Bozidar Matic, Ph.D., President of the Academy of Sciences and Arts of Bosnia and Herzegovina
Professor Zlatko Lagumdzija, Ph.D., Director of the MIT Center at the Faculty of Economics, Sarajevo Assistant
Professor Zeljko Sain, Ph.D., Deputy Director of the MIT Center at the Faculty of Economics, University of Sarajevo

Assistant Professor Fikret Causevic, Ph.D., Research Fellow and Deputy Director of the Economic Institute Sarajevo

Botswana
Botswana Institute for Development Policy Analysis (BIDPA)
Dr N.H. Fidzani, Executive Director
Kedikilwe P. Maroba, Programme Coordinator

Brazil
Fundação Dom Cabral Professor Carlos Arruda, Associate Dean for Development
Fabiana Santos
Movimento Brasil Competitivo (MBC)
José Fernando Mattos, President
Jorge H. S. Lima, Project Coordinator

Bulgaria
Center for Economic Development
Anelia Damianova, Ph.D., Senior Expert

Canada
Institute for Competitiveness and Prosperity
Roger Martin, Dean of the Rotman School of Management, University of Toronto and Chairman of the Institute for Competitiveness and Prosperity
James Milway, Executive Director of the Institute for Competitiveness and Prosperity

Chad
Groupe de Recherches Alternatives et de Monitoring du Projet Pétrole-Tchad-Cameroun (GRAMP-TC)
Professor Gilbert Maoundonodji, Director

Chile
University of Adolfo Ibáñez
Andres Allamand Zavala, Dean of the School of Government
Victoria Hurtado Larrain, Academic Coordinator of the School of Government
Sergio Selman Hasbún, Research Assistant

China
Institute of Economics Systems and Management
National Development and Reform Commission
Dr Liu Fuyuan, Director
Ms Dong Ying, Professorial Fellow
Mr Chen Wei, Research Fellow

Colombia
National Planning Department
María Isabel Agudelo
Fernando J. Estupiñan V.

Croatia
National Competitiveness Council
Mira Lenardic, Secretary General
Ivana Cesljas, Advisor Ruzica Simic, Advisor

Cyprus

Center of Applied Research, Cyprus College

Dr Bambos Papageorgiou

The Cyprus Development Bank Ltd

Maria Markidou-Georgiadou, Manager, International Banking Services Unit and Business Development Manager

Czech Republic

CMC Graduate School of Business

Peter Loewenguth, President Partner Institutes

Denmark

Copenhagen Business School

Dr Heather Alison Hazard, Associate Professor, Program Director, Vice President for International Affairs

Ecuador

Escuela Superior Politecnica del Litoral (ESPOL)

Escuela de Postgrado en Administracion de Empresas (ESPAE)

Virginia Lasio, Acting Director

Karina Astudillo, Project Assistant

Egypt

The Egyptian Center for Economic Studies

Dr Ahmed Galal, Executive Director and Director of Research

Dr Samiha Fawzy, Deputy Director and Lead Economist

Estonia

Estonian Chamber of Commerce and Industry

Siim Raie, Director General

Ethiopia

Ethiopian Economic Association/Ethiopian Economic Policy Research Institute

Berhanu Nega, Director

Kibre Moges, Senior Researcher

Worku Gebeyehu, Researcher

Finland

ETLA—The Research Institute of the Finnish Economy

Pentti Vartia, President

Pekka Ylä-Anttila, Managing Director

Petri Rouvinen, Research Director

France

HEC School of Management, Paris

Bernard Ramanantsoa, Professor, Dean of HEC School of Management

Bertrand Moingeon, Professor, Associate Dean for Executive Education

Gambia

Gambia Economic and Social Development Research Institute (GESDRI)

Makaireh A. Njie, Director

Georgia

Business Initiative for Reforms in Georgia

Irakli Burdiladze, Executive Director

Mamuka Tsereteli, Founding Member of the Board of Directors

Giga Makharadze, Founding Member of the Board of Directors

Germany

Wissenschaftliche Hochschule für Unternehmensführung Koblenz

(WHU)—Otto Beisheim Graduate School of Management

Professor Michael Frenkel

Ghana

Ghana Investment Promotion Centre

Kwasi Abeasi, Chief Executive Officer

Greece

Federation of Greek Industries

Antonis Tortopidis, Coordinator, Research and Analysis

Thanasis Printsipas, Economist, Research and Analysis

Hong Kong SAR

The Hong Kong General Chamber of Commerce

David O'Rear, Chief Economist

Federation of Hong Kong Industries

Alexandra Poon, Director

Hungary

Kopint-Datorg, Economic Research

Dr Éva Palócz, Deputy General Director

Ágnes Nagy, Project Manager

Iceland

ICETEC

Hallgrimur Jonasson, General Director

India

Confederation of Indian Industry

Tarun Das, Chief Mentor

Natraj Srinivasan, Director General

Ajay Khanna, Deputy Director General

Indonesia

LP3E-Kadin Indonesia

Dr Tulus Tambunan

Ireland

Department of Economics, University College Cork

Dr Eleanor Doyle

Bernadette Power

Rosemary Kelleher

Niall O'Sullivan

Israel

Manufacturers' Association of Israel (MAI)

Yoram Blizovsky, Managing Director

Moshe Nahum, Director, Foreign Trade and International Relations Division

Italy

SDA Bocconi

Paola Dubini, Associate Professor, Strategy

Olga E. Annushkina, Assistant Professor, Strategic and Entrepreneural Management Department

Jamaica

Private Sector Organisation of Jamaica (PSOJ)

Greta Bogues, Chief Executive Officer Mona School of Business at the University of the West Indies (MSB)

Gordon Shirley, Professor

Japan

Hitotsubashi University Graduate School of International Corporate Strategy (ICS)

Professor Yoko Ishikura

Jordan

Ministry of Planning, Competitiveness Unit

Naseem Al-Rahahleh, Director

Kenya

Institute for Development Studies, University of Nairobi

Professor Dorothy McCormick, Director

Paul Kamau, Research Fellow

Korea

Korea Development Institute

Dr Cho Byung-Koo, Director of Center for Economic Information (CEI)

Latvia

Institute of Economics, Latvian Academy of Sciences, Riga

Dr Raita Karnite, Director

Lithuania

Statistikos Tyrimai‹Statistical Surveys, Vilnius

Benonas Miksas, Director

Luxembourg

Chamber of Commerce of the Grand Duchy of Luxembourg

Carlo Thelen, Member of the Management Committee

Macedonia, FYR

National Entrepreneurship and Competitiveness Council (NECC)

Svetozar Janevski, Co-chair of the NECC, Managing Director of Skopsko Brewery

Stevce Jakimovski, Co-chair of the NECC, Minister of Economy

Ana Nikovska, Advisor to the NECC, Macedonia Competitiveness Activity

Madagascar

Centre of Economic Studies, University of Antananarivo

Pépé Andrianomanana, Director

Malawi

Malawi Investment Promotion Agency

Alick C. E. Sukasuka, Acting Deputy General Manager

Malaysia

Institute of Strategic and International Studies (ISIS)

Tan Sri Dato' Dr Mohamed Noordin Sopiee, Chairman and Chief Executive Officer

Mali

Groupe de Recherche en Economie Appliquée et Théorique (GREAT)

Massa Coulibaly, Coordinator

Malta

Competitive Malta—Foundation for National Competitiveness

Dr John C. Grech, President

Mr Adrian Said, Chief Coordinator

Mr Wilfred Kenely, Policy & Programs Coordinator

Dr Jennifer Cassingena Harper, International Relations Coordinator

Mauritius

Joint Economic Council of Mauritius

Raj Makoond, Director

Mexico

Ministry of the Economy

Dr Eduardo J. Solis Sanchez, Chief of the Office for the Co-ordination of International Trade and Investment Promotion

Lic. Veronica Orendain De Los Santos, Assistant in the Office for the Co-ordination of International Trade and Investment Promotion

Center for Intellectual Capital and Competitiveness

Dr Rene Villarreal, President

Dra Rocio Ramos de Villarreal, Vice-President

Morocco

Université Hassan II

Fouzi Mourji, Professor of Economics

Mozambique

EconPolicy Research Group, Lda

Dr Peter Coughlin, Partner

Professor Dr Paulo N. Mole, Partner

Namibia

Namibian Economic Policy Research Unit

Dr Christoph Stork, Senior Researcher

Netherlands

Erasmus Strategic Renewal Center, Erasmus University Rotterdam

Professor Frans A. J. van den Bosch

Professor Henk W. Volberda

New Zealand

Business New Zealand

Marcia Dunnett, Manager, Business Services

Nigeria

Nigerian Economic Summit Group (NESG)

Chris E. Onyemenam, Director, Operations & Administration

Dr Felix Ogbera, Associate Director, Research

Norway

Norwegian School of Management BI, Centre for Corporate Citizenship

Dr Torger Reve, Professor

Tore Dirdal Researcher

Pakistan

Pakistan Institute of Development Economics
Dr A. R. Kemal

Paraguay

Centro de Analisis y Difusion de Economia Paraguaya (CADEP)
Fernando Masi, Director
Dionisio Borda, Research Member
Nelson Aguilera Alfred, Research Member

Peru

Centro de Desarrollo Industrial (CDI)—Sociedad Nacional de Industrias
Luis Tenorio, Executive Director
Néstor Asto, Project Director

Philippines

Makati Business Club
Guillermo M. Luz, Executive Director
Marc P. Opulencia, Deputy Director
Michael B. Mundo, Chief Economist

Poland

Warsaw School of Economics
Professor Bogdan Radomski, Associate Professor

Portugal

PROFORUM, Associação Para o Desenvolvimento da Engenharia
Ilídio António de Ayala Serôdio, Member of the Board of Directors

Romania

Group of Applied Economics (GEA)‹Romanian Economic Society(SOREC)
Professor Daniel Daianu, President SOREC
Professor of Economics [END]
Dragos N. Pislaru, Executive Director, Group of Applied Economics
Dr Liviu Voinea, Programme Coordinator, Group of Applied Economics

Russian Federation

Bauman Technical University (BMSTU) & Bauman Innovation
Dr Alexei Prazdnitchnyk, Postdoctoral University Fellow and Director, Russian Regional Competitiveness Survey Project
Institute for Private Sector Development and Socio-Economic Analysis (IPSSA)
Irina Evseyeva
Stockholm School of Economics, Russia
Professor Carl F. Fey, Associate Dean of Research
Dr Igor Dukeov, Research Fellow

Serbia and Montenegro

USAID
Serbia Enterprise Development Project
Andrew Vonnegut, Chief of Party
Jelena Sevo, Advisor, Legal, Regulatory and Government Affairs

Singapore

Economic Development Board
Shirley Chen, Assistant Managing Director, Corporate Services
Chua Kia Chee, Head, Research and Statistics Unit

Slovak Republic

Business Alliance of Slovakia (PAS)
Robert Kicina, Executive Director
Marian Driensky, Project Manager
Institute for Economic and Social Reforms (INEKO)
Eugen Jurzyca, Director

Slovenia

Institute for Economic Research
Dr Peter Stanovnik, Director
Dr Mateja Drnovšek, Faculty of Economics
Professor Aleš Vahčič, Faculty of Economics

South Africa

Business Unity South Africa
Bheki Sibiya, Chief Executive Officer
Friede Dowie, Chief Officer, Strategic Services

Spain

IESE Business School-Anselmo Rubiralta
Center for Globalization and Strategy
Professor Eduardo Ballarín
María Luisa Blázquez, Research Associate

Sri Lanka

Institute of Policy Studies
Indika Siriwardena, Database Manager
The Ceylon Chamber of Commerce
Gayathri Gunaruwan, Consultant Economist, Economic Intelligence Unit
Sri Lanka Institute of Directors
Alikie Ismail, Secretary

Sweden

Stockholm School of Economics, Institute of International Business
Professor Örjan Sölvell

Switzerland

University of St. Gallen
Professor Dr Franz Jaeger, Director, Research Institute for Empirical Economics and Economic Policy

Taiwan

Council for Economic Planning and Development
Economic Research Department
Dr C.Y. Hu, Director
Chung-Chung Shieh, Researcher

Tanzania

Economic and Social Research Foundation
Professor Haidari Amani, Executive Director
John Ulanga, Coordinator, Commissioned Studies Department
Moses Msuya, Research Assistant, Commissioned Studies Department

Thailand

National Economic and Social Development Board

Arkhom Termpittayapaisith, Senior Advisor in Policy and Planning

Trinidad and Tobago

UWI Institute of Business

Dr Rolph Balgobin, Executive Director

Vashti G. Guyadeen, Manager Business Research

Tunisia

Institut Arabe des Chefs d'Entreprises

Faycal Lakhoua, Conseiller

Turkey

TUSIAD—Sabanci University

Competitiveness Forum

Arzu Ilhan Kibris, Vice Director

Uganda

Makerere Institute of Social Research

Delius Asiimwe, Senior Research Fellow

Wlison Asiimwe, Graduate Fellow

Ukraine

CASE—Ukraine, Center for Social and Economic Research

Vladimir Dubrovskiy, Leading Expert

Oleksandr Rohozynsky, Executive Director

United Arab Emirates

Dubai Strategy Forum

Zayed University

UAE Ministry of Economy and Commerce

United Kingdom

Interdisciplinary Institute of Management, London School of Economics and Political Science

Sir Geoffrey Owen, Senior Fellow

United States

Council on Competitiveness

Deborah Wince-Smith, President

Chad Evans, Vice President, Research & Analysis

Uruguay

Universidad ORT

Professor Isidoro Hodara

Venezuela

CONAPRI, National Council for Investment Promotion

María Eugenia Labrador, Special Projects Manager

Camilo Daza, Investor Service Kitys Gil, Communication Assistant

Vietnam

Central Institute for Economic Management (CIEM)

Dr Dinh Van An, President

Phan Thanh Ha, Deputy Director, Department of Macroeconomic Management

Pham Hoang Ha, Senior researcher, Department of Macroeconomic Management

Institute for Economic Research of HCMC

Tran Du Lich, Director

Du Phuoc Tan, Head of the Scientific Management and International Cooperation Department

Doan Nguyen Ngoc Quynh, Researcher of the Scientific Management and International Cooperation Department

Zambia

Institute of Economic and Social Research (INESOR), University of Zambia

Chileshe L. Mulenga, Director

Zimbabwe

University of Zimbabwe

Professor A.M. Hawkins, Graduate School of Management

Bolivia, Costa Rica, Dominican Republic, Ecuador, El Salvador, Guatemala, Honduras, Nicaragua, Panama

Latin American Center for Competitiveness and Sustainable Development (INCAE)

Roberto Artavia, Rector

Arturo Condo, Associate Dean

Marlene de Estrella, Administrative and Financial Director

Estonia, Latvia, Lithuania

Stockholm School of Economics in Riga

Dr Anders Paalzow, Rector

Dr Karlis Kreslins, Associate Professor